Introduction to Rock Mechanics

Introduction to Rock Mechanics

Second Edition

Richard E. Goodman

University of California at Berkeley

WILEY

John Wiley & Sons

New York / Chichester / Brisbane / Toronto / Singapore

Dedicated to the memory of
Daniel G. Moye

Library of Congress Cataloging in Publication Data:

Goodman, Richard E.
  Introduction to rock mechanics/Richard E. Goodman.—2nd ed.
    p.   cm.

  Bibliography: p.
  Includes index.
  ISBN 0-471-81200-5
  1. Rock mechanics.   I. Title.

TA706.G65   1989
624.1'5132—dc19   87-34689
10 9   CIP

Preface to the First Edition

Rock mechanics is a truly interdisciplinary subject, with applications in geology and geophysics, mining, petroleum, and civil engineering. It relates to energy recovery and development, construction of transportation, water resources and defense facilities, prediction of earthquakes, and many other activities of greatest importance. This book introduces specific aspects of this subject most immediately applicable to civil engineering. Civil engineering students, at the advanced undergraduate and beginning graduate level, will find here a selection of concepts, techniques, and applications pertaining to the heart of their field—for example, how to evaluate the support pressure required to prevent squeezing of claystone in tunnels, how to evaluate the optimum angle of a rock cut through a jointed rock mass, and how to determine the bearing capacity of a pier socketed into rock. Students in other fields should also find this work useful because the organization is consistently that of a textbook whose primary objective is to provide the background and technique for solving practical problems. Excellent reference books cover the fundamental bases for the subject well. What has been lacking is a relatively short work to explain how the fundamentals of rock mechanics may be applied in practice.

The book is organized into three parts. Part 1, embracing the first six chapters, provides a survey of the methods for describing rock properties. This includes index properties for engineering classification, rock strength and deformability properties, the properties and behavior of joints, and methods of characterizing the state of initial stress. Modern fracture mechanics has been omitted but some attention is given to anisotropy and time dependency. Part 2, consisting of Chapters 7, 8, and 9, discusses specific applications of rock mechanics for surface and underground excavations and foundations. Part 3 is a series of appendices. One appendix presents derivations of equations, which were omitted from the chapters to highlight usable results. There is also a thorough discussion of stresses in two and three dimensions and instructions in the measurement of strains. Appendix 3 presents a simple scheme for identifying rocks and minerals. It is assumed that the reader has some familiarity with introductory geology; this section distills the terminology of petrology and mineralogy to provide a practical naming scheme sufficient for many purposes in rock mechanics. Part 3 also includes answers to all problems, with elaboration of the methods of solution for a selected set. The problems presented at the ends of each chapter and the worked out solutions in the answers section are a vital part of this book. Most of the problems are not just exercises in filling in values for equations offered in the text, but try to explore new material. I always enjoy learning new material in a practical context and therefore have elected to introduce new ideas in this way.

Although this is largely a presentation of results already published in journals and proceedings, previously unpublished materials are sprinkled through the text, rounding out the subject matter. In almost all such cases, the derivations in the appendix provide complete details.

This book is used for a one-quarter, three-credits course for undergraduates and beginning graduate students at the University of California, Berkeley, Department of Civil Engineering. Attention is riveted to the problems with little time spent on derivations of equations. Appendices 1 and 2 and all materials relating to time dependency are skipped. In a second course, derivations of equations are treated in class and the materials presented here are supplemented with the author's previous book *Methods of Geological Engineering in Discontinuous Rocks* (West Publishing Co.) 1976, as well as with selected references.

I am deeply indebted to Dr. John Bray of Imperial College for illuminating and inspiring contributions from which I have drawn freely. A number of individuals generously loaned photographs and other illustrations. These include K. C. Den Dooven, Ben Kelly, Dr. Wolfgang Wawersik, Professor Tor Brekke, Dr. Dougall MacCreath, Professor Alfonso Alvarez, Dr. Tom Doe, Duncan Wyllie, Professor H. R. Wenk et al., and Professor A. J. Hendron Jr. Many colleagues assisted me in selection of material and criticism of the manuscript. The list includes E. T. Brown, Fred Kulhawy, Tor Brekke, Gregory Korbin, Bezalel Haimson, P. N. Sundaram, William Boyle, K. Jeyapalan, Bernard Amadei, J. David Rogers and Richard Nolting. I am particularly grateful to Professor Kulhawy for acquainting me with much material concerning rock foundations. I am also very appreciative of Cindy Steen's devoted typing.

Richard E. Goodman

Preface

Since the publication of the first edition in 1980 we have developed a geometric approach to rock mechanics called "block theory." This theory is based on the type of data that comes most easily and naturally from a geological investigation, namely the orientations and properties of the joints. Block theory formalizes procedures for selecting the wisest shapes and orientations for excavations in hard jointed rock and is expounded in a book by Gen hua Shi and myself, published in 1985, and in additional articles derived from subsequent research at Berkeley. In preparing this edition my main objective was to incorporate an introduction to the principles of block theory and its application to rock slopes and underground excavations. This has been accomplished in lengthy supplements to Chapters 7 and 8, as well as in a series of problems and answers.

An additional objective in preparing this new edition was to incorporate previously omitted subjects that have since proved to be important in practice, or that have appeared subsequent to initial publication. In the former category are discussions of the Q system of rock classification and the empirical criterion of joint shear strength, both introduced by Barton and co-workers at the Norwegian Geotechnical Institute (NGI). In the latter category are fundamental, new contributions by Indian engineers Jethwa and Dube on the interpretation of extensometer data in squeezing tunnels; analysis of rock bolting using an exponential formulation by Lang and Bischoff; properties of weak rocks brought to light by Dobereiner and deFreitas; representation of the statistical frequency of jointing by Priest and Hudson; an empirical criterion of rock strength by Hoek and Brown; and development of a "block reaction curve" as a model for design of supports in underground openings (analogous to the ground reaction curve concept previously presented in Chapter 7). Additionally, several useful figures presenting derived relationships were updated; these deal with the directions of stresses in the continental United States summarized by Zoback and Zoback, and the relationship between the rock mass rating of Bieniawski, and the "stand-up time" of tunnels.

To present this material, I have elected to develop a series of new problems and worked-out solutions. Thus, to take full advantage of this book you will need to study the problems and answers. The statements of the problems sometimes contain important material not previously presented in the chapters. And, of course, if you can take the time to work them through yourself, you will better understand and appreciate the value of the material.

Today, many workers in rock mechanics tend to use comprehensive numerical modeling to study the complex issues relating to the disposal of nuclear waste, energy storage and conversion, and defense technology. Although these models are powerful, much headway can also be made with simpler approaches by using statics with well-selected free-body diagrams, elegant graphical methods like the stereographic projection, and modest computations facilitated by microcomputers. If there is an overriding purpose in this book, it is to help you see the simple truths before trying to take hold of the big numerical tools.

Richard E. Goodman

Contents

Symbols and Notation

Symbols are defined where they are introduced. Vectors are indicated by bold-face type, for example, **B**, with lowercase boldface letters usually reserved for unit vectors. The summation convention is not used. Matrix notation is used throughout, with ( ) enclosing one- and two-dimensional arrays. Occasionally, { } are used to enclose a column vector. The notation $B(u)$ means that B is a function of u. Dimensions of quantities are sometimes given in brackets, with F = force, L = length, and T = time; for example, the units of stress are given as (FL^{-2}). A dot over a letter or symbol (e.g., $\dot{\sigma}$) usually means differentiation with respect to time. Some of the more commonly used symbols are the following:

$\hat{D}_i$	unit vector parallel to the dip
Δd	change in the length of a diameter of a tunnel or borehole
dev	subscript identifying deviatoric stress components
E	Young's modulus (FL^{-2})
g	acceleration of gravity
G	shear modulus; also, specific gravity
GPa	10^3 MPa
i	angle of the leading edge of an asperity on a joint
I_1, I_2, I_3	invariants of stress
$\hat{I}_{ij}$	unit vector parallel to the line of intersection of planes i and j
k	used for different purposes as defined locally, including conductivity (LT^{-1}) and stiffness coefficients
K	used variously for the bulk modulus, the Fisher distribution parameter, permeability (L^2), $\sigma_{\text{horiz}}/\sigma_{\text{vert}}$, and σ_3/σ_1
l, m, n	direction cosines of a line
ln	natural logarithm
MPa	megapascals (MN/m^2); 1 MPa $\approx$ 145 psi
n, s, t	coordinates perpendicular and parallel to layers (st plane)
n	porosity
$\hat{N}_i$	unit vector perpendicular to layers or joints of one set

p, p_w	pressure, water pressure
p_1, p_2	secondary principal stresses
P	force; also, in Chapter 9, a line load (FL^{-1})
q_f	bearing capacity (FL^{-2})
q_u	unconfined compressive strength
RMR	rock mass rating according to the Geomechanics Classification
S	spacing between joints of a given set
S_i	shear strength intercept according to the Mohr Coulomb relationship ("cohesion")
S_j	shear strength intercept for a joint
T_{MR}	magnitude of the flexural tensile strength ("modulus of rupture")
T_o	magnitude of the tensile strength; uniaxial tensile strength unless indicated otherwise
u, v	displacements parallel to x, y; positive in positive direction of coordinate axis
u_r, v_θ	displacements parallel to r, θ
Δu	shear displacement along a joint; also radial deformation
Δv	normal displacement across a joint
V_l, V_t	longitudinal and transverse stress wave velocities in a bar
V_p, V_s	compressive and shear wave velocities in an infinite medium
$\Delta V/V$	volumetric strain
w	water content, dry weight basis
w_L, w_P	liquid limit and plastic limit
$\mathbf{W}$	weight vector
x, y, z	right-handed Cartesian coordinates
Z	depth below ground surface
γ	weight per unit volume (FL^{-3})
γ_w	unit weight of water
ε, γ	normal and shear strains
η	viscosity $(FL^{-2}T)$
λ	Lamé's constant; also wavelength
μ	friction coefficient $(= \tan \phi)$; also same as η
ν	Poisson's ratio
ρ	mass density $(FL^{-4}T^2)$
σ	normal stress

$\sigma_1, \sigma_2, \sigma_3$	principal stresses; $\sigma_1 > \sigma_2 > \sigma_3$ (compression positive)
$\sigma_{t,B}$	magnitude of the Brazilian (splitting tension) strength
σ_r, σ_θ	radial and tangential normal stresses
σ'	effective stress
τ	shear stress
τ_p, τ_r	peak and residual shear strength
ϕ	friction angle; variously used as internal and surficial friction angles as defined locally
ϕ_μ	friction angle for sliding on a smooth surface ($i = 0$)
ϕ_j	friction angle for a joint
ψ	angle between the direction of σ_1 and the plane of a joint
$\bar{\omega}$	average displacement of a bearing plate

Chapter 1

Introduction

Some knowledge of rock mechanics is vital for civil engineers although it is only since about 1960 that rock mechanics has come to be recognized as a discipline worthy of a special course of lectures in an engineering program. That recognition is an inevitable consequence of new engineering activities in rock, including complex underground installations, deep cuts for spillways, and enormous open pit mines. *Rock mechanics* deals with the properties of rock and the special methodology required for design of rock-related components of engineering schemes. Rock, like soil, is sufficiently distinct from other engineering materials that the process of "design" in rock is really special. In dealing with a reinforced concrete structure, for example, the engineer first calculates the external loads to be applied, prescribes the material on the basis of the strength required (exerting control to insure that strength is guaranteed), and accordingly determines the structural geometry. In rock structures, on the other hand, the applied loads are often less significant than the forces deriving from redistribution of initial stresses. Then, since rock structures like underground openings possess many possible failure modes, the determination of material "strength" requires as much judgment as measurement. Finally, the geometry of the structure is at least partly ordained by geological structure and not completely within the designer's freedoms. For these reasons, rock mechanics includes some aspects not considered in other fields of applied mechanics—geological selection of sites rather than control of material properties, measurement of initial stresses, and analysis, through graphics and model studies, of multiple modes of failure. The subject of rock mechanics is therefore closely allied with geology and geological engineering.

1.1 Fields of Application of Rock Mechanics

Our involvement with rock in the most intimate terms extends backward far into prehistory. Arrowheads, common tools, vessels, fortifications, houses, even tunnels were built of or in rock. Constructions and sculptures, such as the

Abu Simbel Temple in Egypt and the pyramids, testify to a refined technique for selecting, quarrying, cutting, and working rocks. In the eighteenth and nineteenth centuries, great tunnels were driven for mine ventilation and drainage, water supply, canals, and rail transport.

In this century the great sculptures on Mount Rushmore (Figure 1.1) demonstrated to the world the enduring resolve of great figures and well-selected granite alike, even while engineers were turning to other materials. In this age, when materials engineers can concoct alloys and plastics to survive bizarre and demanding special requirements, rock work still occupies the energies of industry and the imagination of engineers; questions concerning the properties and behavior of rock figure prominently in engineering for structures, transportation routes, defense works, and energy supply.

Figure 1.1 Sculpting of Roosevelt and Lincoln in Mount Rushmore. Gutzon Borglum selected the site and adjusted the sculpture to fit its imperfections, even down to the last inch. The weathered rock was removed via controlled blasting with dynamite, the hole spacing and charge becoming progressively finer as the final surface was approached. The last inches were removed by very close drilling and chiseling. (Photo by Charles d'Emery. Reproduced with permission of Lincoln Borglum and K. C. Den Dooven. From *Mount Rushmore, the Story Behind the Scenery*, K. C. Publications (1978).)

Table 1.1 sketches some of the components of engineering works that involve rock mechanics to a significant degree. Of the many occupations of engineers in planning, design, and construction of works, nine have been singled out in this table because they are often significantly dependent upon rock mechanics input: evaluation of geological hazards in quantitative terms, selection and preparation of rock materials, evaluation of cuttability or drillability of rock and design of cutting and drilling tools, layout and selection of types of structures, analysis of rock deformations, analysis of rock stability, supervision and control of blast procedures, design of support systems, and hydraulic fracturing. These activities are pursued in somewhat different styles according to the nature of the engineering work.

Engineering *structures placed on the surface* of the ground normally do not require study of rock properties and behavior unless the structure is very large, or special, or unless the rock has unusual properties. Of course, the engineer is always on the lookout for geological hazards, such as active faults or landslides that might affect siting. The engineering geologist has the responsibility to discover the hazards; rock mechanics can sometimes help reduce the risk. For example, loose sheets of exfoliating granite pose a threat to buildings near the feet of cliffs in Rio de Janeiro. The rock engineer may be called upon to design a bolting system, or a remedial controlled blast. In the case of light structures like private homes, the only rock mechanics input would concern testing the potential swellability of shale foundations. However, in the case of very large buildings, bridges, factories, etc., tests may be required to establish the elastic and delayed settlement of the rock under the applied loads. Over karstic limestone, or mined-out coal seams at depth, considerable investigation and specially designed foundations may be required to insure structural stability.

An aspect of engineering for *tall buildings* that involves rock mechanics is control of blasting so that the vibrations do not damage neighboring structures or irritate local residents (Figure 1.2). In cities, foundations of new buildings may lie extremely close to older structures. Also, temporary excavations may require tieback systems to prevent sliding or raveling of rock blocks.

The most challenging surface structures with respect to rock mechanics are large *dams*, especially arch and buttress types that impose high stresses on rock foundations or abutments, simultaneously with the force and action of water. In addition to concern about active faults in the foundation, the hazards of possible landslides into the reservoir have to be carefully evaluated; very fresh is the memory of the Vajont catastrophe in Italy when a massive slide displaced the water over the high Vajont arch dam and killed more than 2000 people downstream. Rock mechanics is also involved in the choice of materials—rip-rap for protection of embankment slopes against wave erosion, concrete aggregate, various filter materials, and rock fill. Rock testing may be required to determine the durability and strength properties of such materials. Since the different types of dams exert very different stress regimes on the

Table 1.1 *Some Areas of Rock Mechanics Application Activity Involving a Substantial Rock Mechanics Input*

Project	Eval. of Geol. Hazards	Selection of Materials	Eval. of Cuttability, Drillability	Layout and Selection of Types of Works
Surface Structures				
Housing tracts	(2) Landslides, faults			
Bridges, tall buildings, surface power houses	(2) Landslides, faults	(2) Facing stones, concrete aggregate	(1) Drilled shafts for pier foundations	(2) Location of stable site
Dams	(1) Landslides in reservoirs; faults	(1) Rock fill, riprap concrete aggregate		(1) Selection of arch, gravity or embankment
Transportation Routes				
Highways, railways	(1) Landslides	(2) Embankment, base, aggregate, riprap		(1) Direction and slope of cuts
Canals, pipelines	(1) Landslides	(2) Embankment, base, aggregate, riprap		(1) Direction and slope of cuts
Penstocks	(1) Landslides			(1) Surface penstock vs. lined or unlined tunnel
Surface Excavations for Other Purposes				
Quarries and mine pits	(2) Landslides		(1) Taconite deposits and other hard rocks	(1) Slopes; conveyors; buildings
Spillways	(1) Landslides			(2) Side hill vs. tunnel; slopes

(1) Very relevant.
(2) Somewhat relevant.

Analysis of Deformations	Analysis of Stability	Supervision of Blasting	Design of Support System	Hydraulic Fracturing
(2) Rebound in shales				
(2) Reactions for pretensioning; subsidence engineering	(2) If on cliff edge or over old mines	(1) Control near existing building	Tiebacks in temporary excav.	
(1) Vertical and horizontal	(1) Abutment, foundation	(1) Abutment galleries cutoff trench, quarry	(1) Abutments; found.; reservoir slopes	Potential use for cutoffs
(2) Shale rebound; steep, urban cuts	(1) Cut slopes	(1) Perimeter control	(2) Steep cuts in cities	
(2) Shale rebound; steep, urban cuts	(1) Cut slopes	(1) Perimeter control	(2) Steep cuts in cities	
(1) For tunnel penstocks				
(2) To support monitoring programs	(1) Rock slopes	(1) Protection of struct. in and near pit	(2) Protection of struct., portals	
(2) To support monitoring programs	(1) Rock slopes	(1) Protection of struct. in and near cut	(2) For tunnel spillway	

Table 1.1 (*continued*)
Activity Involving a Substantial Rock Mechanics Input

Project	Eval. of Geol. Hazards	Selection of Materials	Eval. of Cuttability, Drillability	Layout and Selection of Types of Works
Dry Underground Excavations				
Caving mines	(1) Faults; air blasts	(2) Yielding supports	(1) Selection of long-wall cutters; moles	(1) Entire layout
Stable mines	(1) Faults; rock bursts		(1) Selection of mining tools	(1) Selection of mining scheme
Tunnels	(1) Faults; rock bursts		(1) Design of mole cutters	(1) Shape, size
Underground chambers	(1) Faults; rock bursts		(2) Bidding excavation costs	(1) Orientation
Defense works	(1) Faults; rock bursts			(1) Choice of depth
Energy Development				
Petroleum	(2) Faults; rock bursts		(1) Improving rates	
Geothermal	(2) Faults; rock bursts		(1) High temperature and salinity effects	
Nuclear power plants	(1) Faults; landslides	(2) Concrete aggregate		(2) Water-tight core
Nuclear waste disposal	(1) Faults	(1) Best rock choice for waste isolation		(1) Retrievability, stability
Energy storage caverns for oil, water, air, LNG	(1) Faults	(2) Special linings		(1) Leakproof curtain
Solution Mining				

(1) Very relevant.
(2) Somewhat relevant.

Analysis of Deformations	Analysis of Stability	Supervision of Blasting	Design of Support System	Hydraulic Fracturing
(1) To support monitoring programs	(1) Airblast avoidance; ore dilution anal.	(1) Avoid premature detonation	(1) Haulageways	(2) Solution mining
(1) To support monitoring programs	(1) Access tunnels, stopes, etc.	(1) Control of perimeter; vibration	(1) Rock bolts, shotcrete	
(1) To support monitoring programs	(1) Roof, wall, invert.	(1) Control of perimeter; vibration	(1) Select. temp. and perm. supports	
(1) Support monitoring; design of details	(1) Roof; pillars; invert	(1) Control of perimeter; vibration	(1) Rock bolts or shotcrete	
(1) Support monitoring; design of details	(1) Under blast loads	(1) Control of perimeter; vibration	(1) Rock bolts or shotcrete	
	(1) Deep holes in shale, evaporites depth of casing			(1) To improve permeability
	(1) Depth of casing			(1) Development of dry hot rock
(2) Rock slope monitoring	(1) Rock slopes; waste disposal		(1) Rock slopes and core shaft	
(1) Support of monitoring	(1) Effect of +200°C	(1) Control of perimeter; vibration	(1) Backfill for canisters	(1) Intermediate level storage
(1) Design and monitoring	(1) Effect of + or −200°C	(1) Control of perimeter; vibration	(1) Long design life	
(1) Monitoring surface subsidence	(1) Surface subsidence			(1) New technique

Figure 1.2 Excavation in rock very close to existing buildings is a frequent problem for construction in cities. (Photo courtesy of A. J. Hendron, Jr. Manhattan schist, Hunter College, New York.)

rock, rock mechanics assists in confirming the type of dam for the site. Then analysis of rock deformations, and of rock stability, form an important part of the engineering design studies.

In the case of concrete dams, deformability values assigned to the rocks of the foundations and abutments, through laboratory and in situ tests, are integrated in model studies or numerical analyses of concrete stresses. The safety of large and small rock wedges under the dam are calculated by statics. If necessary, cable or rock bolt support systems are designed to prestress the rock or the dam/rock contact.

Blasting for rock cleanup has to be engineered to preserve the integrity of the remaining rock and to limit the vibrations of neighboring structures to acceptable levels. At the Grand Coulee Third Powerhouse site, blasting was performed for the headrace channels very close to the existing Grand Coulee dam, without any possibility for lowering the reservoir. Also, a rock "cofferdam" was constructed by leaving a core of solid granite unexcavated until the completion of the powerhouse excavation some years later; this was accomplished by using controlled blasting technique on the upstream and downstream limits of the blast adjacent to the cofferdam.

Transportation engineering also calls upon rock mechanics in many ways. Design of cut slopes for *highways, railways, canals, pipelines,* and *penstocks* may involve testing and analysis of the system of discontinuities. Considerable cost savings are possible if the orientation of the right of way can be adjusted based on the rock mechanics studies, but this is not always practical. The decision to place portions of such routes underground is partly determined by judgments about the rock conditions and relative costs of open cuts and tunnels. Savings can be realized in penstock steel by assigning a portion of the stress to the rock if the penstock is placed in a tunnel; in that case rock tests can determine rock properties for the design. Sometimes penstocks can be left unlined; rock stress measurements may then be required to assure that leakage will not be disastrous. In urban areas, transportation routes at the surface may have to accept subvertical slopes because of the high values of land, and, accordingly, permanently stable slopes will have to be maintained by artificial supports. Considerable testing and analysis of the rock may be justified to provide an interpretational framework for instruments provided to monitor long-term safety.

Surface excavations for other purposes may also demand rock mechanics input in control of blasting, selection of cut slopes and location of safety benches, and provision for support. In the case of *open pit mines*, which rely on economical excavation for profitable operation, considerable study may be warranted in choosing appropriate rock slopes. Statistical methods of dealing with the many variables are being developed to enable the mine planner to determine mining costs in the most useful terms. Since these mines cannot afford generous factors of safety, they often support thorough monitoring of rock deformation and stress. Normally, artificial supports are not provided because the costs would be prohibitive, but rock bolts, retaining structures, drains, and other measures are sometimes required at the sites of power structures and at crushers or conveyor belts within the pit. *Spillway cuts* for dams also can attain impressive dimensions and demand rock mechanics attention (Fig. 1.3). Such cuts assume a value far greater than their cost since failure at an unfortunate time could allow overtopping of the dam; even so, the costs of major spillway cuts can rival the cost of even a large dam and thus such excavations can be considered engineering structures in their own right. Rock mechanics affects the decision on whether to locate spillways in open cut or in tunnels.

Underground excavations call upon the discipline of rock mechanics in many ways. In *mining*, the design of cutters and drills can be tailored to the rock conditions, which are determined by suitable laboratory tests. This also applied to *tunneling* with moles or tunneling machines. A major decision of mining is whether to attempt to maintain the openings while removing the ore, or instead to let the rock deform. The rock condition and state of stress is fundamentally important in reaching this decision correctly. In stable mining methods, the dimensions of pillars, rooms, and other rock components are based upon rock mechanics studies using numerical analysis or applicable theory, and calling into play thorough rock testing programs. In the case of unsta-

Figure 1.3 The flip bucket for the side-hill spillway for Chivor rock-fill dam, Colombia. Note the roadway and access tunnel in the lower left and the drainage tunnel under the flip bucket. (Owner, I.S.A.; Engineer, Ingetec, Ltda.)

ble mining methods, the layout of haulageways and "draw points" is based upon studies aiming to minimize dilution of ore with waste rock and to optimize efficiency.

Underground chambers are now being used for a variety of purposes other than transportation and mining. Some of these applications are demanding new kinds of data and special technology. Storage of liquefied natural gas in underground chambers requires determination of rock properties under conditions of extreme cold and analysis of heat transfer in the rock. Storage of oil and gas in mined chambers (Figure 1.4) requires a leakproof underground environment. Any large underground chamber, regardless of its special requirements, should

Figure 1.4 An underground chamber for storage of petroleum products in Norway. A storage facility consists of a number of such chambers. (Photo courtesy of Tor Brekke.)

be stable essentially without support and this depends upon the state of stress and the pattern and properties of discontinuities. Underground hydroelectric power plants, which offer advantages over surface power plants in mountainous terrain, feature very large machine halls (e.g., 25-m span) and numerous other openings in a complex three-dimensional arrangement (see Fig. 7.1). The orientation and layout of these openings depend almost entirely upon rock mechanics and geological considerations. Blasting, design of supports, and most other engineering aspects of such schemes depend markedly upon rock conditions; therefore rock mechanics is a basic tool. The *military* is interested in underground *openings* to create invulnerable facilities. Rock dynamics has figured prominently in design of such schemes, since the security of the openings must be maintained in the face of enormous ground shock pressures. The military has sponsored special prototype tests to failure that have advanced the knowledge of rock properties and behavior and of rock/structure interactions.

Rock mechanics is also important in the field of *energy development* (in addition to the hydroelectric works already mentioned). In petroleum engineering, design of drilling bits depends upon rock properties; bit wear is one of the major elements of cost. Rock mechanics studies are being directed toward solving the problems associated with deep drilling, to allow recovery from greater depths. In shales, salts, and certain other rocks, depth limitations are created by flowage of the rock and rapid closure of the hole. A laboratory has been built in Salt Lake City (Terra Tek Drilling Laboratory) to allow full-scale simulation of drilling at depths up to 20,000 feet and at temperatures up to 340°C. The petroleum industry pioneered the use of hydraulically induced fractures to increase reservoir yield. Hydraulic fracturing is now a standard reservoir operation. It is also being investigated as a mechanism for exchanging the earth's heat as a source of *geothermal energy* in dry, hot rocks. In the Los Alamos Scientific Laboratory scheme, under full-scale field investigation, a hydraulic fracture circulates cold water into hot rock; the heated water is returned to the surface through a second drill hole intersecting the top of the fracture. In the *nuclear energy* field, in addition to the problems of constructing the surface and/or underground facilities in rock and the elaborate precautions required by licensing agencies to insure that there are no active faults or other geological hazards on site, the industry is burdened with large quantities of highly toxic, long-lived *radioactive wastes*. The current plan is to isolate these wastes in stainless steel canisters by emplacement in specially mined cavities in deposits of rock salt and perhaps in granite, basalt, tuff or other rock types. Salt was selected because of its relatively high heat conductivity together with general water tightness since fractures tend to be absent or healed. The rock will assume temperatures of approximately 200°C after emplacement of the canisters.

New applications for rock mechanics are appearing with great rapidity. Exploration and development of extraterrestrial space, prediction of earthquakes, solution mining, compressed air storage in underground chambers, and other exotic fields are calling on further development of rock technology. Meanwhile, we are still not completely in command of the essential ingredients for rational design in some of the more mundane applications mentioned previously. This is because of the special nature of rock, which renders it different and perhaps more difficult to deal with than other engineering materials.

1.2 The Nature of Rocks

When attempting to formulate mechanical behavior of solids, it is common to assume they are ideally homogeneous, continuous, isotropic (nondirectional in properties), linear, and elastic. Rocks can be nonideal in a number of ways. First, they are seldom truly continuous, because pores or fissures are usual. Interconnected *pores*, approximately equidimensional cavities, are found between the grains of sedimentary rocks. Isolated vugs of other origins are found in volcanic rocks and soluble carbonate rocks. Since the capacity of rocks to store and transmit fluids is largely dependent upon the behavior of these voids, a special theory has been developed, primarily by workers in petroleum engineering, to deal with the deformations, stresses, and water pressures in porous rocks. *Microfissures* are small planar cracks common in hard rocks that have undergone internal deformation; they occur as intracrystalline and crystal boundary cracks. A fissured rock is like a test specimen that has been loaded into the cracking region (i.e., that has been damaged). The behavior of the network of fissures is as important or even more vital with regard to rock properties than the mineralogic composition itself. Collectively, fissures and pores do the following: they create nonlinear load/deformation response, especially at low stress levels; they reduce the tensile strength (especially fissures); they create stress dependency in materials properties; they produce variability and scatter in test results; and they introduce a scale effect into predictions of behavior.

A related nonideality of most rocks is the presence of *macrodiscontinuities*. Regular cracks and fractures are usual at shallow depths beneath the surface and some persist to depths of thousands of meters. Joints, bedding-plane partings, minor faults, and other recurrent planar fractures radically alter the behavior of rock in place from that predictable on the basis of testing intact samples, even though the latter may possess fissures. The mechanics of discontinuous rocks is especially relevant to engineers of surface structures, surface excavations, and shallow underground excavations. Indeed, it was the movement of a block bounded by faults and joints that undermined the Malpasset Arch Dam in 1959 (Figure 1.5).

The effect of a single fracture in a rock mass is to lower the tensile strength nearly to zero in the direction perpendicular to the fracture plane, and to restrict the shear strength in the direction parallel to the fracture plane. If the joints are not randomly distributed (and they almost never are) then the effect is to create pronounced anisotropy of strength, as well as of all other properties in the rock mass. For example, the strength of a foundation loaded obliquely to the bedding may be less than one-half of the strength when the load is applied perpendicular or parallel to the bedding. *Anisotropy* is common in many rocks even without discontinuous structure because of preferred orientations of mineral grains or directional stress history. Foliation and schistosity make schists, slates, and many other metamorphic rocks highly directional in their deformability, strength, and other properties. Bedding makes shales, thin-bedded sandstones and limestones, and other common sedimentary rocks highly anisotropic. Also, even rock specimens apparently free from bedding structures, such as thick-bedded sandstones and limestones, may prove to have directional

Figure 1.5 A view of the left abutment of Malpasset arch dam after its failure. The movement of a
wedge delimited by discontinuity surfaces, one of
which forms the newly exposed rock surface on
the abutment, brought on the rupture of the concrete arch.

properties because they were subjected to unequal principal stresses as they
were gradually transformed from sediment into rock. Finally, any fissured rock
that maintains unequal initial stresses will be anisotropic because its properties
are greatly influenced by the state of stress across the fissures; they are one
material when the fissures are closed, and another when the fissures are opened
or sheared.

We can discuss a "mechanics of rocks" in these chapters but such a
discussion must be broad in scope if it is to have general value because the term
"rock" includes a great variety of material types. Granite can behave in a

brittle, elastic manner, up to confining pressures of hundreds of megapascals[1] (MPa), while carbonate rocks become plastic at moderate pressures and flow like clay. Compaction shales and friable sandstones are weakened by immersion in water. Gypsum and rock salt are inclined to behave plastically at relatively low confining pressures and are highly soluble.

Despite all these problems with rock as an engineering material, it is possible to support engineering decisions with meaningful tests, calculations, and observations. This is the subject of our study.

Sources of Information in Rock Mechanics

BIBLIOGRAPHIES

KWIC Index of Rock Mechanics Literature published before 1969, in two volumes, E. Hoek (Ed.). Produced by Rock Mechanics Information Service, Imperial College, London. Published by AIME, 345 E. 47th Street, New York, NY 10017. A companion volume, Part 2, carrying the bibliography forward from 1969 to 1976 was published by Pergamon Press Ltd, Oxford (1979); J. P. Jenkins and E. T. Brown (Eds.).

Geomechanics Abstracts: see *International Journal of Rock Mechanics and Mining Science*. These are key-worded abstracts of articles published worldwide; issued and bound with the journal.

BOOKS

Attewell, P. B. and Farmer, I. W. (1976) *Principles of Engineering Geology*, Chapman & Hall, London.

Bieniawski, Z. T. (1984) *Rock Mechanics Design in Mining and Tunneling*, Balkema, Rotterdam.

Brady, B. H. G. and Brown, E. T. (1985) *Rock Mechanics for Underground Mining*, Allen & Unwin, London.

Brown, E. T. (Ed.) (1981) *Rock Characterization, Testing, and Monitoring: ISRM Suggested Methods*, Pergamon, Oxford.

Brown, E. T. (Ed.) (1987) *Analytical and Computational Methods in Engineering Rock Mechanics*, Allen & Unwin, London.

Budavari, S. (Ed.) (1983) *Rock Mechanics in Mining Practise*, South African Institute of Mining and Metallurgy, Johannesburg.

Coates, R. E. (1970) *Rock Mechanics Principles*, Mines Branch Monograph 874, revised, CANMET (Canadian Dept. of Energy, Mines and Resources), Ottawa.

Dowding, C. H. (1985) *Blast Vibration Monitoring and Control*, Prentice-Hall, Englewood Cliffs, NJ.

Farmer, I. W. (1983) *Engineering Behaviour of Rocks*, 2d ed., Chapman & Hall, London.

---

[1] One megapascal equals 145 psi.

Goodman, R. E. (1976) *Methods of Geological Engineering in Discontinuous Rocks*, West, St. Paul, MN.
Goodman, R. E. and Shi, G. H. (1985) *Block Theory and Its Application to Rock Engineering*, Prentice-Hall, Englewood Cliffs, NJ.
Hoek, E. and Bray, J. (1981) *Rock Slope Engineering*, 3d ed., Institute of Mining and Metallurgy, London.
Hoek, E. and Brown, E. T. (1980) *Underground Excavations in Rock*, Institute of Mining and Metallurgy, London.
Jaeger, C. (1972) *Rock Mechanics and Engineering*, Cambridge Univ. Press, London.
Jaeger, J. C. and Cook, N. G. W. (1979) *Fundamentals of Rock Mechanics*, 3d ed., Chapman & Hall, London.
Krynine, D. and Judd, W. (1959) *Principles of Engineering Geology and Geotechnics*, McGraw-Hill, New York.
Lama, R. D. and Vutukuri, V. S., with Saluja, S. S. (1974, 1978) *Handbook on Mechanical Properties of Rocks* (in four volumes), Trans Tech Publications, Rockport, MA. Vol. 1 (1974) by Vutukuri, Lama, and Saluja; Vols. 2–4 (1978) by Lama and Vutukuri.
Obert, L. and Duvall, W. (1967) *Rock Mechanics and the Design of Structures in Rocks*, Wiley, New York.
Priest, S. D. (1985) *Hemispherical Projection Methods in Rock Mechanics*, Allen & Unwin, London.
Roberts, A. (1976) *Geotechnology*, Pergamon, Oxford.
Turchaninov, I. A., Iofis, M. A., and Kasparyan, E. V. (1979) *Principles of Rock Mechanics*, Terraspace, Rockville, MD.
Zaruba, Q. and Mencl, V. (1976) *Engineering Geology*, Elsevier, New York.

JOURNALS

Canadian Geotechnical Journal, Canadian National Research Council, Toronto, Canada.
International Journal of Rock Mechanics and Mining Sciences & Geomechanics Abstracts, Pergamon Press, Ltd., Oxford.
Geotechnical Testing Journal, American Society for Testing Materials.
Journal of the Geotechnical Division, Proceedings of the American Society of Civil Engineering (ASCE), New York.
Rock Mechanics, Springer-Verlag, Vienna.
Underground Space, American Underground Association, Pergamon Press, Ltd., Oxford.

PROCEEDINGS

Canadian Rock Mechanics Symposia, Annual; various publishers. Sponsored by the Canadian Advisory Committee on Rock Mechanics.
Congresses of the International Society of Rock Mechanics (ISRM), First—Lisbon (1966); Second—Belgrade (1970); Third—Denver (1974); Fourth—Montreux (1979); Fifth—Melbourne (1983); Sixth—Montreal (1987).

Specialty Conferences and Symposia sponsored by ISRM, Institute of Civil Engineers (London); British Geotechnical Society, AIME, International Congress on Large Dams (ICOLD), and other organizations as cited in the references after each chapter.

Symposia on Rock Mechanics, Annual U. S. Conference; various publishers. Sponsored by the U. S. National Committee on Rock Mechanics.

STANDARDS AND SUGGESTED METHODS

Rock mechanics has not yet advanced to the stage where testing and observational techniques can be rigorously standardized. However, the International Society for Rock Mechanics (ISRM) and the American Society for Testing and Materials (ASTM) have published "designations" and "suggested methods" for laboratory and field testing and for description of rock materials. Several of these are listed with the references at the ends of the appropriate chapters. See Brown (1981) under "BOOKS" above. For up-to-date information about standardization in rock mechanics, communicate directly with ISRM, Commission on Standardization, Laboratorio Nacional de Engenharia Civil, Avenida do Brasil, P-1799, Lisbon, Portugal; and with ASTM, Committee D-18 on Soil and Rock for Engineering Purposes, 1916 Race Street, Philadelphia, PA 19103.

Chapter 2

Classification and Index Properties of Rocks

2.1 Geological Classification of Rocks

Although they were not developed to satisfy the needs of civil engineers, the names geologists are able to attach to rock specimens on the basis of limited observations with a hand lens, or with the eye alone, do often reveal something about rock properties. If you are unfamiliar with the common rock names and how to assign them to an unknown rock, a review of geology is highly recommended. A good way to begin is to study Appendix 3, which explains simplified schemes for classifying and naming the principal rocks and minerals. Appendix 3 also lists the periods of the earth's history, the names of which indicate the age of a rock. A rock's age often, but not infallibly, correlates with its hardness, strength, durability, and other properties.

From a genetic point of view, rocks are usually divided into the three groups: *igneous, metamorphic,* and *sedimentary.* Yet these names are the *results*, not the starting point of classification. Since we are interested in behavioral rather than genetic attributes of rocks, it makes more sense to divide the rocks into the following classes and subclasses:

I. Crystalline Texture

	Examples
A. Soluble carbonates and salts	Limestone, dolomite, marble, rock salt, trona, gypsum
B. Mica or other planar minerals in continuous bands	Mica schist, chlorite schist, graphite schist
C. Banded silicate minerals without continuous mica sheets	Gneiss
D. Randomly oriented and distributed silicate minerals of uniform grain size	Granite, diorite, gabbro, syenite
E. Randomly oriented and distributed silicate minerals in a background of very fine grain and with vugs	Basalt, rhyolite, other volcanic rocks
F. Highly sheared rocks	Serpentinite, mylonite

II. Clastic Texture

	Examples
A. Stably cemented	Silica-cemented sandstone and limonite sandstones
B. With slightly soluble cement	Calcite-cemented sandstone and conglomerate
C. With highly soluble cement	Gypsum-cemented sandstones and conglomerates
D. Incompletely or weakly cemented	Friable sandstones, tuff
E. Uncemented	Clay-bound sandstones

III. Very Fine-Grained Rocks

	Examples
A. Isotropic, hard rocks	Hornfels, some basalts
B. Anisotropic on a macro scale but microscopically isotropic hard rocks	Cemented shales, flagstones
C. Microscopically anisotropic hard rocks	Slate, phyllite
D. Soft, soil-like rocks	Compaction shale, chalk, marl

IV. Organic Rocks

Examples

A. Soft coal   Lignite and bituminous coal
B. Hard coal
C. "Oil shale"
D. Bituminous shale
E. Tar sand

Crystalline rocks are constructed of tightly interlocked crystals of silicate minerals or carbonate, sulfate, or other salts (Figure 2.1*a*). Unweathered crystalline silicates like fresh granite are usually elastic and strong with brittle failure characteristics at pressures throughout the usual range for civil engineering works. However, if the crystals are separated by grain boundary cracks (fissures), such rocks may deform nonlinearly and "plastically" (irreversibly). Carbonates and crystalline salt rocks may also be strong and brittle but will become plastic at modest confining pressures due to intracrystalline gliding. Also, they are soluble in water. Mica and other sheet minerals like serpentine, talc, chlorite, and graphite reduce the strength of rocks due to easy sliding along the cleavage surfaces. Mica schists and related rocks are highly anisotropic rocks with low strength in directions along the schistosity (Figure 2.1*b*) except when the schistosity has been deformed through refolding. Volcanic rocks like basalts may present numerous small holes (vugs); otherwise, they behave similarly to granitic rocks (Figure 2.2*c*). Serpentinites, because they tend to be pervasively sheared on hidden surfaces within almost any hand specimen, are highly variable and often poor in their engineering properties.

The clastic rocks, composed of pieces of various rock types and assorted mineral grains, owe their properties chiefly to the cement or binder that holds the fragments together. Some are stably and tightly cemented and behave in a brittle, elastic manner. Others are reduced to sediment upon more soaking in water. In the clastic rock group, the geological names are not very useful for rock mechanics because the name doesn't indicate the nature of the cement. However, a full geological description can often suggest the properties of the cement; for example, a *friable* sandstone, where grains can be liberated by rubbing, is obviously incompletely or weakly cemented at best.

Shales are a group of rocks primarily composed of silt and clay that vary widely in durability, strength, deformability, and toughness. Cemented shales can be hard and strong. Many so-called "compaction shales" and "mudstones," however, are just compacted clay soils without durable binder, and have the attributes of hard soils rather than of rocks: they may exhibit volume change upon wetting or drying together with extreme variation in properties with variations in moisture content. Unlike soils, which quickly lose strength when kept moist at their natural water content, compaction shales remain

Figure 2.1 Photomicrographs of thin sections of rocks, viewed in polarized, transmitted light (courtesy of Professor H. R. Wenk). (*a*) Tightly interlocked fabric of a crystalline rock—diabase (×27).

Figure 2.1 Photomicrographs of thin sections or rocks, viewed in polarized, transmitted light (courtesy of Professor H. R. Wenk). (*b*) Highly anisotrophic fabric of a quartz mylonite (×20).

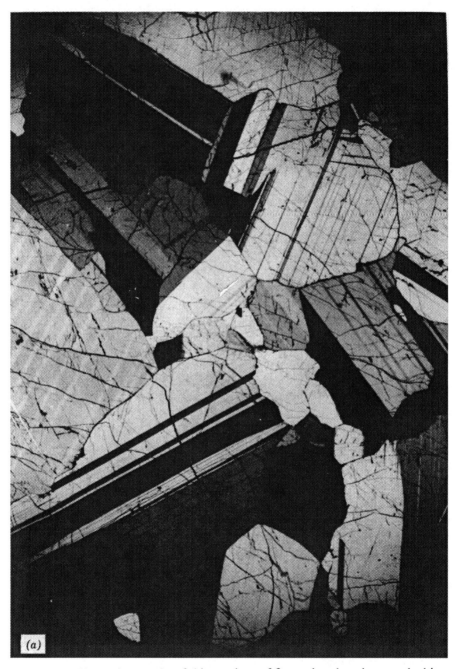

Figure 2.2 Photomicrographs of thin sections of fissured rocks, photographed in transmitted, polarized light (courtesy of H. R. Wenk). (*a*) Anorthosite with many intracrystalline and some intercrystalline fractures (×6.5).

Figure 2.2 Photomicrographs of thin sections of fissured rocks, photographed in transmitted, polarized light (courtesy of H. R. Wenk). (*b*) Gabbro with regular fissures oriented across the cleavage (×7).

Figure 2.2 Photomicrographs of thin sections of fissured rocks, photographed in transmitted, polarized light (courtesy of H. R. Wenk). (*c*) Volcanic rock (trachyte) with fissured sanidine phenocrysts (×30).

intact for some time. However, when dried and then immersed in water, they gradually decrease in density and strength over days, weeks, or longer. Chalk is a highly porous clastic carbonate rock that is elastic and brittle at low pressures, but plastic at moderate pressures.

Organic rocks include viscous, plastic, and elastic types. Hard coal and oil shale are strong, elastic rocks; however, the former may be fissured. Soft coal is highly fissued and may contain hydrocarbon gases under pressure in the pores. Tar sand may behave like a viscous liquid at high pressure or temperature; it also may contain gas under pressure.

We see that the rock family is large and "nonexclusive." Some of the simple laboratory tests and measurements enumerated below will help to decide what kind of material you are dealing with in any specific case.

2.2 Index Properties of Rock Systems

Because of the vast range in properties of rocks, which reflects varieties of structures, fabrics, and components, we rely on a number of basic measurements to describe rocks quantitatively. Certain properties that are relatively easy to measure are valuable in this regard and may be designated *index properties* for rock specimens. *Porosity* identifies the relative proportion of solids and voids; *density* adds information about the mineralogic or grain constituents. The *sonic velocity* together with a petrographic description evaluate the degree of fissuring. *Permeability* evaluates the relative interconnection of the pores; *durability* indicates the tendency for eventual breakdown of components or structures, with degradation of rock quality. Finally, *strength* determines the present competency of the rock fabric to bind the components together. These attributes need to be evaluated for engineering classification of rock, and together they permit one to draw useful correlations with experience for practical applications. However, the behavior of rock specimens under changing stress, temperature, fluid pressure, and time includes many other facets that are not represented by the above list of index properties. Therefore, characterization of a series of indexes in the laboratory is not a substitute for careful and detailed testing in other areas of special concern.

A list of index properties related to laboratory specimens of rock can help classify it for applications related primarily to the behavior of the *rock* itself as opposed to the *rock mass* with the interactions among its system of discontinuities. A little reflection on the spectrum of applications of rock mechanics will yield some that do involve mainly rock specimen characteristics, for example, drillability, cuttability, aggregate selection, and rip-rap evaluation. Most applications involving excavation at the surface or underground, on the other hand, test the system of discontinuities as much as or more than the nature of the rock itself. In these instances, the classification of the rock mass for engineering purposes reflects not only laboratory tests but structural and environmental characteristics of the rock mass in the field. We consider engineering classification of rock masses later in this chapter.

2.3 Porosity

The *porosity* of a rock, indicated by the dimensionless quantity n, is a fraction expressing the proportion of void space to total space in the rock.

$$n = \frac{v_p}{v_t} \qquad (2.1)$$

where v_p is the volume of pores in total volume v_t. In sedimentary rocks, formed by the accumulation of grains, rock fragments, or shells, the porosity varies from close to 0 to as much as 90% ($n = 0.90$) with 15% as a typical value for an average sandstone. In these rocks, porosity generally decreases with age, and with depth below the surface, other things being equal. Table 2.1 illustrates these tendencies for a number of sedimentary rocks: a typical Cambrian sandstone had a porosity of 11% while a Cretaceous sandstone contained 34% pores. The effect of depth is most striking in the rocks derived from compaction[1] of clay as shown in Table 2.1. A Pennsylvanian age shale from Oklahoma encountered at depth of 1000, 3000, and 5000 feet had porosities of 16%, 7%, and 4%, respectively. Chalk is among the most porous of all rocks with porosities in some instances of more than 50%. These rocks are formed of the hollow skeletons of microscopic animals—coccoliths. Some volcanic rocks (e.g., pumice) can also present very high porosity due to the preservation of the sites of volcanic gas bubbles; in volcanic rocks, the system of pores is not always well connected.

In crystalline limestones and evaporites, and most igneous and metamorphic rocks, a large proportion of the pore space belongs to planar cracks termed *fissures* (Figure 2.2). A relatively small porosity due to fissures affects the properties of the rock to the same degree as a much larger percentage of subspherical pore space and, as noted in the previous chapter, creates stress dependency in a number of physical properties. In the igneous rocks, porosity is usually less than 1 or 2% unless weathering has taken hold. As weathering

---

[1] *Compaction* is a term used by geologists and petroleum engineers to describe processes by which a sediment is densified. Soils engineers reserve this term for processes of densification involving the expulsion of air from the voids. *Consolidation* refers to the expulsion of water from the voids of a clay, in soil mechanics usage, whereas geologists and petroleum engineers use *consolidation* for processes of lithification.

Table 2.1 *Porosities of Some Typical Rocks Showing Effects of Age and Depth*[a]

Rock	Age	Depth	Porosity (%)
Mount Simon sandstone	Cambrian	13,000 ft	0.7
Nugget sandstone (Utah)	Jurassic		1.9
Potsdam sandstone	Cambrian	Surface	11.0
Pottsville sandstone	Pennsylvanian		2.9
Berea sandstone	Mississippian	0–2000 ft	14.0
Keuper sandstone (England)	Triassic	Surface	22.0
Navajo sandstone	Jurassic	Surface	15.5
Sandstone, Montana	Cretaceous	Surface	34.0
Beekmantown dolomite	Ordovician	10,500 ft	0.4
Black River limestone	Ordovician	Surface	0.46
Niagara dolomite	Silurian	Surface	2.9
Limestone, Great Britain	Carboniferous	Surface	5.7
Chalk, Great Britain	Cretaceous	Surface	28.8
Solenhofen limestone		Surface	4.8
Salem limestone	Mississippian	Surface	13.2
Bedford limestone	Mississippian	Surface	12.0
Bermuda limestone	Recent	Surface	43.0
Shale	Pre-Cambrian	Surface	1.6
Shale, Oklahoma	Pennsylvanian	1000 ft	17.0
Shale, Oklahoma	Pennsylvanian	3000 ft	7.0
Shale, Oklahoma	Pennsylvanian	5000 ft	4.0
Shale	Cretaceous	600 ft	33.5
Shale	Cretaceous	2500 ft	25.4
Shale	Cretaceous	3500 ft	21.1
Shale	Cretaceous	6100 ft	7.6
Mudstone, Japan	Upper Tertiary	Near surface	22–32
Granite, fresh		Surface	0 to 1
Granite, weathered			1–5
Decomposed granite (Saprolyte)			20.0
Marble			0.3
Marble			1.1
Bedded tuff			40.0
Welded tuff			14.0
Cedar City tonalite			7.0
Frederick diabase			0.1
San Marcos gabbro			0.2

[a] Data selected from Clark (1966) and Brace and Riley (1972).

progresses, the porosity tends to increase to 20% or more. As a result, measurement of porosity can serve as an accurate index to rock quality in such rocks. In several projects in granitic rocks the National Civil Engineering Laboratory of Portugal was able to classify the rock for the purposes of engineering design mainly on the basis of a quick porosity measurement, obtained from the water content of the rock after immersion for 24 hours at a standard temperature and pressure (Hamrol, 1961). Among unweathered rocks, there is also a general correlation between porosity and mechanical properties such as unconfined compressive strength and modulus of elasticity; but such relationships are usually marked by enormous scatter. In the case of weak sandstones (having saturated compressive strength less than 20 MPa) Dobereiner and de Freitas (1986) have demonstrated good correlations of density, modulus of elasticity, and compressive strength with the saturated moisture content. The moisture content of a saturated specimen is linked with its porosity by Equation 2.5. Saturation can be approached by soaking a specimen in water while it is subjected to a laboratory vacuum.

Porosity can be measured in rock specimens by a variety of techniques. Since it is the pore space that governs the quantity of oil contained in a saturated petroleum reservoir, accurate methods for porosity determination in sandstones have been developed by the oil industry. However, these methods are not always suitable for measurements in hard rocks with porosities of less than several percent. Porosities can be determined from the following calculations.

1. Measured density.
2. Measured water content after saturation in water.
3. Mercury content after saturation with mercury using a pressure injector.
4. Measured solid volume and pore air volume using Boyle's law.

These are considered further below.

2.4 Density

The density or "unit weight" of a rock, γ, is its specific weight (FL^{-3}),[2] for example, pounds per cubic foot or kilonewtons per cubic meter. The *specific gravity* of a solid, G, is the ratio between its density and the unit weight of water γ_w; the latter is approximately equal to 1 g-force/cm³ (9.8 kN/m³ or approximately 0.01 MN/m³).[3] Rock with a specific gravity of 2.6 has a density

---

[2] The terms in parenthesis indicate the dimensions of the preceding quantity. F, L, T indicate force, length, and time, respectively.

[3] At 20°C, the unit weight of water is 0.998 g/cm³ × 980 cm/s² = 978 dynes/cm³ or = 0.998 g-force/cm³.

of approximately 26 kN/m³. In the English system, the density of water is 62.4 pounds per cubic foot. (Mass density ρ equals γ/g.)

It was stated previously that the porosity of a rock can be calculated from knowledge of its weight density. This assumes that the specific gravity of the grains or crystals is known; grain specific gravity can be determined by grinding the rock and adapting methods used in soils laboratories. If the percentages of different minerals can be estimated under a binocular microscope, or from a thin section, the specific gravity of the solid part of a rock can then be calculated as the weighted average of the specific gravities of the component grains and crystals:

$$G = \sum_{i=1}^{n} G_i V_i \tag{2.2}$$

where G_i is the specific gravity of component i, and V_i is its volume percentage in the solid part of the rock. The specific gravities of a number of common rock-forming minerals are listed in Table 2.2. The relation between porosity and dry density γ_{dry} is

$$\gamma_{dry} = G\gamma_w(1 - n) \tag{2.3}$$

Table 2.2 Specific Gravities of Common Minerals[a]

Mineral	G
Halite	2.1–2.6
Gypsum	2.3–2.4
Serpentine	2.3–2.6
Orthoclase	2.5–2.6
Chalcedony	2.6–2.64
Quartz	2.65
Plagioclase	2.6–2.8
Chlorite and illite	2.6–3.0
Calcite	2.7
Muscovite	2.7–3.0
Biotite	2.8–3.1
Dolomite	2.8–3.1
Anhydrite	2.9–3.0
Pyroxene	3.2–3.6
Olivine	3.2–3.6
Barite	4.3–4.6
Magnetite	4.4–5.2
Pyrite	4.9–5.2
Galena	7.4–7.6

[a] A. N. Winchell (1942).

The dry density is related to the wet density by the relationship

$$\gamma_{dry} = \frac{\gamma_{wet}}{1 + w} \tag{2.4}$$

where w is the water content of the rock (dry weight basis).

Water content and porosity are related by

$$n = \frac{w \cdot G}{1 + w \cdot G} \tag{2.5}$$

If the pores of the rock are filled with mercury, and the mercury content is determined to be w_{Hg} (as a proportion of the dry weight of the rock before mercury injection), the porosity can be calculated more accurately as follows:

$$n = \frac{w_{Hg} \cdot G/G_{Hg}}{1 + (w_{Hg} \cdot G/G_{Hg})} \tag{2.6}$$

The specific gravity of mercury (G_{Hg}) equals 13.546.

The densities of some common rocks are given in Table 2.3. These figures are only sample values, of course, since special factors can cause wide variations in individual formations.

Rocks exhibit a far greater range in density values than do soils. Knowledge of rock density can be important to engineering and mining practice. For example, the density of a rock governs the stresses it will experience when acting as a beam spanning an underground opening; unusually high density in a roof rock implies a shortened limiting safe span. A concrete aggregate with higher than average density can mean a smaller volume of concrete required for a gravity retaining wall or dam. Lighter than average aggregate can mean lower stresses in a concrete roof structure. In oil shale deposits, the density indicates the value of the mineral commodity because the oil yield correlates directly with the unit weight; this is true because oil shale is a mixture of a relatively light constituent (kerogen) and a relatively heavy constituent (dolomite). In coal deposits, the density correlates with the ash content and with the previous depth of cover, accordingly with the strength and elasticity of the rock. It is easy to measure the density of a rock; simply saw off the ends of a dried drill core, calculate its volume from the dimensions, and weight it. In view of the possible significance of variations from the norm, density should therefore be measured routinely in rock investigations.

2.5 Hydraulic Permeability and Conductivity

Measurement of the permeability of a rock sample may have direct bearing on a practical problem, for example, pumping water, oil, or gas into or out of a

Table 2.3 *Dry Densities of Some Typical Rocks*[a]

Rock	Dry (g/cm³)	Dry (kN/m³)	Dry (lb/ft³)
Nepheline syenite	2.7	26.5	169
Syenite	2.6	25.5	162
Granite	2.65	26.0	165
Diorite	2.85	27.9	178
Gabbro	3.0	29.4	187
Gypsum	2.3	22.5	144
Rock salt	2.1	20.6	131
Coal	0.7–2.0		
(density varies with the ash content)			
Oil shale	1.6–2.7		
(density varies with the kerogen content, and therefore with the oil yield in gallons per ton)			
30 gal/ton rock	2.13	21.0	133
Dense limestone	2.7	20.9	168
Marble	2.75	27.0	172
Shale, Oklahoma[b]			
1000 ft depth	2.25	22.1	140
3000 ft depth	2.52	24.7	157
5000 ft depth	2.62	25.7	163
Quartz, mica schist	2.82	27.6	176
Amphibolite	2.99	29.3	187
Rhyolite	2.37	23.2	148
Basalt	2.77	27.1	173

[a] Data from Clark (1966), Davis and De Weist (1966), and other sources.
[b] This is the Pennsylvanian age shale listed in Table 2.1.

porous formation, disposing of brine wastes in porous formations, storing fluids in mined caverns for energy conversion, assessing the water tightness of a reservoir, dewatering a deep chamber, or predicting water inflows into a tunnel. In many instances the system of discontinuities will radically modify the permeability values of the rock in the field as compared to that in the lab, so that some sort of in situ pumping test will be required for an acceptable forecast of formation permeabilities. Our motivation for selecting permeability as an index property of rock is that it conveys information about the degree of interconnection between the pores or fissures—a basic part of the rock framework. Furthermore, the variation of permeability with change in normal stress, especially as the sense of the stress is varied from compression to tension, evaluates the degree of fissuring of the rock, since flat cracks are greatly affected by normal stress whereas spherical pores are not. Also, the degree to which the permeability changes by changing the permeant from air to water expresses interaction between the water and the minerals or binder of the rock and can detect subtle but fundamental flaws in the integrity of the rock; this promising aspect of permeability as an index has not been fully researched.

Most rocks obey Darcy's law. For many applications in civil engineering practice, which may involve water at about 20°C, it is common to write Darcy's law in the form

$$q_x = k \frac{dh}{dx} A \qquad (2.7)$$

where q_x is the flow rate (L^3T^{-1}) in the x direction
h is the hydraulic head with dimension L
A is the cross-sectional area normal to x (dimension L^2)

The coefficient k is termed *the hydraulic conductivity*; it has dimensions of velocity (e.g., centimeters per second or feet per minute). When temperature will vary considerably from 20°C or when other fluids are to be considered, a more useful form of Darcy's law is

$$q_x = \frac{K}{\mu} \frac{dp}{dx} A \qquad (2.8)$$

in which p is the fluid pressure (equal to $\gamma_w h$) with dimensions of FL^{-2} and μ is the viscosity of the permeant with dimensions $FL^{-2}T$. For water at 20°C, μ = 2.098×10^{-5} lb s/ft^2 = 1.005×10^{-3} N s/m^2 and γ = 62.4 lb/ft^3 = 9.80 kN/m^3.

When Darcy's law is written this way, the coefficient K is independent of the properties of the fluid. Its dimensions are those of area (e.g., square centimeters). K is termed *the hydraulic permeability*.

A common permeability unit is the darcy: 1 darcy equals 9.86×10^{-9} cm^2. Table 2.4 gives typical values of conductivities calculated for the properties of water at 20°C; 1 darcy corresponds approximately to a conductivity value of 10^{-3} cm/s.

Permeability can be determined in the laboratory by measuring the time for a calibrated volume of fluid to pass through the specimen when a constant air pressure acts over the surface of the fluid. An alternative method is to generate radial flow in a hollow cylindrical specimen, prepared by drilling a coaxial central hole in a drill core. When the flow is from the outer circumference toward the center, a compressive body force is generated, whereas when the flow is from the central hole toward the outside, a tensile body force is set up. Consequently, rocks that owe their permeability partly to the presence of a network of fissures demonstrate a profound difference in permeability values according to the direction of flow. A radial permeability test was devised by Bernaix (1969) in testing the foundation rock of the Malpasset Dam after the failure. The permeability of the mica schist from that site varied over as much as 50,000 times as the conditions were changed from radially outward flow with

Table 2.4 *Conductivities of Typical Rocks*[a]

	k (cm/s) for Rock with Water (20°C) as Permeant	
Rock	Lab	Field
Sandstone	3×10^{-3} to 8×10^{-8}	1×10^{-3} to 3×10^{-8}
Navajo sandstone	2×10^{-3}	
Berea sandstone	4×10^{-5}	
Greywacke	3.2×10^{-8}	
Shale	10^{-9} to 5×10^{-13}	10^{-8} to 10^{-11}
Pierre shale	5×10^{-12}	2×10^{-9} to 5×10^{-11}
Limestone, dolomite	10^{-5} to 10^{-13}	10^{-3} to 10^{-7}
Salem limestone	2×10^{-6}	
Basalt	10^{-12}	10^{-2} to 10^{-7}
Granite	10^{-7} to 10^{-11}	10^{-4} to 10^{-9}
Schist	10^{-8}	2×10^{-7}
Fissured schist	1×10^{-4} to 3×10^{-4}	

[a] Data from Brace (1978), Davis and De Wiest (1966), and Serafim (1968).

ΔP of 1 bar, to radially inward flow with ΔP of 50 bars. The hydraulic conductivity (velocity units) from a radial flow test can be approximated by

$$k = \frac{q \ln(R_2/R_1)}{2\pi L \Delta h} \tag{2.9}$$

where q is the volume rate of flow

L is the length of the specimen

R_2 and R_1 are the outer and inner radii of the specimen

Δh is the head difference across the flow region corresponding to ΔP

An advantage of the radial permeability test, in addition to its capability to distinguish flow in fissures from flow in pores, is the fact that very large flow gradients can be generated, allowing permeability measurement in the millidarcy region. For rocks considerably less permeable than that, for example, granites with permeability in the region 10^{-9} darcy and below, Brace et al. (1968) devised a transient flow test.

Dense rocks like granite, basalt, schist, and crystalline limestone usually exhibit very small permeability as laboratory specimens, yet field tests in such rocks may show significant permeability as observed in Table 2.4. The reason for this discrepancy is usually attributed to regular sets of open joints and fractures throughout the rock mass. Snow (1965) showed that it is useful to idealize the rock mass as a system of parallel smooth plates, all flow running between the plates. When there are three mutually perpendicular sets of fractures with parallel walls, all with identical aperture and spacing and ideally smooth, the conductivity of the rock mass is theoretically expressed by

$$k = \frac{\gamma_w}{6\mu} \left(\frac{e^3}{S}\right) \tag{2.10}$$

where S is the *spacing* between fractures and e is the fracture *aperture* (interwall separation). It is seldom feasible to calculate the rock permeability from a description of the fractures, although Rocha and Franciss (1977) have shown how this can be done by using oriented, continuous core samples and correcting the data with results from a few pumping tests. Equation 2.10 is useful, however, for calculating the hypothetical fracture aperture e, that gives the same permeability value as measured in the field (corresponding to an assigned fracture spacing S). The aperture and spacing of the fractures then provide quantitative indexes of rock mass quality.

2.6 Strength

The value of having an index to rock strength is self-evident. The problem is that strength determinations on rock usually require careful test setup and specimen preparation, and the results are highly sensitive to the method and style of loading. An index is useful only if the properties are reproducible from one laboratory to another and can be measured inexpensively. Such a strength index is now available using the point load test, described by Broch and Franklin (1972). In this test, a rock is loaded between hardened steel cones, causing failure by the development of tensile cracks parallel to the axis of loading. The test is an outgrowth of experiments with compression of irregular pieces of rock in which it was found that the shape and size effects were relatively small and could be accounted for, and in which the failure was usually by induced tension. In the Broch and Franklin apparatus, which is commercially available, the *point load strength* is

$$I_s = \frac{P}{D^2} \tag{2.11}$$

where P is the load at rupture, and D is the distance between the point loads. Tests are done on pieces of drill core at least 1.4 times as long as the diameter. In practice there is a strength/size effect so a correction must be made to reduce results to a common size. Point load strength is found to fall by a factor of 2 to 3 as one proceeds from cores with diameter of 10 mm to diameters of 70 mm; therefore, size standardization is required. The *point load index* is reported as the point load strength of a 50-mm core. (Size correction charts are

Table 2.5 *Typical Point Load Index Values*[a]

Material	Point Load Strength Index (MPa)
Tertiary sandstone and claystone	0.05–1
Coal	0.2–2
Limestone	0.25–8
Mudstone, shale	0.2–8
Volcanic flow rocks	3.0–15
Dolomite	6.0–11

[a] Data from Broch and Franklin (1972) and other sources.

given by Broch and Franklin.) A frequently cited correlation between point load index and unconfined compression strength is

$$q_u = 24 I_{s(50)} \tag{2.12}$$

where q_u is the unconfined compressive strength of cylinders with a length to diameter ratio of 2 to 1, and $I_{s(50)}$ is the point load strength corrected to a diameter of 50 mm. However, as shown in Table 3.1, this relationship can be severely inaccurate for weak rocks and it should be checked by special calibration studies wherever such a correlation is important in practice.

The point load strength test is quick and simple, and it can be done in the field at the site of drilling. The cores are broken but not destroyed, since the fractures produced tend to be clean, single breaks that can be distinguished from preexisting fractures sampled by the drilling operation. Point load test results can be shown on the drill log, along with other geotechnical information, and repetition of tests after the core has dried out can establish the effect of natural water conditions on strength. Values of the point load index are given for a number of typical rocks in Table 2.5.

2.7 Slaking and Durability

Durability of rocks is fundamentally important for all applications. Changes in the properties of rocks are produced by exfoliation, hydration, decrepitation (slaking), solution, oxidation, abrasion, and other processes. In some shales and some volcanic rocks, radical deterioration in rock quality occurs rapidly after a new surface is uncovered. Fortunately, such changes usually act imperceptibly through the body of the rock and only the immediate surface is degraded in tens of years. At any rate, some index to the degree of alterability of rock is required. Since the paths to rock destruction devised by nature are many and varied, no test can reproduce expectable service conditions for more than a few special situations. Thus an index to alteration is useful mainly in offering a relative ranking of rock durability.

One good index test is the *slake durability* test proposed by Franklin and Chandra (1972). The apparatus consists of a drum 140 mm in diameter and 100 mm long with sieve mesh forming the cylindrical walls (2 mm opening); about 500 g of rock is broken into 10 lumps and loaded inside the drum, which is turned at 20 revolutions per minute in a water bath. After 10 min of this slow rotation, the percentage of rock retained inside the drum, on a dry weight basis, is reported as the *slake durability index* (I_d). Gamble (1971) proposed using a second 10-min cycle after drying. Values of the slake durability index for representative shales and claystones tested by Gamble varied over the whole range from 0 to 100%. There was no discernible connection between durability and geological age but durability increased linearly with density and inversely with natural water content. Based upon his results, Gamble proposed a classification of slake durability (Table 2.6).

Morgenstern and Eigenbrod (1974) expressed the durability of shales and claystones in terms of the rate and amount of strength reduction resulting from soaking. They showed that noncemented claystone or shale immersed in water tends to absorb water and soften until it reaches its *liquid limit*. The latter can be determined by a standard procedure described in ASTM designation D423-54T after disaggregating the rock by shaving it with a knife and mixing the shavings with water in a food blender. Materials with high liquid limits are more severely disrupted by slaking than those with low liquid limits. Classes of amounts of slaking were therefore defined in terms of the value of the liquid limit as presented in Table 2.7. The *rate* at which slaking occurs is independent of the liquid limit but can be indexed by the rate of water content change following soaking. The rate of slaking was classified in terms of the *change in liquidity index* (ΔI_L) following immersion in water for 2 h; ΔI_L is defined as

$$\Delta I_L = \frac{\Delta w}{w_L - w_P} \tag{2.13}$$

Table 2.6 *Gamble's Slake Durability Classification*

Group Name	% Retained after One 10-min Cycle (Dry Weight Basis)	% Retained after Two 10-min Cycles (Dry Weight Basis)
Very high durability	>99	>98
High durability	98–99	95–98
Medium high durability	95–98	85–95
Medium durability	85–95	60–85
Low durability	60–85	30–60
Very low durability	<60	<30

Table 2.7 Description of Rate and Amount of Slaking[a]

Amount of Slaking	Liquid Limit (%)
Very low	<20
Low	20–50
Medium	50–90
High	90–140
Very high	>140

Rate of Slaking	Change in Liquidity Index after Soaking 2 h
Slow	<0.75
Fast	0.75–1.25
Very fast	>1.25

[a] After Morgenstern and Eigenbrod (1974).

where Δw is the change in water content of the rock or soil after soaking for 2 h
on filter paper in a funnel
w_P is the water content at the plastic limit
w_L is the water content at the liquid limit

All the water contents are expressed as a percentage of the dry weight. These indexes and procedures for determining them are described in most textbooks on soil mechanics (e.g., Sowers and Sowers, cited in Chapter 9).

2.8 Sonic Velocity as an Index to Degree of Fissuring

Measurement of the velocity of sound waves in a core specimen is relatively simple and apparatus is available for this purpose. The most popular method pulses one end of the rock with a piezoelectric crystal and receives the vibrations with a second crystal at the other end. The travel time is determined by measuring the phase difference with an oscilloscope equipped with a variable delay line. It is also possible to resonate the rock with a vibrator and then calculate its sonic velocity from the resonant frequency, known dimensions, and density. Both longitudinal and transverse shear wave velocities can be determined. However, the index test described here requires the determination of only the longitudinal velocity V_l, which proves the easier to measure. ASTM Designation D2845-69 (1976) describes laboratory determination of pulse velocities and ultrasonic elastic constants of rock.

   Theoretically, the velocity with which stress waves are transmitted through rock depends exclusively upon their elastic properties and their density (as explored in Chapter 6). In practice, a network of fissures in the specimen superimposes an overriding effect. This being the case, the sonic velocity can serve to index the degree of fissuring within rock specimens.

Fourmaintraux (1976) proposed the following procedure. First calculate the longitudinal wave velocity (V_l^*) that the speciment would have if it lacked pores or fissures. If the mineral composition is known, V_l^* can be calculated from

$$\frac{1}{V_l^*} = \sum_i \frac{C_i}{V_{l,i}} \tag{2.14}$$

where $V_{l,i}$ is the longitudinal wave velocity in mineral constituent i, which has volume proportion C_i in the rock. Average velocities of longitudinal waves in rock-forming minerals are given in Table 2.8. Table 2.9 lists typical values of V_l^* for a few rock types.

Now measure the actual velocity of longitudinal waves in the rock specimen and form the ratio V_l/V_l^*. As a quality index define

$$IQ\% = \frac{V_l}{V_l^*} \times 100\% \tag{2.15}$$

Experiments by Fourmaintraux established that IQ is affected by pores (spherical holes) according to

$$IQ\% = 100 - 1.6n_p\% \tag{2.16}$$

where $n_p\%$ is the porosity of nonfissured rock expressed as a percentage. However, if there is even a small fraction of flat cracks (fissures), Equation 2.16 breaks down.

Table 2.8 *Longitudinal Velocities of Minerals*

Mineral	V_l (m/s)
Quartz	6050
Olivine	8400
Augite	7200
Amphibole	7200
Muscovite	5800
Orthoclase	5800
Plagioclase	6250
Calcite	6600
Dolomite	7500
Magnetite	7400
Gypsum	5200
Epidote	7450
Pyrite	8000

From Fourmaintraux (1976).

Table 2.9 *Typical Values of V_l^* for Rocks*

Rock	V_l^* (m/s)
Gabbro	7000
Basalt	6500–7000
Limestone	6000–6500
Dolomite	6500–7000
Sandstone and quartzite	6000
Granitic rocks	5500–6000

ª From Fourmaintraux (1976).

For example, a sandstone with n_p equals 10% had IQ equal to 84%. After heating the rock to a high temperature that produced an additional increment of flat cracks amounting to 2% pore space ($n_p = 10\%$, $n = 12\%$), IQ fell to 52%. (Heating opens grain boundary cracks in minerals with different coefficients of thermal expansion in different directions, in this case quartz.)

Because of this extreme sensitivity of IQ to fissuring and based upon laboratory measurements and microscopic observations of fissures, Fourmaintraux proposed plotting IQ versus porosity (Figure 2.3) as a basis for describing the degree of fissuring of a rock specimen. Entering the figure with known porosity

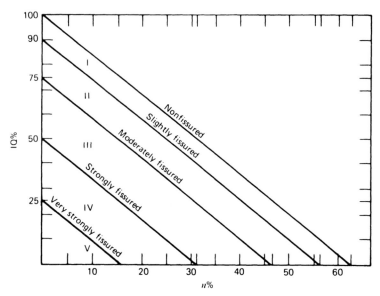

Figure 2.3 Classification scheme for fissuring in rock specimens. (After Fourmaintraux 1976.)

and calculated IQ defines a point in one of the five fields: (I) nonfissured to slightly fissured, (II) slightly to moderately fissured, (III) moderately to strongly fissured, (IV) strongly to very strongly fissured, and (V) extremely fissured. Although it would be better to determine the length, distribution, and extent of fissures by direct microscopic techniques, this necessitates tools and procedures that are not generally available. On the other hand, using Figure 2.3, the degree of fissuring can be appreciated and named readily and inexpensively in almost any rock mechanics laboratory.

2.9 Other Physical Properties

Many other physical properties are important to specific engineering tasks in rock. The hardness of rock affects drillability. Elasticity and stress-strain coefficients are basic to engineering for dams and pressure tunnels. The thermal properties—heat conductivity and heat capacity and the coefficient of linear expansion—affect storage of hot and cold fluids in caverns and geothermal energy recovery. The following chapters consider some of these rock specimen attributes further. As noted previously, an overriding influence on rock behavior in many instances stems from the characteristics of the discontinuities, including joints, bedding, foliation, and fractures. This is addressed by a meaningful system of rock classification that attempts to overlay index properties of rocks and of discontinuities.

2.10 Classification of Rock Masses for Engineering Purposes

It is not always convenient to make a definitive test in support of engineering decision involving rock, and sometimes it is not even possible. Frequently, experience and judgment are strained in trying to find answers to design decisions involving rock qualities. Where there are particular and recurrent needs for quantitative values from rock, useful index tests are used routinely as in evaluating the need for continued grouting below a dam, deepening a pier shaft before filling it with concrete, or establishing the thickness of shotcrete lining in a newly excavated stretch of a rock tunnel. Thus it is not surprising that numerous schemes have been devised to guide judgment through standardized procedures and descriptions. Three especially well-received classification systems, originally advanced for tunneling, are those developed by Barton, Lien, and Lunde (1974), Bieniawski (1974, 1984), and Wickham, Tiedemann, and Skinner (1974).

Bieniawski's *Geomechanics Classification* system provides a general rock mass rating (RMR) increasing with rock quality from 0 to 100. It is based upon five universal parameters: strength of the rock, drill core quality, groundwater conditions, joint and fracture spacing, and joint characteristics. A sixth parameter, orientation of joints, is entered differently for specific application in tunneling, mining, and foundations. Increments of rock mass rating corresponding to each parameter are summed to determine RMR.

The *strength* of the rock can be evaluated using a laboratory compression test on prepared core, as discussed in the next chapter. But for rock classification purposes, it is satisfactory to determine compressive strength approximately using the point load test described previously on intact pieces of drill core. To simplify class boundaries, Bieniawski revised Equation 2.12 to $q_u = 25I_s$. The rock mass rating increment corresponding to compressive strength values are listed in Table 2.10.

Drill core quality is rated according to the rock quality designation (RQD) introduced by Deere (1963). Although the RQD is widely used as a sole parameter for classification of rock quality, it is preferable to combine it with other parameters accounting for rock strength, joint character, and environmental factors as done here, since the RQD alone ignores these features. The RQD of a rock is evaluated by determining the percentage recovery of core in lengths greater than twice its diameter. The index was first applied solely to NX core, usually 2.125 in. in diameter, the percentage core recovery being modified to reject from the "recovered" category any fragments less than 4 in. in length. The rock mass rating increments corresponding to five bands of RQD values are given in Table 2.11.

The *spacing* of joints is also evaluated from drill core, if available. It is assumed that the rock mass contains three sets of joints in general and the spacing entered in Table 2.12 to determine the rating increment should reflect that joint set considered to be most critical for the particular application. If the

Table 2.10 *Rock Mass Rating Increments for Compressive Strength of the Rock*

Point Load Index (MPa)	OR	Unconfined Compressive Strength (MPa)	Rating
>10		>250	15
4–10		100–250	12
2–4		50–100	7
1–2		25–50	4
Don't use		10–25	2
Don't use		3–10	1
Don't use		<3	0

Table 2.11 *Rock Mass Rating Increments for Drill Core Quality*

RQD (%)	Rating
90–100	20
75–90	17
50–75	13
25–50	8
<25	3

rock mass has fewer sets of joints, the rating may be established more favorably than indicated in this table. The *condition* of joints is also examined with respect to the joint sets most likely to influence the work. In general, the descriptions of joint surface roughness and coating material should be weighted toward the smoothest and weakest joint set. Joint condition ratings are given in Table 2.13. Further discussion of the influence of joint roughness and spacing on the properties of rocks is presented in Chapter 5.

Groundwater can strongly influence rock mass behavior so the geomechanics classification attempts to include a groundwater rating term as given in Table 2.14. If an exploratory adit or pilot tunnel is available, measurements of water inflows or joint water pressures may be used to determine the rating increment directly. The drill core and drilling log can be used in lieu of such

Table 2.12 *Increments of Rock Mass Rating for Spacing of Joints of Most Influential Set*

Joint Spacing (m)	Rating
>2.0	20
0.6–2.0	15
0.2–.6	10
0.06–0.2	8
<0.06	5

Table 2.13 *Rock Mass Rating*
Increments for Joint Condition

Description	Rating
Very rough surfaces of limited extent; hard wall rock	30
Slightly rough surfaces; aperture less than 1 mm; hard wall rock	25
Slightly rough surfaces; aperture less than 1 mm; soft wall rock	20
Smooth surfaces, OR gouge filling 1–5 mm thick, OR aperture of 1–5 mm; joints extend more than several meters	10
Open joints filled with more than 5 mm of gouge, OR open more than 5 mm; joints extend more than several meters	0

information to assign the rock to one of four categories from which the rating increment is assigned—completely dry, moist, water under moderate pressure, or severe water problems.

Since the orientation of the joints relative to the work can have an influence on the behavior of the rock, Bieniawski recommended adjusting the sum of the first five rating numbers to account for favorable or unfavorable orientations, according to Table 2.15. No points are subtracted for very favorable orientations of joints, up to 12 points are deducted for unfavorable orientations of joints in tunnels, and up to 25 for unfavorable orientations in foundations. It is difficult to apply these corrections by universal charts because a given orienta-

Table 2.14 *Increments of Rock Mass Rating*
Due to Groundwater Condition

Inflow per 10 m Tunnel Length (L/min)	OR	Joint Water Pressure Divided by Major Principal Stress	OR	General Condition	Rating
None		0		Completely dry	15
<10		0.0–0.1		Damp	10
10–25		0.1–0.2		Wet	7
25–125		0.2–0.5		Dripping	4
>125		>0.5		Flowing	0

Table 2.15 *Adjustment in RMR for Joint Orientations*

Assessment of Influence of Orientation on the Work	Rating Increment for Tunnels	Rating Increment for Foundations
Very favorable	0	0
Favorable	−2	−2
Fair	−5	−7
Unfavorable	−10	−15
Very unfavorable	−12	−25

tion may be favorable or unfavorable depending upon the groundwater and joint conditions. Thus, applying Table 2.14 requires advice from an engineering geologist familiar with the particular rock formations and the works in question. The orientation of joint sets cannot be found from normal, routine drilling of rock masses but can be determined from drill core with special tools or procedures, as reviewed by Goodman (1976) (work cited in Chapter 1). Logging of the borehole using a television or camera downhole will reveal orientations of joints, and absolute orientations will also be obtained from logging shafts and adits.

For applications in mining, involving assessments of caveability, drillability, blasting, and supports, Laubscher and Taylor (1976) modified Tables 2.10 to 2.15 and introduced factors to adjust for blasting practice, rock stress, and weathering. They also presented a table to find joint spacing ratings given the separate spacings of all joint sets. The overall RMR rating of a rock mass places the rock in one of the five categories defined in Table 2.16. Specific applications of the rock mass rating are presented in later chapters.

Table 2.16 *Geomechanics Classification of Rock Masses*

Class	Description of Rock Mass	RMR Sum of Rating Increments from Tables 2.9–2.14
I	Very good rock	81–100
II	Good rock	61–80
III	Fair rock	41–60
IV	Poor rock	21–40
V	Very poor rock	0–20

The Q system by Barton, Lien, and Lunde (1974) (also called the NGI system) combines six parameters in a multiplicative function:

$$Q = (RQD/J_n) \times (J_r/J_a) \times (J_w/SRF) \qquad (2.16)$$

where RQD is the Rock Quality Designation
 J_n relates to the number of joint sets
 J_r relates to the roughness of the most important joints
 J_a relates to the wall rock condition and/or filling material
 J_w relates to the water flow characteristics of the rock
 SRF relates to looseness and stress conditions.

The first term of Equation 2.16 is a measure of the sizes of joint blocks, the second factor expresses the shear strength of the block surfaces, and the last factor evaluates the important environmental conditions influencing the behavior of the rock mass. Numerical values are assigned to each parameter of the Q system according to detailed descriptions to be found in the article by Barton et al., which are abbreviated in Table 2.17. Table 2.18 assigns qualitative classes to the rock according to the overall value of Q.

The Q system and the RMR system include somewhat different parameters and therefore cannot be strictly correlated. Equation 2.17 is an approximate connecting relationship proposed by Bianiawski, based upon a study of a large number of case histories (standard deviation = 9.4).

$$RMR = 9 \log Q + 44 \qquad (2.17)$$

Table 2.17 *Values of the Parameters in the* Q *System*

Number of Sets of Discontinuities	J_n
Massive	0.5
One set	2.0
Two sets	4.0
Three sets	9.0
Four or more sets	15.0
Crushed rock	20.0
Roughness of Discontinuities	J_r*
Noncontinuous joints	4.0
Rough, wavy	3.0
Smooth, wavy	2.0
Rough, planar	1.5
Smooth, planar	1.0
Slick, planar	0.5
"Filled" discontinuities	1.0

*Add 1.0 if mean joint spacing exceeds 3 m

Filling and Wall Rock Alteration	J_a
Essentially unfilled	
Healed	0.75
Staining only; no alteration	1.0
Silty or sandy coatings	3.0
Clay coatings	4.0
Filled	
Sand or crushed rock filling	4.0
Stiff clay filling <5 mm thick	6.0
Soft clay filling <5 mm thick	8.0
Swelling clay filling <5 mm thick	12.0
Stiff clay filling >5 mm thick	10.0
Soft clay filling >5 mm thick	15.0
Swelling clay filling >5 mm thick	20.0
Water Conditions	J_w
Dry	1.0
Medium water inflow	0.66
Large inflow with unfilled joints	0.5
Large inflow with filled joints that wash out	0.33
High transient inflow	0.2–0.1
High continuous inflow	0.1–0.05
Stress Reduction Class	**SRF***
Loose rock with clay-filled discontinuities	10.0
Loose rock with open discontinuities	5.0
Rock at shallow depth (<50 m) with clay-filled discontinuities	2.5
Rock with tight, unfilled discontinuities under medium stress	1.0

* Barton et al. also define SRF values corresponding to degrees of bursting, squeezing, and swelling rock conditions.

The use of engineering classification systems for rock is still somewhat controversial. Proponents point to the opportunities they offer for empiricism in design of tunnels, mines, and other works in rock. Furthermore, an attempt to fill out the tables of values required by these schemes disciplines the observer and produces a careful, thorough scrutiny of the rock mass. On the other hand, these classifications tend to promote generalizations that in some cases are

Table 2.18 *After Barton, Lien, and Lunde (1974)*

Q	Rock Mass Quality for Tunneling
<0.01	Exceptionally poor
0.01– 0.1	Extremely poor
0.1 – 1.0	Very poor
1.0 – 4.0	Poor
4.0 – 10.0	Fair
10.0 – 40.0	Good
40.0 –100.0	Very good
100.0 –400.0	Extremely good
>400.0	Exceptionally good

inadequate to describe the full range of specifics of real rocks. Whichever argument prevails in a particular case, there can be no doubt that classification systems are proving valuable to many in various aspects of applied rock mechanics.

References

Aastrup, A. and Sallstrom, S. (1964) Further Treatment of Problematic Rock Foundations at Bergeforsen Dam. *Proc. Eighth Cong. on Large Dams,* Edinburgh, p. 627.

Barton, N. (1976) Recent experiences with the Q-system of tunnel support design, *Proceedings of Symposium on Exploration for Rock Engineering* (Balkema, Rotterdam), Vol. 1, pp. 107–118.

Barton, N., Lien, R., and Lunde, J. (1974) Engineering classification of rock masses for the design of tunnel support, *Rock Mech.* **6:** 189–236.

Bernaix, J. (1969) New Laboratory methods of studying the mechanical properties of rock, *Int. J. Rock Mech. Min. Sci.* **6:** 43–90.

Bieniawski, Z. T. (1974) Geomechanics classification of rock masses and its application in tunneling, *Proc. 3rd Cong. ISRM* (Denver), Vol. 2A, p. 27.

Bieniawski, Z. T. (1976) Rock mass classifications in rock engineering, *Proceedings of Symposium on Exploration for Rock Engineering* (Balkema, Rotterdam), Vol. 1, pp. 97–106.

Bieniawski, Z. T. (1984) *Rock Mechanics Design in Mining and Tunneling,* Balkema, Rotterdam.

Brace, W. F. and Riley, D. K. (1972) Static uniaxial deformation of 15 rocks to 30 kb, *Int. J. Rock Mech. Mining Sci.* **9:** 271–288.

Brace, W. F., Walsh, J. B., and Frangos, W. T. (1968) Permeability of granite under high pressure, *J. Geoph. Res.* **73:** 2225–2236.

Broch, E. and Franklin, J. A. (1972) The point load strength test, *Int. J. Rock Mech. Mining Sci.* **9**: 669–697.

Clark, S. P. (Ed.) (1966) *Handbook of Physical Constants,* Geological Society of America, Memoir 97.

Daly, R. A., Manger, G. I., and Clark, S. P., Jr. (1966) Density of rocks. In S. P. Clark, Ed., *Handbook of Physical Constants,* rev. ed., Geological Society of America, Memoir 97, pp. 19–26.

Davis, S. N. and DeWiest, R. J. M. (1966) *Hydrogeology,* Wiley, New York.

Deere, D. U. (1963) Technical description of rock cores for engineering purposes, *Rock Mech. Eng. Geol.* **1**: 18.

Dobereiner, L. and de Freitas, M. H. (1986) Geotechnical properties of weak sandstones, *Geotechnique* **36**: 79–94.

Fourmaintraux, D. (1976) Characterization of rocks; laboratory tests, Chapter IV in *La Mécanique des roches appliquée aux ouvrages du génie civil* by Marc Panet et al. Ecole Nationale des Ponts et Chaussées, Paris.

Franklin, J. A. and Chandra, R. (1972) The slake durability index, *Int. J. Rock Mech. Min. Sci.* **9**: 325–342.

Franklin, J. A., Vogler, U. W., Szlavin, J., Edmond, J. M., and Bieniawski, Z. T. (1979) Suggested methods for determining water content, porosity, density, absorption and related properties and swelling and slake durability index properties for ISRM Commission on Standardization of Laboratory and Field Tests, *Int. J. Rock Mech. Min. Sci.* **16**: 141–156.

Gamble, J. C. (1971) Durability-plasticity classification of shales and other argillaceous rocks, Ph. D. thesis, University of Illinois.

Hamrol, A. (1961) A quantitative classification of the weathering and weatherability of rocks, *Proceedings, 5th International Conference on Soil Mechanics and Foundation Engineering* (Paris), Vol. 2, p. 771.

Kulhawy, F. (1975) Stress deformation properties of rock and rock discontinuities, *Eng. Geol.* **9**: 327–350.

Laubscher, D. H. and Taylor, H. W. (1976) The importance of geomechanics classification of jointed rock masses in mining operations, *Proceedings of Symposium on Exploration for Rock Engineering* (Johannesburg), Vol. 1, pp. 119–135.

Morgenstern, N. R. and Eigenbrod, K. D. (1974) Classification of argillaceous soils and rocks, *J. Geotech. Eng. Div.* (ASCE) **100** (GT 10): 1137–1158.

Müller-Salzburg, L. (1963, 1978) *Der Felsbau,* Vols. 1 and 3, (In German), Ferdinand-Enke, Stuttgart.

Nakano, R. (1979) Geotechnical properties of mudstone of Neogene Tertiary in Japan, *Proceedings of International Symposium on Soil Mechanics in Perspective* (Oaxaca, Mexico), March, Session 2 (International Society of Soil Mechanics and Foundation Engineering).

Rocha, M. and Franciss, F. (1977) Determination of permeability in anisotropic rock masses from integral samples, *Rock Mech.* **9**: 67–94.

Rummel, F. and Van Heerden, W. L. (1978) Suggested methods for determining sound velocity, for ISRM Commission on Standardization of Laboratory and Field Tests, *Int. J. Rock Mech. Min. Sci.* **15**: 53–58.

Rzhevsky, V. and Novik, G. (1971) *The Physics of Rocks,* Mir, Moscow.

Snow, D. T. (1965) A parallel plate model of fractured permeable media, Ph.D. thesis, University of California, Berkeley.

Snow, D. T. (1968) Rock fracture spacings, openings, and porosities, *J. Soil Mech. Foundations Div.* (ASCE) **94** (SM 1): 73–92.

Techter, D. and Olsen, E. (1970) *Stereogram Books of Rocks, Minerals & Gems,* Hubbard, Scientific. Northbrook, IL.

Underwood, L. B. (1967) Classification and identification of shales, *J. Soil Mech. Foundations Div.* (ASCE) **93** (SM 6): 97–116.

Wickham, G. E., Tiedemann, H. R., and Skinner, E. H. (1974) Ground support prediction model—RSR concept, *Proc. 2nd RETC Conf.* (AIME), pp. 691–707.

Winchell, A. N. (1942) *Elements of Mineralogy,* Prentice-Hall, Englewood Cliffs, NJ.

Problems

1. A shale of Cretaceous age is composed of 60% illite, 20% chlorite, and 20% pyrite. The porosity values at different depths are as follows: n equals 33.5% at 600 ft; 25.4% at 2500 ft; 21.1% at 3500 ft, and 9.6% at 6100 ft. Estimate the vertical stress at 6000 ft depth in this shale (assuming a continuous thickness of shale from the surface to depth 6000 ft and saturation with water).

2. Three samples of rock were subjected to diametral point load tests. The pressure gage readings at rupture were 250, 700, and 1800 psi. If the ram area was 2.07 in.², and the diameter of the cores tested was 54 mm, calculate an estimate for the unconfined compressive strength of each rock. (Ignore a size correction.)

3. A sandstone core composed of quartz and feldspar grains with calcite cement is 82 mm in diameter and 169 mm long. On saturation in water, its wet weight is 21.42 N; after oven drying its weight is 20.31 N. Calculate its wet unit weight, its dry unit weight, and its porosity.

4. Another core specimen from the same formation as the rock of Problem 3, displays large voids. Its wet unit weight is 128 lb/ft³. Assuming its specific gravity is the same as for the rock in Problem 3, estimate its porosity.

5. A granitic rock is composed of a mixture of 30% quartz, 40% plagioclase, and 30% augite. Its porosity is 3.0% and its longitudinal wave velocity measured in the laboratory is 3200 m/s. Describe its state of fissuring.

6. A sandstone with porosity of 15% is composed of a mixture of 70% quartz grains and 30% pyrite grains. Determine its dry density in pounds per cubic foot and meganewtons per cubic meter.

7. Determine the water content of the above rock when it is saturated with water.

8. A rock is injected with mercury by subjecting it to a high pressure. Derive a formula expressing its porosity in terms of the measured mercury content, the specific gravity of mercury, and the specific gravity of the component minerals.

9. If a rock has a permeability of 1 millidarcy, how much water will flow through it per unit of time and area under a gradient of unity? (The water temperature is 20°C.)

10. What will be the vertical stress in the ground at a depth of 5000 ft in the Pennsylvanian age shale whose porosity is given in Table 2.1 and whose density is given in Table 2.3 (Oklahoma shale). (Integrate the varying density depth relation.) Express your answer in psi and MPa.

11. A rock mass has field conductivity of 10^{-5} cm/s. Assuming the rock itself is impervious and three orthogonal sets of smooth fractures recur with spacing 1 m, calculate the aperture (e) of the fractures.

12. Derive a formula expressing the conductivity k (cm/s) of a rock mass with orthogonal fractures characterized by identical spacing S and aperture e if the fractures are filled with soil having permeability k_f (cm/s).

13. A moist rock mass is characterized by the following parameters: joint water pressure is nil; the point load index = 3 MPa; the joint spacing = 0.5 m; and RQD = 55%. Prepare a table of rock mass rating versus joint condition using the terminology of Table 2.16 for the former and Table 2.13 for the latter.

14. An orthogonally jointed rock mass has a field permeability of 55.0 darcies. The mean joint spacing is 0.50 m. Calculate the corresponding average aperture of the fractures.

15. A frequently used estimator of rock mass hydraulic conductivity is the water loss coefficient (C) determined with "pump-in" tests. A section of an exploratory borehole is isolated by packers, and the pressure is brought to an elevated level (Δp) above the initial water pressure in the middle of the test section, while the flow rate (q) into the hole is monitored. For steady state flow, a rate of water loss of 1 "lugeon" corresponds to $q = 1$ L/min per meter length of the test section at a pressure difference (Δp) of 10 atmospheres (≈ 1 MPa) applied at the test section. How many lugeons of water loss corresponds to a flow of 4.0 gal/min in a 10-foot-long test section under a differential pressure (Δp) of 55 psi?

16. A rock mass has initial unit weight equal to γ and, after loosening, it

assumes unit weight γ_1. A coefficient of loosening (n) was defined by Müller (1978) as

$$n = \frac{\gamma - \gamma_1}{\gamma}$$

(a) A jointed sedimentary rock mass assumes a value of $n = 0.35$ after loosening and 0.08 after recompaction. Calculate the corresponding values of γ_1. ($\gamma = 27$ kN/m³)

(b) Crystalline igneous rocks like granite, gneiss, and diabase have a range of values of $n = 0.35$ to 0.50 after loosening and 0.08 to 0.25 after recompaction. Compute the corresponding values of the unit weight (γ_1).

Rock Strength and Failure Criteria

Whenever we place an engineered structure against rock, we ask the following two questions: Will the *stresses in the rock* reach the maximum levels that are tolerable, with consequent local or gross rock failure? Will the *displacements of the rock* under the loads to be applied produce such large strains in the structure that they cause its damage or destruction? This chapter discusses the first question. Assuming that we can estimate the initial stresses in the rock mass and that we can predict how these stresses will be modified by the construction and operation of the engineering work, how may we discover if the rock will flow, yield, crush, crack, buckle, or otherwise give way in service? For this we utilize "criteria of failure"—equations that link the limiting combinations of stress components separating acceptable from inadmissible conditions. Before we can propose meaningful criteria, however, we should examine how rocks usually fail, that is, whether in bending, shearing, crushing, or otherwise.

3.1 Modes of Rock Failure

The varieties of load configurations in practice are such that no single mode of rock failure predominates. In fact, flexure, shear, tension, and compression can each prove most critical in particular instances. *Flexure* refers to failure by bending, with development and propagation of tensile cracks. This may tend to occur in the layers above a mine roof (Figure 3.1a). As the "immediate roof" detaches from the rock above, under gravity, a gap forms and a beam of rock sags downward under its own weight. As the beam begins to crack, its neutral axis advances upward; eventually, the cracks extend right through the beam, after which sections of rocks may come loose and fall. Flexural failure can also

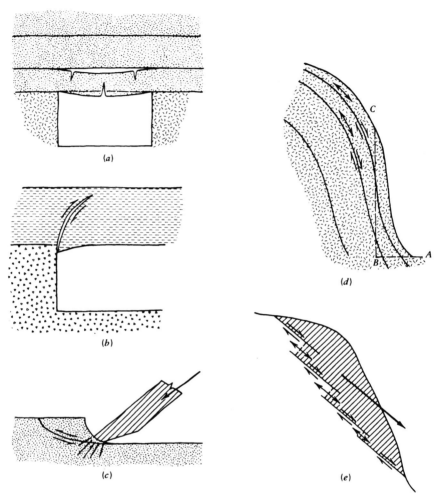

Figure 3.1 Examples of failure modes involving breakage of rock. (a) Flexure. (b) Shear. (c) Crushing and tensile cracking, followed by shear. (d and e) Direct tension.

occur in rock slopes with steeply dipping layers as the layers overturn toward the free space ("toppling failure").

Shear failure refers to formation of a surface of rupture where the shear stresses have become critical, followed by release of the shear stress as the rock suffers a displacement along the rupture surface. This is common in slopes cut in weak, soil-like rocks such as weathered clay shales and crushed rock of fault zones. It may occur in a mine with stiff ore and a softer, weaker roof or floor; the shear stresses in the roof or pillar base can allow the pillar to "punch" relatively upward into the roof (Figure 3.1*b*) or downward into the floor. Rock cutters employing "drag bits" or "picks" owe their cutting action partly to shear along fractures caused by compression under the edge of the bit (Figure 3.1*c*). The vibration of such cutters as they advance reflects the periodic formation and removal of rock chips.

Direct tension is occasionally set up in rock layers resting on convex upward slope surfaces (e.g., in sheeted granites (Figure 3.1*d*)) and in sedimentary rocks on the flank of an anticline. The base of the slope has layers inclined more steeply than friction will allow and the balance of support for the weight of the layers is the tensile pull from the stable part of the slope above. Direct tension also is the mechanism of failure in rock slopes with nonconnected, short joint planes; the formation of tension cracks severs the rock bridges and allows a complete block of rock to translate downward en masse (Figure 3.1*e*). When rock breaks in tension, the surface of rupture is rather rough and free from crushed rock particles and fragments. With shear failure, on the contrary, the surface of failure is slick and there is much powder from crushing and communition of rock. Direct tension failure also occurs when the circumference of a borehole or a tunnel is stretched owing to internal water or gas pressure. The former situation arises when a pressure tunnel is operated at excessive pressure and when a drill hole is "hydraulically fractured" by pumping water to a high pressure in a section isolated by "packers." Detonation of an explosive agent in a borehole will raise gas pressure against the wall to millions of pounds per square inch; tensile failure then creates a series of radial cracks beyond the immediate periphery of the borehole, which may be crushed or in extreme cases actually melted. Some extension joints in bedrock are believed to have arisen from circumferential strain accompanying large amounts of uplift over broad geographic belts ("epeirogeny").

Crushing or *compression failure* occurs in intensely shortened volumes or rock penetrated by a stiff punch. Examination of processes of crushing shows it to be a highly complex mode, including formation of tensile cracks and their growth and interaction through flexure and shear. When the particles and slivers formed by cracking are not free to move away from the zone of compression, they become finely comminuted. This happens under some drill bits and under disk cutters of boring machines. In a mine pillar, overextraction of ore can lead to pillar failure by splitting and shear, although the destruction of the load-carrying capacity of the pillar through growth and coalescence of cracks is sometimes spoken of as "compression failure."

It may be appreciated that the actual destruction of a load-carrying rock mass is rather complex and involves one or more of the modes mentioned. It is no wonder then that no single method of testing rock has been advanced to the exclusion of others. In fact, the theory of failure makes use of a variety of laboratory and field testing techniques adapted to the special nature of the problem at hand.

3.2 Common Laboratory Strength Tests

To characterize the strength of rock specimens, unconfined and confined compression tests, shear tests, and direct and indirect tension tests are used widely. Other test configurations are preferred for special applications and a great variety of procedures has been investigated. We review here the important features of the most widely used tests—unconfined compression, triaxial compression, splitting tension ("Brazilian tests"), beam bending, and ring shear. Figure 3.2 shows rock preparation equipment required to prepare specimens for such tests.

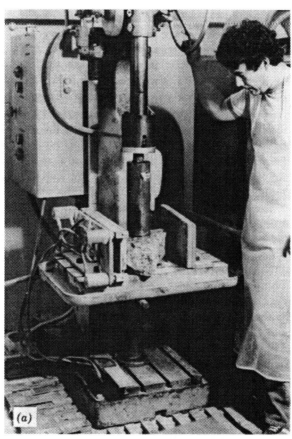

(a)

Figure 3.2 Equipment for preparing rock specimens for laboratory tests. (*a*) A drill press modified for feed under constant pressure and equipped with a vise to retain arbitrary blocks during drilling. (The drill press was devised by Quentin Gorton.)

Figure 3.2 Equipment for preparing rock specimens for laboratory tests. (*b*) A diamond saw. (*c*) A surface grinder adapted from a milling machine by adding a diamond wheel and water bath.

Unconfined compression (Figure 3.3a) is the most frequently used strength test for rocks, yet it is not simple to perform properly and results can vary by a factor of more than two as procedures are varied. The test specimen should be a rock cylinder of length-to-width ratio in the range 2 to 2.5 with flat, smooth, and parallel ends cut perpendicularly to the cylinder axis. Procedures are recommended in ASTM designation D2938-71a and by Bieniawski and Bernede (1979). Capping of the ends with sulfur or plaster to specified smoothness is thought to introduce artificial end restraints that overly strengthen the rock. However, introduction of Teflon pads to reduce friction between the ends and the loading surfaces can cause outward extrusion forces producing a premature splitting failure, especially in the harder rocks. When mine pillars are studied, it is sometimes preferable to machine the compression specimen from a large cylinder to achieve loading through rock of the upper and lower regions into the more slender central region. In the standard laboratory compression test, however, cores obtained during site exploration are usually trimmed and compressed between the crosshead and platen of a testing machine. The compres-

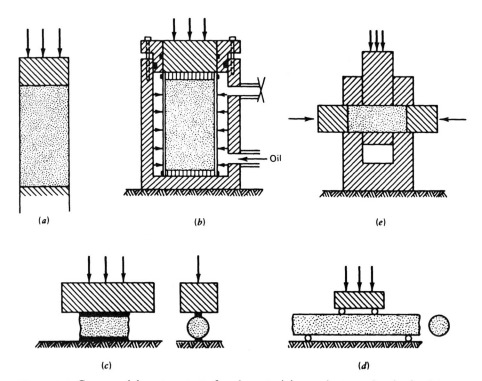

Figure 3.3 Common laboratory tests for characterizing rock strength criteria. (*a*) Unconfined compression. (*b*) Triaxial compression. (*c*) Splitting tension (Brazilian). (*d*) Four-point flexure. (*e*) Ring shear.

sive strength q_u is expressed as the ratio of peak load P to initial cross-sectional area A:

$$q_u = \frac{P}{A} \tag{3.1}$$

Representative values of q_u are listed in Table 3.1.

Triaxial compression (Figure 3.3*b*) refers to a test with simultaneous compression of a rock cylinder and application of axisymmetric confining pressure. Recommended procedures are described in ASTM designation D2664-67 (1974) and in an ISRM Committee report by Vogler and Kovari (1978).

Table 3.1 *Unconfined Compressive Strength* (q$_u$) *and Ratio of Compressive to Indirect Tensile Strength* (q$_u$/T$_0$) *for Specimens of Representative Rocks*

Description[a]	q_u		q_u/T_0[b]	Reference[c]
	MPa	psi		
Berea sandstone	73.8	10,700	63.0	5
Navajo sandstone	214.0	31,030	26.3	5
Tensleep sandstone	72.4	10,500		1
Hackensack siltstone	122.7	17,800	41.5	5
Monticello Dam s.s. (greywacke)	79.3	11,500		4
Solenhofen limestone	245.0	35,500	61.3	5
Bedford limestone	51.0	7,400	32.3	5
Tavernalle limestone	97.9	14,200	25.0	5
Oneota dolomite	86.9	12,600	19.7	5
Lockport dolomite	90.3	13,100	29.8	5
Flaming Gorge shale	35.2	5,100	167.6	3
Micaceous shale	75.2	10,900	36.3	2
Dworshak Dam gneiss				
45° to foliation	162.0	23,500	23.5	5
Quartz mica schist ⊥ schistocity	55.2	8,000	100.4	5
Baraboo quartzite	320.0	46,400	29.1	5
Taconic marble	62.0	8,990	53.0	5
Cherokee marble	66.9	9,700	37.4	5
Nevada Test Site granite	141.1	20,500	12.1	7
Pikes Peak granite	226.0	32,800	19.0	5
Cedar City tonalite	101.5	14,700	15.9	6
Palisades diabase	241.0	34,950	21.1	5
Nevada Test Site basalt	148.0	21,500	11.3	7
John Day basalt	355.0	51,500	24.5	5
Nevada Test Site tuff	11.3	1,650	10.0	7

[a] Description of rocks listed in Table 3.1:

   Berea sandstone, from Amherst, Ohio; fine grained, slightly porous; cemented. *Navajo sandstone,* from Glen Canyon Dam site, Arizona; friable, fine to medium grained. (Both sandstones are

Table Footnote (*continued*)

predominately composed of quartz grains.) *Tensleep sandstone*, Pennsylvanian-age sandstone from Alcova Powerhouse, Wyoming, (near Casper); calcite cemented; medium grained. *Hackensack siltstone*, New Jersey; from Triassic Newark Series; cemented with hematite; argillaceous. *Monticello Dam greywacke*, Cretaceous sandstone from the Monticello dam foundation, California; medium to coarse grained, cemented feldspar, quartz, and other components; some feldspars altered to mica. *Solenhofen limestone*, from Bavaria; very fine, interlocked crystalline texture. *Bedford limestone*, Indiana; slightly porous, oolitic, bioclastic limestone. *Tavernalle limestone*, from Carthage, Missouri; fine grained, cemented and interlocked crystalline limestone with fossils. *Oneota dolomite*, Kasota, Minnesota; fine-grained interlocking granular texture with mottled appearance due to disseminated calcite veins. *Lockport dolomite*, Niagara Falls, New York; very fine-grained cemented granular texture grading to interlocking crystalline texture; some grains. *Flaming Gorge shale*, from Flaming Gorge damsite, Utah, Wyoming border. *Micaceous shale*, from the Jonathan mine, Ohio; the clay mineral is kaolinite. *Dworshak dam gneiss*, from Orofino, Idaho; fine to medium-grained granodiorite gneiss with prominent foliation. *Quartz mica schist* with crenulated schistocity; origin unknown. *Baraboo quartzite*, from Wisconsin; fine-grained, brittle, massive Pre-Cambrian quartzite with tightly interlocking crystalline texture. *Taconic white marble*, Rutland, Vermont; uniform, fine-grained massive marble, with sugary texture. *Cherokee marble*, from Tate, Georgia; medium- to coarse-grained massive marble with tightly interlocking crystalline texture. *Nevada Test Site* "granite," granodiorite from Piledriver Experiment; coarse-grained. *Pikes Peak granite*, Colorado Springs, Colorado; fine- to medium-grained dense; interlocked crystalline texture. *Cedar City tonalite*, somewhat weathered quartz monzonite, with porosity of 4.9%, from Cedar City, Utah. *Palasades diabase*, from West Nyack, New York; medium-grained. *Nevada Test Site basalt*, from Buckboard Mesa; fine, olivine basalt. *John Day basalt*, from John Day dam site, Arlington, Oregon. *Nevada Test Site tuff*, from "Red Hot" experiment; welded volcanic ash; porosity 19.8%.

[b] Tensile strengths were determined by point load tests for all entries corresponding to reference 5; determined by Brazilian test for entries corresponding to references 6 and 7. The point load tensile strength T_0 in megapascals was calculated from the load at failure (F), in meganewtons for point loading across the rock core diameter (d), in meters; $T_0 = 6.62 \ 10^{-3} \ F/d^2$ (Reichmuth, 1963).

[c] References for Table 3.1:

General

Kulhawy, F. (1975) cited in references at the end of this chapter.
Lama, R. D. and Vutukuri, V. S., cited in references in Chapter 1.

Specific

1. Balmer, G. G. (1953) Physical properties of some typical foundation rocks, U. S. Bureau of Reclamation Concrete Lab Report SP-39.
2. Blair, B. E. (1956) Physical properties of mine rock, Part IV, U. S. Bureau of Mines Rep. Inv. 5244.
3. Brandon, T. R. (1974) Rock mechanic properties of typical foundation rocks, U. S. Bureau of Reclamation Rep. REC-ERC 74-10.
4. Judd, W. R. (1969) Statistical methods to compile and correlate rock properties, Purdue University, Department of Civil Engineering.
5. Miller, R. P. (1965) Engineering classification and index properties for intact rock, Ph.D. Thesis, University of Illinois.
6. Saucier, K. L. (1969) Properties of Cedar City tonalite, U. S. Army Corps of Engineers, WES Misc. Paper C-69-9.
7. Stowe, R. L. (1969) Strength and deformation properties of granite, basalt, limestone, and tuff, U. S. Army Corps of Engineers, WES Misc. Paper C-69-1.

At the peak load, the stress conditions are $\sigma_1 = P/A$ and $\sigma_3 = p$, where P is the highest load supportable parallel to the cylinder axis, and p is the pressure in the confining medium. The confinement effect, that is, the strengthening of the rock by the application of confining pressure p, is realized only if the rock is enclosed in an impervious jacket. The confining fluid is normally hydraulic oil and the jacket is oil-resistant rubber (e.g., polyurethane); for tests of short duration, bicycle inner tube is suitable. Most rocks show a considerable strengthening effect due to confining pressure and it has become routine to conduct triaxial compression tests on rocks.

Many varieties of triaxial cells are in use in rock mechanics laboratories and several types are available from commercial suppliers. Figure 3.4a shows two cells used at the University of California, Berkeley. The one on the left was designed by Owen Olsen for the U. S. Bureau of Reclamation. It provides extra room for inserting instruments and gages and is easily adapted for pore pressure and other special measurements; however, the diameter of the piston is considerably larger than the diameter of the specimen, with the result that a large uplift force from the confining pressure must be reacted by the axial loading machine. The chamber on the right, based on a design by Fritz Rummel, avoids this problem. The rock specimen, with strain gages attached, will be jacketed before insertion in the triaxial chamber. Figure 3.4b shows a high-pressure, high-temperature triaxial test facility at the TerraTek Laboratory, Salt Lake City, Utah. This computer-controlled apparatus can supply confining

Figure 3.4 Equipment for triaxial compression tests. (a) Two types of cells used at Berkeley.

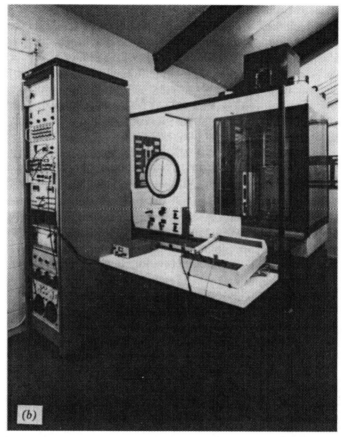

Figure 3.4 Equipment for triaxial compression tests. (*b*) A
high-pressure, high-temperature facility at TerraTek, Salt Lake
City, Utah.

pressures up to 200 MPa to specimens as large as 10 cm in diameter at temperatures as high as 200°C (5-cm-diameter specimens can be heated up to 535°C).

The usual procedure for conducting a triaxial compression test is first to
apply the confining pressure all round the cylinder (i.e., $\sigma_1 = \sigma_3 = p$) and then
to apply the axial load $\sigma_1 - p$ as the lateral pressure is held constant. In this
case, the triaxial compression experiment can be interpreted as the superposition of a uniaxial compression test on an initial state of all-round compression.
However, the actual path of loading in service may be quite different; since
some rocks demonstrate strong path effects it may then be desirable to follow
different procedures. For example, the stresses in the rock at the front of a
traveling plane wave are applied simultaneously in all directions. With com-

puter or manual feedback control, it is possible to follow almost any prescribed path of loading, although, as will be shown later, not all paths can result in fracture under load. For the best results and a clear interpretation of the effects of load, both the axial shortening, and the lateral expansion of the specimen should be monitored during loading as discussed later.

The *Brazilian test*, described for cylindrical concrete specimens in ASTM designation C496-71,[1] is convenient for gaining an estimate of the tensile strength of rock. It has been found that a rock core about as long as its diameter will split along the diameter and parallel to the cylinder axis when loaded on its side in a compression machine (Figure 3.2c). The reason for this can be demonstrated by examining the stress inside a disk loaded at opposite sides of a diametral plane. In such a configuration the horizontal stresses perpendicular to the loaded diameter are uniform and tensile with magnitude

$$\sigma_{t,B} = \frac{2P}{\pi dt} \tag{3.2}$$

where P is the compression load, d is the cylinder diameter, and t is the thickness of the disk (the length of the cylinder). It is much easier to perform this type of test than to arrange the precise alignment and end preparation required for a direct tensile test.

The "Brazilian tensile strength" is estimated from the test result by reporting the value of $\sigma_{t,B}$ corresponding to the peak compression load. It should be understood, however, that the actual cause of failure may also reflect the action of the vertical stress along the vertical diameter in concert with the horizontal tension; the vertical stress is nonuniform increasing from a compressive stress of three times $\sigma_{t,B}$ at the center of the disk to progressively higher values as the ends are approached. According to the Griffith theory of failure, the critical point ought to be the center where the ratio of compression to tension is 3. With a principal stress ratio of 3, failure ought to result from the application of the tensile stress alone, without any complication from the simultaneous compression parallel to the eventual rupture plane. In fact, the Brazilian test has been found to give a tensile strength higher than that of the direct tension test, probably owing to the effect of fissures. Short fissures weaken a direct tension specimen more severely than they weaken a splitting tension specimen. The ratio of Brazilian to direct tensile strength has been found to vary from unity to more than ten as the length of preexisting fissures grows larger (Tourenq and Denis, 1970).

A *flexural test* causes failure of a rock beam by bending. Like the Brazilian test, flexural tests also can be run on rock cores lacking machined ends. Four-point flexural loading (Figure 3.3d), with the bottom of the core supported on

---

[1] Standard Method of Test for Splitting Tensile Strength of Cylindrical Concrete Specimens, ASTM Committee C-9 on Concrete and Concrete Aggregate.

points near the ends and the top of the core loaded from above at the third points, produces uniform moment in the central third of the specimen and gives better reproducibility of results than three-point loading in which the upper load is central. The flexural strength or "modulus of rupture" is the maximum tensile stress on the bottom of the rock corresponding to peak load. It is calculated from simple beam theory assuming elastic conditions throughout. The flexural strength is found to be two to three times as great as the direct tensile strength. For four-point bending of cylindrical rock specimens, with loads applied at $L/3$ from each end and reactions at the ends, the modulus of rupture (MR) is:

$$T_{MR} = \frac{16 P_{max} L}{3 \pi d^3} \qquad (3.3)$$

where P_{max} is the maximum load, L is the length between load reactions on the lower surface, and d is the diameter of the core.

The *ring shear test* (Figure 3.3e) provides a relatively simple method to test intact rock strength as a function of confining pressure (Lundborg, 1966). In contrast to compression tests, core specimens for the ring shear test do not require perfectly square and smooth ends. As with the triaxial test, the results permit an appreciation of the rate of increase of strength with confining pressure. The latter is provided by the load parallel to the axis of the core. Two sets of complex fracture surfaces form along the two planes of imposed shear as the load is applied to the plunger.

If P is the peak load on the plunger, the peak shear stress (τ_p) is called the "shear strength" and is calculated by

$$\tau_p = \frac{P}{2A} \qquad (3.4)$$

where A is the area across the core sample.

3.3 Stress-Strain Behavior in Compression

STRESS AND STRAIN

In discussing the deformations of rock undergoing compression from various directions, it proves useful to divide the stresses into two parts. *Nondeviatoric stresses* (σ_{mean}) are compressions equally applied in all directions, that is, a hydrostatic state of stress. *Deviatoric* stresses (σ_{dev}) are the normal and shear stresses that remain after subtracting a hydrostatic stress, equal to the mean normal stress, from each normal stress component. In the triaxial compression experiment, for example, the principal stresses are $\sigma_1 = P/A$ and $\sigma_2 = \sigma_3 = p$.

The nondeviatoric stress is given by $\frac{1}{3}(\sigma_1 + 2p)$ all around while the deviatoric stress is then what remains: $\sigma_{1,\text{dev}} = \frac{2}{3}(\sigma_1 - p)$ and $\sigma_{3,\text{dev}} = \sigma_{2,\text{dev}} = -\frac{1}{3}(\sigma_1 - p)$. There is strong motivation for doing this: deviatoric stress produces distortion and destruction of rocks while nondeviatoric stresses generally do not (as discussed in the next section). In the triaxial test, the initial pressuring is nondeviatoric; subsequently, both deviatoric and nondeviatoric stresses are raised simultaneously.

Normal strains in a triaxial compression specimen can be measured with surface-bonded electric resistance strain gages. A gage parallel to the specimen axis records the longitudinal strain $\varepsilon_{\text{axial}} = \Delta l / l$, while a strain gage affixed to the rock surface in the circumferential direction yields the lateral strain $\varepsilon_{\text{lateral}} = \Delta d / d$, where d is the diameter of the rock and l is its length (see Figure 3.4a). Assuming that the strain gage readings are zeroed after the confining pressure has been applied, we can write

$$\varepsilon_{\text{lateral}} = -\nu\varepsilon_{\text{axial}} \tag{3.5}$$

in which the constant of proportionality ν is called *Poisson's ratio*. In fact, proportionality is maintained only in the restricted range of loading during which there is no initiation and growth of cracks. For linearly elastic and isotropic rocks, ν must lie in the range 0 to 0.5 and is often assumed equal to 0.25. Because a rock expands laterally as its shortens axially (Figure 3.5), a negative sign is introduced to define Poisson's ratio as a positive quantity. For strains of less than several %, the volume change per unit of volume, $\Delta V/V$, is

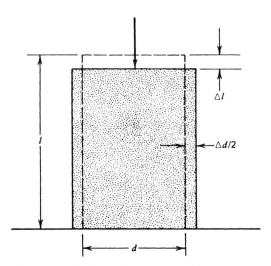

Figure 3.5 Deformations in compression tests.

closely approximated by the algebraic sum of the three normal strains. In the triaxial compression experiment then,

$$\frac{\Delta V}{V} = \varepsilon_{axial} + 2\varepsilon_{lateral}$$

or

$$\frac{\Delta V}{V} = \varepsilon_{axial}(1 - 2\nu) \tag{3.6}$$

Volumetric strain produced either by deviatoric or nondeviatoric stresses can be measured indirectly using surface strain gages and applying Equation 3.5 or directly by monitoring the flow of oil into or out of the confining vessel as the confining pressure is held constant by a servomechanism.

HYDROSTATIC COMPRESSION

Applying a nondeviatoric stress to a rock produces a volume decrease and eventually changes the rock fabric permanently, as pores are crushed. However, it cannot produce a peak load response; that is, the rock can always accept an added increment of load, apparently for as high a pressure as one can generate. Tests have been conducted into the megabar region (millions of psi) producing phase changes in the solid. The pressure, volumetric strain curve is generally concave upward as shown in Figure 3.6 with four distinct regions. In the first, which may be the principal region for many good rocks in civil engineering service, preexisting fissures are closed and the minerals are slightly compressed. When the load is removed, most of the fissures remain closed and there is a net deformation or "per-def." The fissure porosity is related to the per-def.

After most of the fissures have closed, further compression produces bulk rock compression, consisting of pore deformation and grain compression at an approximately linear rate. The slope of the pressure-volumetric strain curve in this region is called the *bulk modulus*,[2] K. In porous rocks like sandstone, chalk, and clastic limestone, the pores begin to collapse due to stress concentrations around them; in well-cemented rocks, this may not occur until reaching a pressure of the order of 1 kbar (100 MPa or 14,500 psi), but in poorly or weakly cemented rocks, pore crushing can occur at much lower pressures. Finally, when all the pores have been closed, the only compressible elements remaining are the grains themselves and the bulk modulus becomes progressively higher. Nonporous rocks do not demonstrate pore "crush up" but show uniformly concave-upward deformation curves to 300 kbar or higher. Pore

---

[2] The compressibility C is $1/K$.

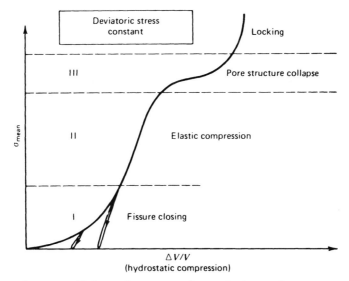

Figure 3.6 Volumetric compression under increasing mean stress, with constant deviatoric stress.

crushing is destructive in very porous rocks like chalk and pumice, which are converted to a cohesionless sediment on removal from the test chamber.

DEVIATORIC COMPRESSION

Applying deviatoric stress produces strikingly different results as shown in Figure 3.7. With initial application of the deviatoric stress, fissures and some pores begin to close, producing an inelastic, concave-upward stress-strain section. In most rocks, this is followed by linear relationships between axial stress and axial strain and between axial stress and lateral strain. At point *B* (Figure 3.7*a*), the rate of lateral strain begins to increase relative to the rate of axial strain (Poisson's ratio increases) as new cracks begin to form inside the most critically stressed portions of the specimen—usually near the sides of the mid-section. A microphone attached to the rock will begin to pick up "rock noise" as new cracks form and old ones extend parallel to the direction of σ_1. In the region between stress *B* and stress *C*, cracks are considered to be "stable" meaning that with each increment of stress they grow to a finite length, and then stop growing. After point *C*, cracks that form propagate to the edges of the specimen and a system of intersecting, coalescing cracks is developed, which eventually form a semicontinuous rupture surface termed a "fault." Figures 3.7*c* and *d*, from Wawersik and Brace (1971), show this development. Bieniawski (1967a, b) suggested that point *C* corresponds to the yield point in the axial stress-axial strain curve. The peak load, point D, is the usual object of failure criteria. However, the rock may not fail when the load reaches this point, as is discussed later. In a stiff loading system, it is possible to continue to shorten the specimen, as long as stress is reduced simultaneously. If the volumetric strain is plotted against the deviatoric stress as in Figure 3.7b, it is seen that the attainment of the crack initiation stress (B) is marked by a beginning of an increase in volume associated with sliding and buckling of rock slivers between cracks and opening of new cracks. At a stress level corresponding to stress point C, the specimen may have a bulk volume larger than at the start of the test. This increase in volume associated with cracking is termed *dilatancy*.

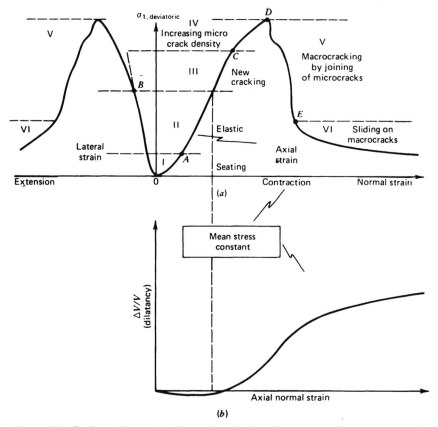

Figure 3.7 Deformation under increasing deviatoric stress, with constant mean stress (hypothetical curves). (*a*) Axial and lateral normal strain with increasing deviatoric axial stress. (*b*) Volumetric strain with increasing axial normal strain (dilatancy).

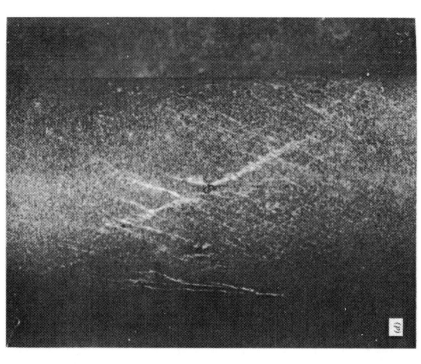

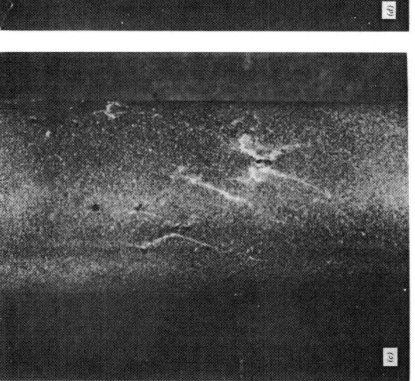

Figure 3.7 Deformation under increasing deviatoric stress, with constant mean stress (hypothetical curves). (c) at the peak axial stress; (d) just after the peak showing linkup of fractures to form a rupture surface. c and d Fractures formed in diabase during triaxial compression at 500 psi, reproduced from Wawersik and Brace (1971) with permission.

EFFECT OF CONFINING PRESSURE

Most rocks are significantly strengthened by confinement. This is especially striking in a highly fissured rock, can be imagined as a mosaic of perfectly matching pieces. Sliding along the fissures is possible if the rock is free to displace normal to the average surface of rupture, as shown in Figure 3.8. But under confinement, the normal displacement required to move along such a jagged rupture path requires additional energy input. Thus it is not uncommon for a fissured rock to achieve an increase in strength by 10 times the amount of

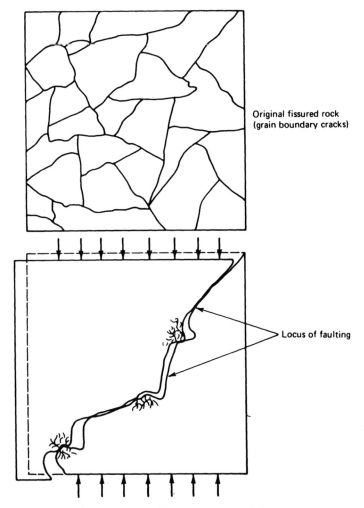

Original fissured rock
(grain boundary cracks)

Locus of faulting

Figure 3.8 Dilatancy caused by roughness of the rupture surface.

a small increment in mean stress. This is one reason why rock bolts are so effective in strengthening tunnels in weathered rocks.

As mean pressure is increased, the rapid decline in load carrying capacity after the peak load (point *D* in Figure 3.7) becomes gradually less striking until, at a value of the mean pressure known as the *brittle-to-ductile transition pressure*, the rock behaves fully plastically (Figure 3.9). That is, after point *D* continued deformation of the rock is possible without any decrease in stress.

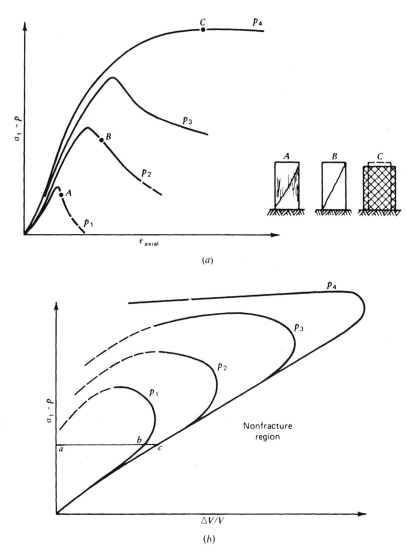

Figure 3.9 Behavior in triaxial compression. (*a*) Transition from brittle-to-ductile behavior. (*b*) Volumetric compression and dilatancy.

("Stress-hardening" behavior is sometimes observed at even higher pressures, meaning that there is actually a strengthening of the rock as it deforms without any "peak stress.") Figure 3.10 shows triaxial test data for a crystalline rock (norite) and a clastic rock (sandstone), both of which demonstrate loss of brittleness with confining pressure.

The brittle-to-ductile transition occurs at pressures far beyond the region of interest in most civil engineering applications. However, in evaporite rocks and soft clay shales, plastic behavior can be exhibited at engineering service loads. Table 3.2 lists some values of the transition pressure. Without confining pressure, most rocks tested past point D of Figure 3.7 will form one or more fractures parallel to the axis of loading (Figure 3.9a). When the ends are not smooth, the rock will sometimes split neatly in two, parallel to the axis, like a Brazilian specimen. As the confining pressure is raised, the failed specimen demonstrates faulting, with an inclined surface of rupture traversing the entire specimen. In soft rocks, this may occur even with unconfined specimens. If the specimen is too short, continued deformation past the faulting region will drive the edges of the fault blocks into the testing machine platens, producing complex fracturing in these regions and possibly apparent strain-hardening behavior. At pressures above the brittle-to-ductile transition, there is no failure per se, but the deformed specimen is found to contain parallel inclined lines that are the loci of intersection of inclined rupture surfaces and the surfaces of the specimen. Examination of the deformed rock will show intracrystalline twin gliding, intercrystal slip, and rupture.

The effect of confining pressure is also expressed in changing volumetric strain response as shown for a series of triaxial compression tests in Figure 3.9b. At successively higher confining pressures, the volumetric strain curves shift smoothly upward and to the right. These curves are the algebraic sum of hydrostatic compression under increasing mean stress (e.g., distance ac) and dilatancy under increasing deviatoric stress (cb). The response shown in Figure 3.9b applies when the ratio of σ_3 to σ_1 is sufficiently small. When this ratio is

Table 3.2 Brittle-to-Ductile Transition Pressures for Rocks (At Room Temperature)

Rock Type	Gage Pressure	
	(MPa)	(psi)
Rock salt	0	0
Chalk	<10	<1500
Compaction shale	0–20	0–3000
Limestone	20–100	3000–15,000
Sandstone	>100	>15,000
Granite	≥100	≥15,000

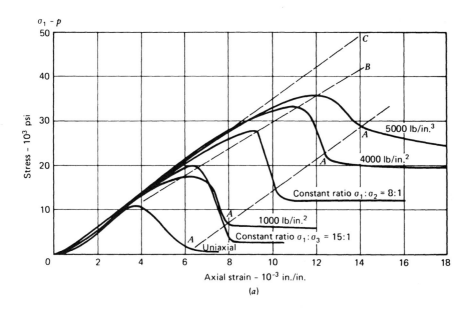

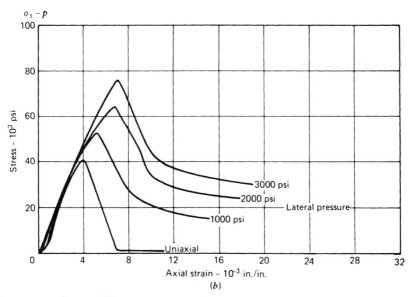

Figure 3.10 Stress difference versus axial strain curves as a function of confining pressure in triaxial compression experiments on sandstone (*a*) and norite (*b*). [From Bieniawski (1972).]

larger than a value of the order of 0.2, fracture does not occur and dilatancy is suppressed (as discussed in Section 3.8). In the usual triaxial test procedure, the principal stress ratio is decreased progressively during application of the deviator stress, until fracture occurs. But in practice, loading may occur such that the principal stress ratio remains fixed or increases.

3.4 The Meaning of "Rock Strength"

The word "failure" connotes an almost total loss of integrity in a sample of rock; in an engineering context, it usually implies loss of ability to perform the intended function. Obviously phenomena that constitute failure will depend on the function—varying from loss of a commodity in storage to structural collapse, property damage, and death. Even in a single specimen of rock, however, the concept of "failure" is unclear, for a total loss of cohesion occurs or does not occur in a single given rock according to the way it is loaded. The reason for such apparently fickle response is that the destruction of a rock by load is partly dependent on the loading system and is not a true rock property. For purposes of engineering design, it is useful to map *peak stress* values (i.e., the stresses corresponding to point D of Figure 3.7), and the criteria of failure discussed later will relate to such points. However, the compression test does not have to end in rupture at that point, but may proceed all the way to point E or beyond if the loading system is very stiff. The rock will exhibit what has been called "a complete stress-strain curve" if tested in a stiff system because the system responds to gradual deterioration in load-carrying capacity through automatic reduction in the applied load.

A testing machine is a reaction frame in which a screw or a hydraulic cylinder is operated to load a specimen. A screw-powered machine is characterized in Figure 3.11a. The rock specimen is fitted between a test table and an upper platen that are connected by stiff screws parallel to the axis of the specimen. A motor below the table turns a gear that causes the screws to turn thus bringing the upper platen up or down. If the screws are turned so that the rock specimen feels a load and then the motor is switched off, any subsequent movement of the upper platen relative to the table must alter the load at a rate given by the stiffness of the testing machine k_m. The family of lines marked A through J in Figure 3.11b describes the machine stiffness at different platen positions. Operation of the testing machine to build up load on a specimen corresponds to moving across the family of curves as shown in the figure. At the point of peak load of the rock, assuming we turn the motor off, the specimen will fail or not depending upon the relative values of k_r and k_m where k_r is the slope of the post peak portion of the complete stress-strain curve. For example, rock 1, Figure 3.11b, will continue to deform without sudden rupture as the testing machine is continuously shortened, whereas even with the ma-

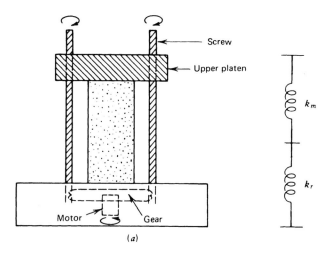

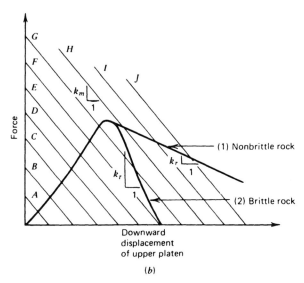

Figure 3.11 Influence of testing machine stiffness on failure. (*a*) A testing machine and its representation by a freebody. (*b*) Stable and unstable samples.

chine turned off, rock 2 will fail because the machine as it "unstretches" cannot reduce the load sufficiently to satisfy the unloading requirements of the rock. However, if the motor were quickly switched to run backward and the upper platen were moved upward, it would be possible to follow the complete stress-strain curve of rock 2 as the system returned to curves G, F, etc. In a servocontrolled testing machine, this can be done automatically, the motor responding to the commands of an electrical circuit processing the signal of a deformation transducer attached to the specimen. Even without such a machine, it is possible to determine the complete stress-strain curve of a brittle rock by quick manual response, switching back and forth from loading to unloading modes. Figure 3.12 shows an actual record from such an operation with a specimen of coarse-grained marble (Chino marble, California). When the rock was loaded in the usual way in a 160,000-lb-capacity screw testing machine, the peak load was followed by a violent rupture that reduced the specimen to rock powder. By quickly unloading whenever the *x-y* plotter record revealed the onset of yielding, a series of hysteresis loops was created, the envelope to which estimates the right side of the complete stress-strain curve. At the end of the test, the specimen displayed a continuous fault but was still integral. Since the stiffness of the rock is proportional to its cross-sectional area, such a test is relatively easier to perform using small specimens. A convincing demonstration of the influence of the loading system stiffness on the mode of failure can be achieved by running two tests with varying machine stiffness while the rock stiffness is held constant. This can be achieved by adding a spring in series with the rock for one of the two tests.

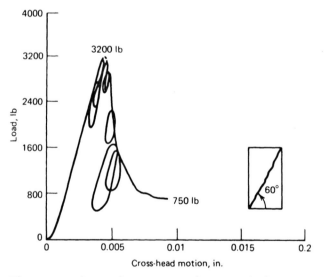

Figure 3.12 A complete stress-strain curve obtained by load cycling on a moderately stiff testing machine. The specimen was a coarse marble cylinder, 0.8 in. in diameter and 1.45 in. long.

3.5 Applications of the Complete Stress-Strain Curve

Normally when stresses become high enough to cause fractures in the wall of a tunnel or mine, rock simply spalls off, producing a destressed zone that drives the flow of stress away from the opening.

In a well-designed mine, the roof load will find somewhere else to go when a pillar collapses. But if a room and pillar mine were made with very wide rooms, the loss of one pillar might be insufferable.

These varying behavior modes in practice are understandable in terms of the complete stress-strain curve concept. In the mine with very wide rooms, the deflection of the roof due to the removal of one pillar can be calculated by assuming the roof span to be two rooms and one pillar wide. Using beam formulas or numerical model methods, the ratio of peak pillar load to the increment of roof deflection caused by removal of the pillar defines the *system stiffness*. If this stiffness is greater in magnitude than the slope of the postpeak part of the complete force-displacement curve, the mine can survive the failure of a pillar.

The complete stress-strain curve can also be used to preduct failure of rock as a result of *creep*. As shown in Figure 3.13, the locus of a creep test in the stress-strain graph is a horizontal line. If the initial stress in the rock is close to the peak load, creep will terminate in rupture when accumulated strain is such as to intersect the falling part of the complete stress-strain curve. A creep test started at *A* will terminate in rupture at point *B* after a relatively short time. A

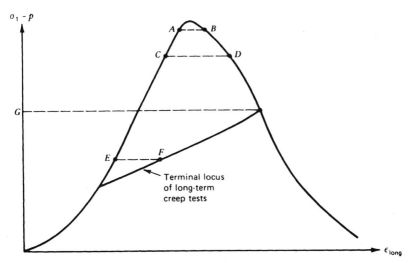

Figure 3.13 Creep in relation to the complete stress-strain curve.

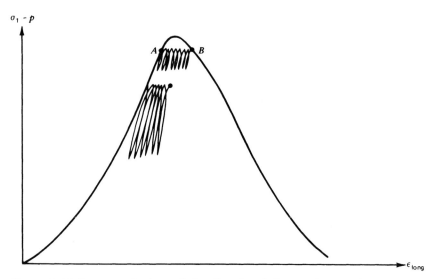

Figure 3.14 Response to dynamic loading, in relation to the complete stress-strain curve.

creep test begun at C will terminate in rupture at D after a much longer time. And a creep test initiated at E below critical stress level G will approach point F without rupture after a long time (compare with Figure 6.16).

A similar concept applies to *cyclic loading* beneath the peak load level, as shown in Figure 3.14. Cycles of loading and unloading produce "hysteresis loops" as energy is consumed in sliding on cracks and fissures inside the rock volume. Multiple load cycles begun at point A such that the peak load is never surpassed will cause a migration of the envelope of hysteresis loops which terminate in rupture at point B.[3]

3.6 The Mohr-Coulomb Failure Criterion

We have noted that the peak stress of rock undergoing deviatoric loading will increase if the rock is confined. The variation of peak stress σ_1 with confining pressure σ_3 is known as a *criterion of failure*. The simplest and best-known criterion of failure for rocks is the Mohr-Coulomb criterion; as shown in Figure 3.15, this consists of a linear envelope touching all Mohr's circles representing

[3] This suggests a method of relieving stored energy along faults near the rupture point by cyclic loading. Professor B. Haimson of the University of Wisconsin proposed this be done by cyclic pumping of water from wells in the fault zone. The effect of water pressure is considered in Section 3.7.

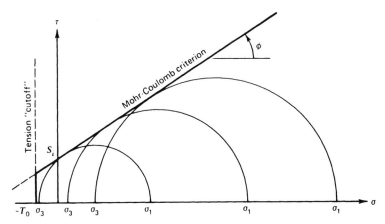

Figure 3.15 The Mohr-Coulomb failure criterion with a tension cutoff.

critical combinations of principal stresses. Stated in terms of normal and shear stresses on the plane represented by the point of tangency of a Mohr circle with the envelope

$$\tau_p = S_i + \sigma \tan \phi \tag{3.7}$$

ϕ is called the angle of internal friction, for like a friction angle for sliding between surfaces, it describes the rate of increase of peak strength with normal stress. τ_p is the peak shear stress, or *shear strength*.

The Mohr-Coulomb criterion is also used to represent the "residual" strength, that is, the minimum strength reached by a material subjected to deformation beyond the peak, as shown in Figure 3.10. In this case, the subscript r may be used with each of the terms of Equation 3.7 to identify them as parameters of residual strength. The residual shear strength ($S_{i,r}$ may approach zero while the residual friction angle ϕ_r will usually lie between zero and the peak friction angle. However, in compaction shales rich in montmorillonite, like the Cretaceous Bearpaw and Pierre shales, values as low as 4–6° are reported, even in "drained" test arrangements that prevent pore water pressure accumulation during deformation (Townsend and Gilbert, 1974).

Equation 3.7 has the following physical interpretation. "Failure" occurs when the applied shear stress less the frictional resistance associated with the normal stress on the failure plane becomes equal to a constant of the rock, S_i. Since it would not be reasonable to admit a frictional resistance in the presence of a tensile normal stress, this equation then loses its physical validity when the value of σ crosses into the tensile region; σ represents the normal stress on the plane of failure. The minimum principal stress σ_3 may be tensile as long as σ remains compressive. Other theories of failure (e.g., the Griffith theory) are more exact in the tensile region. However, the Mohr-Coulomb theory has the merit of simplicity and will be retained here by extrapolating the Mohr-Coulomb line into the tensile region up to the point where σ_3 becomes equal to the uniaxial tensile strength $-T_0$. The minor principal stress can never be less than $-T_0$.

Respecting the last as a constraint on the criterion of failure is, in effect, recognizing a "tension cutoff" superimposed on the Mohr-Coulomb criterion of failure as shown in Figure 3.15. The actual envelope of critical Mohr's circles with one principal stress negative will lie beneath the Mohr-Coulomb criterion with the superimposed tension cutoff as indicated in Figure 3.16, so it is necessary to reduce the tensile strength T_0 and the shear strength intercept S_i when applying this simplified failure criterion in any practical situation.

In terms of the principal stresses at peak load conditions, the Mohr-Coulomb criterion can be written

$$\sigma_{1,p} = q_u + \sigma_3 \tan^2 \left(45 + \frac{\phi}{2} \right) \tag{3.8}$$

where $\sigma_{1,p}$ is the major principal stress corresponding to the peak of the stress-strain curve, and q_u is the unconfined compressive strength. The change of variables leads to the following relationship between shear strength intercept S_i and unconfined compressive strength q_u

$$q_u = 2S_i \tan \left(45 + \frac{\phi}{2} \right) \tag{3.9}$$

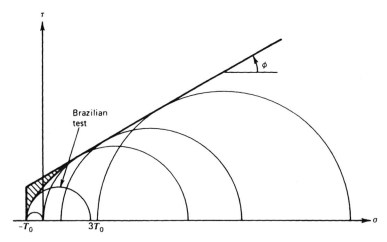

Figure 3.16 Comparison of empirical envelope and Mohr-Coulomb criterion in the tensile region. Inside the ruled region, the Mohr-Coulomb criterion with tension cutoff overestimates the strength.

Table 3.3 Representative Values for Shear Strength Intercept
(S_i) and Angle of Internal Friction (ϕ) for Selected Rocks[a]

Description	Porosity (%)	S_i (MPa)	ϕ	Range of Confining Pressure (MPa)	Reference[b]
Berea sandstone	18.2	27.2	27.8	0–200	4
Bartlesville sandstone		8.0	37.2	0–203	3
Pottsville sandstone	14.0	14.9	45.2	0–68.9	8
Repetto siltstone	5.6	34.7	32.1	0–200	4
Muddy shale	4.7	38.4	14.4	0–200	4
Stockton shale		0.34	22.0	0.8–4.1	2
Edmonton bentonitic shale	44.0	0.3	7.5	0.1–3.1	9
(water content 30%)					
Sioux quartzite		70.6	48.0	0–203	3
Texas slate; loaded					
30° to cleavage		26.2	21.0	34.5–276	6
90° to cleavage		70.3	26.9	34.5–276	6
Georgia marble	0.3	21.2	25.3	5.6–68.9	8
Wolf Camp limestone		23.6	34.8	0–203	3
Indiana limestone	19.4	6.72	42.0	0–9.6	8
Hasmark dolomite	3.5	22.8	35.5	0.8–5.9	4
Chalk	40.0	0	31.5	10–90	1
Blaine anhydrite		43.4	29.4	0–203	3
Inada biotite granite	0.4	55.2	47.7	0.1–98	7
Stone Mountain granite	0.2	55.1	51.0	0–68.9	8
Nevada Test Site basalt	4.6	66.2	31.0	3.4–34.5	10
Schistose gneiss					
90° to schistocity	0.5	46.9	28.0	0–69	2
30° to schistocity	1.9	14.8	27.6	0–69	2

[a] Data from Kulhawy (1975) (Ref. 5).

[b] 1. Dayre, M., Dessene, J. L., and Wack, B. (1970) *Proc. 2nd Congress of ISRM*, Belgrade, Vol. 1, pp. 373–381.

2. DeKlotz, E., Heck, W. J., and Neff, T. L. (1964) First Interim Report, *MRD Lab Report 64/493*, U. S. Army Corps of Engineers, Missouri River Division.

3. Handin, J. and Hager, R. V. (1957) *Bull. A.A.P.G.* **41**: 1–50.

4. Handin, J., Hager, R. V., Friedman, M., and Feather, J. N. (1963) *Bull. A.A.P.G.* **47**: 717–755.

5. Kulhawy, F. (1975) *Eng. Geol.* **9**: 327–350.

6. McLamore, R. T. (1966) Strength-deformation characteristics of anisotropic sedimentary rocks, Ph.D. Thesis, University of Texas, Austin.

7. Mogi, K. (1964) *Bull. Earthquake Res. Inst.*, Tokyo, Vol. 42, Part 3, pp. 491–514.

8. Schwartz, A. E. (1964) *Proc. 6th Symp. on Rock Mech.*, Rolla, Missouri, pp. 109–151.

9. Sinclair, S. R. and Brooker, E. W. (1967) *Proc. Geotech. Conf. on Shear Strength Properties of Natural Soils and Rocks*, Oslo, Vol. 1, pp. 295–299.

10. Stowe, R. L. (1969) U. S. Army Corps of Engineers Waterways Experiment Station. Vicksburg, *Misc. Paper C-69-1*.

The maximum tension criterion must be superimposed on Equation 3.8, that is, failure is presumed to occur because of tensile stress whenever σ_3 becomes equal to $-T_0$, regardless of the value of σ_1.

Typical values of the peak shear strength intercept S_i and the peak angle of internal friction ϕ for a representative set of rock specimens are listed in Table 3.3. The ratio of unconfined compressive to tensile strength a_u/T_0 for a sampling of rock types is given in Table 3.1.

3.7 The Effect of Water

Some rocks are weakened by the addition of water, the effect being a chemical deterioration of the cement or clay binder. A friable sandstone may, typically, lose 15% of its strength by mere saturation. In extreme cases, such as montmorillonitic clay shales, saturation is totally destructive. In most cases, however, it is the effect of pore and fissure water pressure that exerts the greatest influence on rock strength. If drainage is impeded during loading, the pores or fissures will compress the contained water, raising its pressure.

Development of pore pressure and consequent loss in strength of a Pennsylvanian shale tested in triaxial compression is shown in Figure 3.17. Two separate test results are presented in this diagram: the circles represent triaxial compression of a saturated specimen under conditions such that excess pore pressures could drain away rather than accumulate ("drained conditions"); the triangles represent a saturated shale specimen tested without drainage, so that excess pore pressures that develop must accumulate ("undrained conditions"). The curve of differential axial stress versus axial strain for the drained test displays a peak and then a descending tail as depicted in Figure 3.7a. Since the mean stress increases simultaneously with the axial stress in a triaxial test, the curve of volumetric strain shown in Figure 3.17 is the sum of hydrostatic compression (Figure 3.6) and dilatancy behavior (Figure 3.7b). Initially, the volume decreases by hydrostatic compression until the specimen begins to dilate, whereupon the rate of volume decrease slows, eventually becoming negative, meaning the volume increases on subsequent load increments. In the undrained test, the tendency for volume change cannot be fully realized because the water filling the voids undergoes compression rather than drainage. As a result, the water pressure p_w inside the pores begins to increase. This dramatically lowers the peak stress and flattens the postpeak curve.

Many investigators have confirmed the validity of Terzaghi's *effective stress law* for rocks, which states that a pressure of p_w in the pore water of a rock will cause the same reduction in peak normal stress as caused by a reduction of the confining pressure by an amount equal to p_w. We can make use of this result by introducing the term *effective stress* σ' defined by

$$\sigma' = \sigma - p_w \tag{3.10}$$

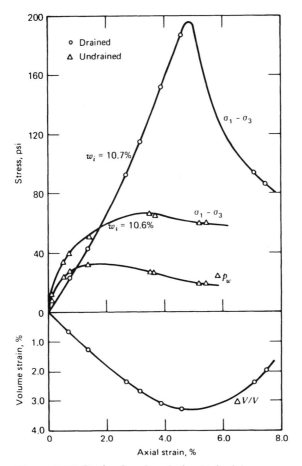

Figure 3.17 Drained and undrained triaxial compression test results for a shale of Pennsylvanian age; w_i is the initial water content; p_w is the pore water pressure. (From Mesri and Gibala, 1972.)

Differential stress $(\sigma_1 - \sigma_3)$ is unaffected by water pressure since $\sigma'_1 - \sigma'_3 = (\sigma_1 - p_w) - (\sigma_3 - p_w) = \sigma_1 - \sigma_3$.

The effect of water pressure can be input in the failure criterion simply by restating the conditions for failure in terms of effective stresses. In a test on a dry rock, there is no difference between normal stresses and effective normal stresses. For a saturated rock, rewrite Equation 3.8 in terms of effective stress by introducing ' on the normal stress terms:

$$\sigma'_{1,p} = q_u + \sigma'_3 \tan^2 \left(45 + \frac{\phi}{2}\right) \tag{3.11}$$

or

$$\sigma'_{1,p} - \sigma'_3 = q_u + \sigma'_3 \left[\tan^2 \left(45 + \frac{\phi}{2} \right) - 1 \right] \qquad (3.12)$$

Since the differential stress is unaffected by pore pressure, Equation 3.12 may also be written

$$\sigma_{1,p} - \sigma_3 = q_u + (\sigma_3 - p_w) \left[\tan^2 \left(45 + \frac{\phi}{2} \right) - 1 \right]$$

Solving for p_w, we can calculate the water pressure in the pores or fissures of a rock required to initiate failure from an initial state of stress defined by σ_1 and σ_3:

$$p_w = \sigma_3 - \frac{(\sigma_1 - \sigma_3) - q_u}{\tan^2(45 + \phi/2) - 1} \qquad (3.13)$$

Figure 3.18 portrays this condition graphically. The buildup of water pressure in the rock near a reservoir or in an aquifer can cause rock failure and earthquakes, if the rock is initially stressed near the limit. However, earthquakes induced by reservoir construction and by pumping water into deep aquifers are believed to originate from rupture along preexisting faults in determined orientations. The mechanism is similar but the equations contain the influence of the relative directions of initial stress as discussed in Chapter 5 (compare with Equation 5.9).

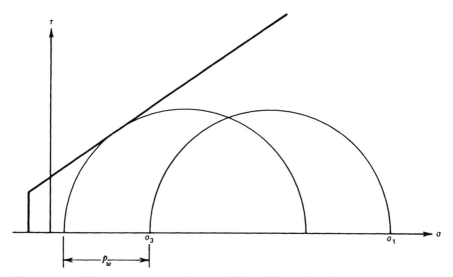

Figure 3.18 Water pressure required to initiate failure of an intact rock from a given initial state of stress.

3.8 The Influence of the Principal Stress Ratio on Failure

In the usual triaxial compression experiment, the rock is seated with a hydrostatic stress, that is, with a principal stress ratio $K = \sigma_3/\sigma_1$ equal to unity. Thereafter, as σ_1 is raised, the value of K is reduced until cracking occurs, and eventually peak strength is reached. This may not be a realistic loading path for all situations, and it may be desirable to consider rock behavior when the principal stress ratio is fixed at some value. In examining the conditions resulting from excavating an underground chamber in a rock mass, for example, the directions and relative magnitudes of principal stresses can be found throughout the region of influence of the opening. Changing the assumption about the magnitudes of the initial stresses will increase or decrease these stresses but will not alter any value of K as long as the rock behaves elastically. Therefore, there is some merit in expressing the criterion of failure in terms of the principal stress ratio, as discussed by Hoek (1968). When this is done, it is easily seen that there is a value of K above which failure cannot occur, and this can be verified by tests. In terms of the Mohr-Coulomb criterion of failure, dividing both sides of Equation 3.8 by $\sigma_{1,P}$ and introducing $K = \sigma_3/\sigma_1$ yields

$$\sigma_{1,p} = \frac{q_u}{1 - K \tan^2(45 + \phi/2)} \qquad (3.14)$$

from which we can see that the peak major principal stress becomes large when K approaches $\cot^2(45 + \phi/2)$. For example, for $\phi = 45°$, failure cannot occur above a principal stress ratio $K = 0.17$.

3.9 Empirical Criteria of Failure

While the Mohr-Coulomb criterion is easy to work with and affords a useful formula for manipulation in practical situations, a more precise criterion of failure can be determined for any rock by fitting an envelope to Mohr's circles representing values of the principal stresses at peak conditions in laboratory tests. As shown in Figure 3.19, this envelope will frequently curve downward. Jaeger and Cook (1976)[4] and Hoek (1968) demonstrated that the failure envelopes for most rocks lie between a straight line and a parabola. The Griffith theory of failure predicts a parabola in the tensile stress region. This theory is premised on the presence of randomly oriented fissures in the rock that act to create local stress concentrations, facilitating new crack initiation. However, the Griffith theory has no physical basis in the region where both principal

[4] See references, Chapter 1.

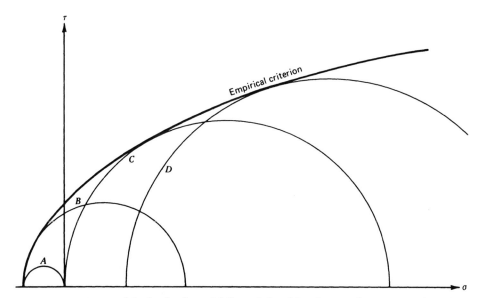

Figure 3.19 An empirical criterion of failure defined by the envelope to a series of Mohr's circles: A, direct tension; B, Brazilian; C, unconfined compression; D, triaxial compression.

stresses are compressive. In practice, empirical curve fitting is the best procedure for producing a criterion of failure tailored to any given rock type. See, for example, Herget and Unrug (1976). A satisfactory formula for many purposes will be afforded by the union of a tension cutoff, $\sigma_3 = -T_0$, and a power law (Bieniawski, 1974):

$$\frac{\sigma_{1,p}}{q_u} = 1 + N \left(\frac{\sigma_3}{q_u}\right)^M \tag{3.15}$$

The constants N and M will be determined by fitting a curve to the family of points

$$\left(\frac{\sigma_3}{q_u}, \frac{\sigma_{1,p}}{q_u} - 1\right)$$

Another approach fits data from the ring shear test (Figure 3.3e) to find an empirical equation for a Mohr envelope of intact rock (Lundborg, 1966). The peak shear strength (τ_p) (Equation 3.4) is plotted against σ to define the strength envelope, Figure 3.20a. Lundborg found that such data define a curved envelope with intercept S_i and asymptote S_f, fit by

$$\tau_p = S_i + \frac{\mu'\sigma}{1 + \dfrac{\mu'\sigma}{S_f - S_i}} \tag{3.16}$$

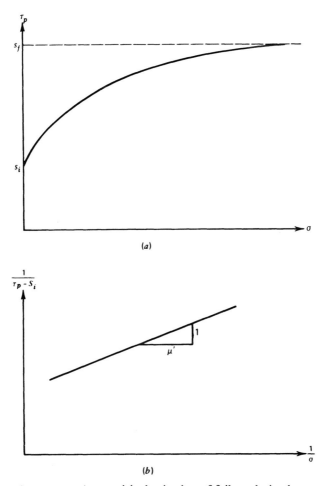

Figure 3.20 An empirical criterion of failure derived from the ring shear test. (After Lundborg, 1966.) (a) A plot of equation 3.16; (b) graphical determination of μ'.

(The symbol μ' has been used in place of Lundborg's μ to distinguish this coefficient from the Mohr-Coulomb coefficient of internal friction, $\mu = \tan \phi$.) Equation 3.16 can àlso be written

$$\frac{1}{\tau_p - S_i} = \frac{1}{\mu'\sigma} + \frac{1}{S_f - S_i} \qquad (3.17)$$

so that μ' is determined as the inverse of the slope of the line obtained by plotting $(\tau_p - S_i)^{-1}$ as ordinate against $(\sigma)^{-1}$ as abscissa (Figure 3.20b). Table 3.4 lists typical values of Lundborg's parameters. Strengths determined by ring

Table 3.4 Some Values of
Constants for Lundborg's
Strength Equation

	μ'	S_i (MPa)	S_f (MPa)
Granite	2.0	60	970
Pegmatite	2.5	50	1170
Quartzite	2.0	60	610
Slate	1.8	30	570
Limestone	1.2	30	870

shear tend to be slightly higher than corresponding strengths determined by triaxial tests.

3.10 The Effect of Size on Strength

Rocks are composed of crystals and grains in a fabric that includes cracks and fissures; understandably, rather large samples are required to obtain statistically complete collections of all the components that influence strength. When the size of a specimen is so small that relatively few cracks are present, failure is forced to involve new crack growth, whereas a rock mass loaded through a larger volume in the field may present preexisting cracks in critical locations. Thus rock strength is size dependent. Coal, altered granitic rocks, shale, and other rocks with networks of fissures exhibit the greatest degree of size dependency, the ratio of field to laboratory strengths sometimes attaining values of 10 or more.

A few definitive studies have been made of size effect in compressive strength over a broad spectrum of specimen sizes. Bieniawski (1968) reported tests on prismatic in situ coal specimens up to $1.6 \times 1.6 \times 1$ m, prepared by cutting coal from a pillar; the specimens were then capped with strong concrete and loaded by hydraulic jacks. Jahns (1966) reported results of similar tests on cubical specimens of calcareous iron ore; the specimens were prepared by means of slot cutting with overlapping drill holes. Jahns recommended a specimen size such that 10 discontinuities intersect any edge. Larger specimens are more expensive without bringing additional size reduction, while smaller specimens yield unnaturally high strengths. Available data are too sparse to accept Jahn's recommendation for all rock types but it does appear that there is generally a size such that larger specimens suffer no further decrease in

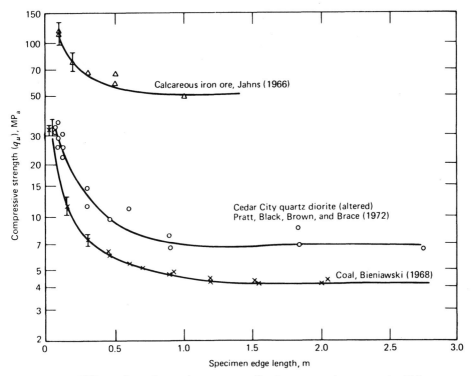

Figure 3.21 Effect of specimen size on unconfined compressive strength. (After Bieniawski and Van Heerden, 1975.)

strength. Figure 3.21 demonstrates this pattern of behavior in a summary of the tests on coal and iron ore, as well as tests on an altered and fissured quartz diorite by Pratt et al. (1972). This clever series of tests included specimens of equilateral triangular cross section 6 ft (1.83 m) on edge, and 9 ft (2.74 m) long, loaded via stainless steel flat jacks in a vertical slot at one end. Figure 3.22a shows a specimen being freed by drilling a slot inclined at 60° and Figure 3.22b shows the surface of the specimen, with completed slots, jacks in place on one end, and extensometers positioned for strain measurements on the surface. The quartz diorite tested displayed a large size effect because it contains highly fractured plagioclase and amphibole phenocrysts in a finer-grained ground mass with disseminated clay; the porosity of this rock is 8–10%.

The influence of size on shear and tension tests is less well documented but undoubtedly as severe for rocks that contain discontinuities. The subject of scale effect will be considered further in Chapter 7 in the context of underground openings.

Figure 3.22 Large uniaxial compression tests conducted in-situ by TerraTek on Cedar City Quartz Diorite. (*a*) Drilling a line of 1–1/2-inch diameter holes plunging 60° to create an inclined slot forming one side of the triangular prism "specimen." (*b*) A view of the test site showing flat jacks at one end and extensometers for relative displacement measurement during loading. (Courtesy of H. Pratt.)

3.11 Anisotropic Rocks

Variation of compressive strength according to the direction of the principal stresses is termed "strength anisotropy." Strong anisotropy is characteristic of rocks composed of parallel arrangements of flat minerals like mica, chlorite, and clay, or long minerals like hornblende. Thus the metamorphic rocks, especially schist and slate, are often markedly directional in their behavior. For example, Donath (1964) found the ratio of minimum to maximum unconfined compressive strength of Martinsburg slate to be equal to 0.17. Anisotropy also occurs in regularly interlayered mixtures of different components, as in banded gneisses, sandstone/shale alternations, or chert/shale alternations. In all such rocks, strength varies continuously with direction and demonstrates pronounced minima when the planes of symmetry of the rock structure are oblique to the major principal stress.

Rock masses cut by sets of joints also display strength anisotropy, except where the joint planes lie within about 30° of being normal to the major principal stress direction. The theory of strength for jointed rocks is discussed in Chapter 5.

Strength anisotropy can be evaluated best by systematic laboratory testing of specimens drilled in different directions from an oriented block sample. Triaxial compression tests at a set of confining pressures for each given orientation then determine the parameters S_i and ϕ as functions of orientation. Expanding on a theory introduced by Jaeger (1960), McLamore (1966) proposed that both S_i and ϕ could be described as continuous functions of direction according to

$$S_i = S_1 - S_2[\cos 2(\psi - \psi_{min,s})]^n \qquad (3.18)$$

and

$$\tan \phi = T_1 - T_2[\cos 2(\psi - \psi_{min,\phi})]^m \qquad (3.19)$$

where S_1, S_2, T_1, T_2, m, and n are constants
    ψ is the angle between the direction of the cleavage (or schistocity, bedding or symmetry plane) and the direction σ_1
    $\psi_{min,s}$ and $\psi_{min,\phi}$ are the values of ψ corresponding to minima in S_i and ϕ, respectively

For a slate, McLamore determined that friction and shear strength intercept minima occur at different values of ψ, respectively 50 and 30°. The strength parameters for the slate are

$$S_i = 65.0 - 38.6[\cos 2(\psi - 30)]^3 \text{ (MPa)} \qquad (3.18a)$$

and

$$\tan \phi = 0.600 - 0.280 \cos 2(\psi - 50) \qquad (3.19a)$$

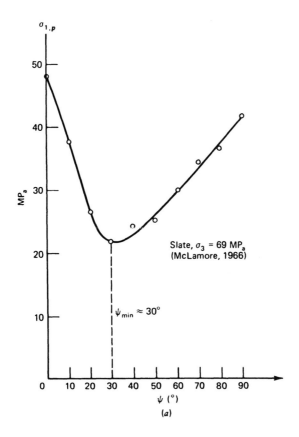

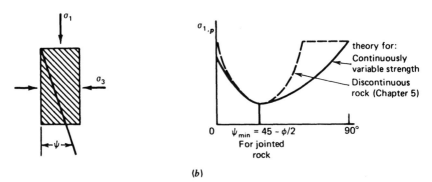

Figure 3.23 Strength anisotropy in triaxial compression.

In general, the entire range of ψ from 0 to 90° cannot be well fit with one set of constants since the theory (Equations 3.18 and 3.19) would then predict strength at $\psi = 0°$ to be less than the strength at $\psi = 90°$; in fact, the strength when loading is parallel to slaty cleavage, schistosity, or bedding is usually higher than the strength when the loading is perpendicular to the planes of weakness within the rock. (Compare Figures 3.23*a* and *b*.) For oil shale, a repetitive layering of marlstone and kerogen, McLamore used one set of constants for the region $0° \leq \psi < 30°$ and a second set of constants for $30° \leq \psi \leq 90°$.

The variation of the friction angle with direction proves generally less severe than the variation of the shear strength intercept. As a simplification, assume $n = 1$, $\psi_{min,s} = 30°$, and ϕ independent of direction ($m = 0$). Then the strength anisotropy can be evaluated from compressive tests run at $\psi = 30°$ and $\psi = 75°$ (see Problem 12).

References

Bieniawski, Z. T. (1967a) Stability concept of brittle fracture propagation in rock, *Eng. Geol.* **2:** 149–162.

Bieniawski, Z. T. (1967b) Mechanism of brittle fracture of rock, *Int. J. Rock Mech. Min. Sci.* **4:** 395–430.

Bieniawski, Z. T. (1968) The effect of specimen size on compressive strength of coal, *Int. J. Rock Mech. Min. Sci.* **5:** 325–335.

Bieniawski, Z. T. (1972) Propagation of brittle fracture in rock, *Proceedings, 10th Symposium on Rock Mechanics* (AIME), pp. 409–427.

Bieniawski, Z. T. (1974) Estimating the strength of rock materials, *J. South African Inst. Min. Metall.* **74:** 312–320.

Bieniawski, Z. T. and Bernede, M. J. (1979) Suggested methods for determining the uniaxial compressive strength and deformability of rock materials, for ISRM Commission on Standardization of Laboratory and Field Tests, *Int. J. Rock Mech. Min. Sci.* **16** (2).

Bieniawski, Z. T. and Hawkes, I. (1978) Suggested methods for determining tensile strength of rock materials, for ISRM Commission on Standardization of Lab and Field Tests, *Int. J. Rock Mech. Min. Sci.* **15:** 99–104.

Bieniawski, Z. T. and Van Heerden, W. L. (1975) The significance of in-situ tests on large rock specimens, *Int. J. Rock Mech. Min. Sci.* **12:** 101–113.

Broch, E. (1974) The influence of water on some rock properties, *Proc. 3rd Cong. ISRM* (Denver), Vol. II A, pp. 33–38.

Brown, E. T., Richards, L. W., and Barr, M. V. (1977) Shear strength characteristics of Delabole slates, *Proceedings, Conference on Rock Engineering* (British Geotechnical Society, Vol. 1, pp. 33–51.

Cook, N. G. W. and Hodgson, K. (1965) Some detailed stress-strain curves for Rock, *J. Geophys. Res.* **70:** 2883–2888.

Donath, F. A. (1964) Strength variation and deformational behavior in anisotropic rocks. In W. Judd (Ed.), *State of Stress in the Earth's Crust,* Elsevier, New York, pp. 281–300.

Fairhurst, C. (1964) On the validity of the Brazilian test for brittle materials, *Int. J. Rock Mech. Min. Sci.* **1:** 535–546.

Haimson, B. C. (1974) Mechanical behavior of rock under cyclic loading, *Proc. 3rd Cong. ISRM* (Denver), Vol. II A, pp. 373–378.

Hallbauer, D. K., Wagner, H., and Cook, N. G. W. (1973) Some observations concerning the microscopic and mechanical behavior of quartzite specimens in stiff triaxial compression tests, *Int. J. Rock Mech. Min. Sci.* **10:** 713–726.

Heard, H. C. (1967) The influence of environment on the brittle failure of rocks, *Proceedings, 8th Symposium on Rock Mechanics* (AIME), pp. 82–93.

Herget, G. and Unrug, K. (1976) In situ rock strength from triaxial testing, *Int. J. Rock Mech. Min. Sci.* **13:** 299–302.

Heuze, F. E. (1980) Scale effects in the determination of rock mass strength and deformability, *Rock Mech.* **12** (3).

Hoek, E. (1968) Brittle failure of rock. In K. Stagg and O. Zienkiewicz (Eds.), *Rock Mechanics in Engineering Practice,* Wiley, New York.

Hoek, E. and Brown, E. T. (1980). Empirical strength criterion for rock masses, *J. Geotech. Eng.* ASCE **106:** 1013–1035.

Hoek, E. and Franklin, J. A. (1968) Sample triaxial cell for field or laboratory testing of rock, *Trans. Section A Inst. Min. Metal.* **77:** A22–A26.

Hudson, J. A., Crouch, S. L., and Fairhurst, C. (1972) Soft, stiff and servo-controlled testing machines: A review with reference to rock failure, *Engl. Geol.* **6:** 155–189.

Hustrulid, W. and Robinson, F. (1972) A simple stiff machine for testing rock in compression, *Proceedings, 14th Symposium on Rock Mechanics* (ASCE), pp. 61–84.

Hustrulid, W. A. (1976) A review of coal pillar strength formulas, *Rock Mech.* **8:** 115–145.

Jaeger, J. C. (1960) Shear failure of anisotropic rocks, *Geol. Mag.* **97:** 65–72.

Jahns, H. (1966) Measuring the strength of rock in-situ at an increasing scale, *Proc. 1st Cong. ISRM* (Lisbon), Vol. 1, pp. 477–482 (in German).

Kulhawy, F. H. (1975) Stress-deformation properties of rock and rock discontinuities, *Eng. Geol.* **9:** 327–350.

Lundborg, N. (1966) Triaxial shear strength of some Swedish rocks and ores, *Proc. 1st Cong. ISRM* (Lisbon), Vol. 1, pp. 251–255.

Maurer, W. C. (1965) Shear failure of rock under compression, *Soc. Petrol. Eng.* **5:** 167–176.

McLamore, R. T. (1966) Strength-deformation characteristics of anisotropic sedimentary rocks, Ph.D. thesis, University of Texas, Austin.

Mesri, G. and Gibala, R. (1972) Engineering properties of a Pennsylvanian shale, *Proceedings, 13th Symposium on Rock Mechanics* (ASCE), pp. 57–75.

Pratt, H. R., Black, A. D., Brown, W. D., and Brace, W. F. (1972) The effect of specimen size on the mechanical properties of unjointed diorite, *Int. J. Rock Mech. Min. Sci.* **9:** 513–530.

Reichmuth, D. R. (1963) Correlation of force-displacement data with physical properties of rock for percussive drilling systems, *Proceedings, 5th Symposium on Rock Mechanics,* p. 33.

Robinson, L. H. Jr. (1959) Effects of pore pressures on failure characteristics of sedimentary rocks, *Trans. AIME* **216:** 26–32.

Tourenq, C. and Denis, A. (1970) The tensile strength of rocks, *Lab de Ponts et Chaussées—Paris, Research Report 4* (in French).

Townsend, F. C. and Gilbert, P. A. (1974) Engineering properties of clay shales, Corps of Engineers, WES Tech, Report S-7i-6.

Vogler, U. W. and Kovari, K. (1978) Suggested methods for determining the strength of rock materials in triaxial compression, for ISRM Commission on Standardization of Laboratory and Field Tests, *Int. J. Rock Mech. Min. Sci.* **15:** 47–52.

Wawersik, W. R. (1972) Time dependent rock behavior in uniaxial compression, *Proceedings, 14th Symposium on Rock Mechanics* (ASCE), pp. 85–106.

Wawersik, W. R. and Brace, W. F. (1971) Post failure behavior of a granite and a diabase, *Rock Mech.* **3:** 61–85.

Wawersik, W. R. and Fairhurst, C. (1970) A study of brittle rock fracture in laboratory compression experiments, *Int. J. Rock Mech. Min. Sci.* **7:** 561–575.

Yudhbir, Lemanza, W., and Prinzl, F. (1983) *Proc. 5th Cong. ISRM* (Melbourne), pp. B1–B8.

Problems

1. In a series of triaxial compression tests on a sandstone, the following represent the stresses at peak load conditions:

Test	σ_3(MPa)	σ_1(MPa)
1	1.0	9.2
2	5.0	28.0
3	9.5	48.7
4	15.0	74.0

   Determine values of S_i and ϕ that best fit the data.

2. The initial state of stress at a point in the ground, in sandstone of Problem 1, is

   $$\sigma_3 = 1300 \text{ psi}$$
   $$\sigma_1 = 5000 \text{ psi}$$

   The pore water pressure (p_w) will be raised by the construction of a reservoir. What value of p_w will cause fracture of the sandstone in situ (assume fracture occurs at peak stress).

3. For the rock of Problems 1 and 2, what is the value of the principal stress ratio (σ_3/σ_1) above which "failure" cannot occur?

4. In a plane wave front, the pressures in the plane of the wave front are $\nu/(1 - \nu)$ times the pressure normal to the wave front. What is the least value of Poisson's ratio (ν) such that compressive or shear failure does not occur as the wave front sweeps through the rock? (See Problem 3.)

5. Triaxial compression tests of porous rock yield S_i equal to 1.0 MPa and ϕ equals 35°. Calculate the unconfined compressive strength and estimate the tensile strength for this rock.

6. In an area underlain by rock of Problem 5, a porous limestone, the in situ stresses at a given point near a reservoir site are $\sigma_1 = 12$ MPa, and $\sigma_3 = 4$ MPa. How deep a reservoir must be built before the pore pressure increase from a corresponding rise in the groundwater levels could fracture the rock? (Express your answer in MPa and psi units for the required increase in water pressure and meters or feet for the reservoir depth.)

7. If a bilinear failure criterion is used with a tension cutoff superimposed on the Mohr-Coulomb criterion, at what value of σ (expressed in terms of T_0, S_i, and ϕ) are shear failure and tensile failure criteria satisfied simultaneously?

8. (a) For the sandstone of Figure 3.10, determine the value of ϕ_p and S_{ip} that best fits the peak strengths given. (The subscript p denotes "peak.")
   (b) Find best fit values of ϕ_r and S_{ir} describing the residual strength of the sandstone.
   (c) Find values of M and N to express the peak strength of the sandstone in Figure 3.10 according to Equation 3.15.
   (d) Find ϕ_p and S_{ip} for the norite of Figure 3.10.
   (e) Find ϕ_r and S_{ir} for the norite.
   (f) Find M and N for the norite.

9. Derive an expression for the modulus of rupture T_{MR} for a test with three-point bending of a core sample (circular cross section).

10. Show that the volume change per unit of volume ($\Delta V/V$) of a rock core undergoing compression is approximately equal to the sum of the three normal strains.

11. (a) Plot Mohr envelopes of strength for (1) $\psi = 0°$, (2) $\psi = 30°$, (3) $\psi = 60°$, and (4) $\psi = 90°$ for the slate whose strength anisotropy is described by Equation 3.18a and 3.19a.
    (b) Plot the peak compressive strength of the slate as a function of ψ for (1) $\sigma_3 = 0$ and (2) $\sigma_3 = 30$ MPa.

12. A set of unconfined compressive strength tests on slate specimens oriented with $\psi = 30°$ and $\psi = 75°$ yields strength values $q_{u,30}$ and $q_{u,75}$, respectively. Show that the directional strength of the rock can be approximated by

$$\sigma_{1,P} = \sigma_3 \tan^2 \left(45 + \frac{\phi}{2}\right) + q_{u,75} - (q_{u,75} - q_{u,30})\cos 2(\psi - 30)$$

13. It has been observed that the degree of anisotropy, as expressed by the ratio of maximum to minimum strength for all directions through a rock, decreases as the confining pressure increases. What explanation can you offer?

14. The four-point loading configuration for the modulus-of-rupture test is desirable because it generates uniform moment with zero shear between the inner load points. Can you find another beam testing configuration such that the central portion of the rock beam receives zero moment with constant shear?

15. (a) Derive an empirical expression similar to Equation 3.15 appropriate for residual strength.
    (b) Find values for the constants M and N fitting the residual strength data for the sandstone in Figure 3.10.
    (c) Do the same for the residual strength of the norite.

16. E. Hoek and E. T. Brown (1980) introduced the empirical criterion of failure for rocks:

$$\frac{\sigma_{1,p}}{q_u} = \frac{\sigma_3}{q_u} + \left(m \frac{\sigma_3}{q_u} + s\right)^{1/2}$$

where m and s are constants;

$$s = \left(\frac{q_u \text{ rock mass}}{q_u \text{ rock substance}}\right)^2$$

(a) Compare this with Equation 3.15 for the case $m = 0$ and $s = 1$.
(b) Hoek and Brown, in studying many sets of data, found the following approximate values: $m = 7$ for carbonate rocks, $m = 10$ for argillaceous rocks, $m \simeq 15$ for sandstone and quartzite, $m = 17$ for volcanic rocks, and $m = 25$ for granitic and other intrusive rocks. Compare the peak compressive strengths as a function of confining pressure for a marble, a rhyolite, and a granite, all having $q_u = 100$ MPa. What is the physical meaning of parameter m?

17. Equation 3.15 was generalized by Yudhbir et al. (1983) by replacing the unity on the right side by the constant A. Permitting A to vary between 1 and 0 offers a continuous variation of rock mass quality in the criterion of failure. They propose linking A to Barton's Q (see Chapter 2) according to the relation $A = 0.0176Q^M$.
    (a) Find a corresponding relationship between A and RMR.
    (b) Based on the answer to part (a), use Equation 3.15 to express the peak major principal stress as a function of confining pressure for a weak sandstone with $M = 0.65$, $N = 5$, $q_u = 2.0$ MPa, and RMR = 50.

Chapter 4

Initial Stresses in Rocks and Their Measurement

Any undisturbed mass of rock in situ contains nonzero stress components due to weight of overlying materials, confinement, and past stress history. Near the surface in mountainous regions the in situ stress may approach zero at some points or lie close to the rock strength at others. In the former case, rocks may fall from surface and underground excavations because joints are open and weak; in the latter case, disturbance of the stress field by tunneling or perhaps even surface excavation may trigger violent release of stored energy. This chapter concerns determination of the magnitude and direction of the initial stresses at the site of a work.

4.1 Influence of the Initial Stresses

It is often possible to estimate the order of magnitude of stresses and their directions, but one can never be certain of the margin of error without backup measurements. Application of such measurements is fairly common in mining practice, but since stress measurements tend to be expensive they are not routine for civil engineering applications. There are several civil engineering situations, however, when knowledge of the state of stress can be helpful or lack of knowledge might prove so costly that a significant stress measurement program is warranted. For example, when choosing the *orientation* for a cavern, one hopes to avoid aligning the long dimension perpendicular to the greatest principal stress. If the initial stresses are very high, the *shape* will have to be selected largely to minimize stress concentrations. Knowledge of rock stresses also aids in *layout* of complex underground works. An underground powerhouse for example, consists of a three-dimensional array of openings including a machine hall, a transformer gallery, low-voltage lead shafts, pressure tunnels, surge shafts, rock traps, access tunnels, ventilation tunnels, muck hauling tunnels, penstocks, draft tubes, and other openings. Cracks that initiate at one opening must not run into another (Figure 4.1a). Since cracks tend to extend in the plane perpendicular to σ_3 knowledge of the direction of the stresses permits choosing a *layout* to reduce this risk. Pressure tunnels and penstocks can be constructed and operated in rock without any *lining* if virgin stress is greater than the internal water pressure, so for such applications stress measurement might permit large cost savings. When *displacement instruments* are installed in an underground or surface excavation, to monitor the rock performance during construction and service, stress measurements beforehand provide a framework for analysis of the data and enhance their value. When making large surface excavations with *presplitting* techniques, economies will be realized if the excavation is oriented perpendicular to σ_3 (Figures 4.1b,c). With underground storage of fluids in reservoir rocks, knowledge of the initial state of stress will help evaluate the potential hazard of *triggering an earthquake*. These are a few examples of situations in which a knowledge of the state of stress can be integrated in engineering design. In a more general sense, however, the state of stress can be considered a basic rock attribute whose magnitudes and directions affect the overall rock strength, permeability, deformability, and other important rock mass characteristics. Thus it is rarely irrelevant to know the initial stress state when dealing with rock in situ.

Sometimes initial stresses are so high that engineering activities can trigger rock failure. Whenever the major stress in the region of an excavation is more than about 25% of the unconfined compressive strength, new cracking can be expected as result of construction no matter how carefully it is performed. This derives from two observations: (1) the maximum stress concentration around an underground opening cannot be less than 2; and (2) cracking occurs in an unconfined compression specimen when the stress reaches about half of the unconfined compressive strength. Close to steep valley sides, where the angle from the excavation to the mountain top is greater than 25°, data show that rock stress problems tend to occur in Norwegian fjord country whenever the weight of rock cover is greater than about $0.15q_u$ (Brekke and Selmer-Olsen, 1966; Brekke, 1970). Such stress problems can vary from slabbing and overbreak of rock on the tunnel wall nearest the valley side, to isolated violent detachment of rocks from the walls or even destructive bursts. Conditions for rock bursts are found underground in deep mines, as in the Canadian Kirkland Lake District, the South African gold mines, and the Idaho Coeur d'Alene district, where mining is pursued at depths of as much as 11,000 ft. In civil engineering work, in addition to the valleyside stress problem noted, railroad and road tunnels under high mountains, such as, the Mont Blanc Tunnel in the Alps, have encountered severe rock stress problems. In shales and other rocks

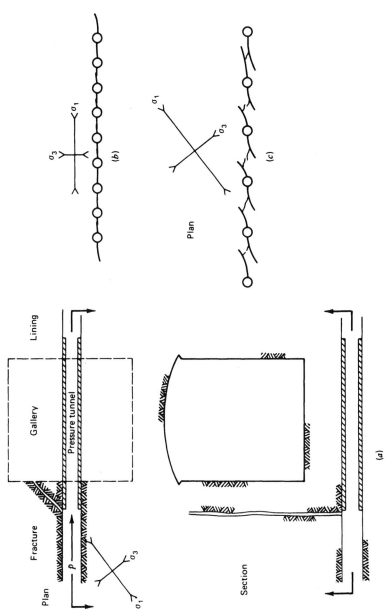

Figure 4.1 Some examples of the influence of stress direction on practice.

with low values of q_u, conditions for rock failure due to concentration of initial stress may lead to slow compression ("squeeze") and destruction of tunnel supports rather than violent collapse, but difficulties can still be significant. The "stand-up time" of a tunnel, that is, the maximum duration for erecting supports, is closely related to the ratio of maximum initial stress to q_u.

4.2 Estimating the Initial Stresses

VERTICAL STRESS

It is generally safe to assume that the vertical normal stress is equal to the weight of the overlying rock, 0.027 MPa/m or 1.2 psi/ft on the average. Near horizontal ground, the principal stress directions are vertical and horizontal. It is often assumed that they are also vertical and horizontal at depth (Figure 4.2a); however, this is just an assumption to reduce the number of unknowns, an assumption that finds reinforcement in Anderson's observations that normal and reverse faults often dip at 60 and 30°, respectively (see Jaeger and Cook, 1976). The simplifying assumption that the principal stresses are vertical and horizontal has been widely adopted in practice. Of course, this breaks down at shallow depths beneath hilly terrain, because the ground surface, lacking normal and shear stresses, always forms a trajectory of principal stress (Figure 4.2). Beneath a valley side, one principal stress is normal to the slope and equals zero, while the other two principal stresses lie in the plane of the slope

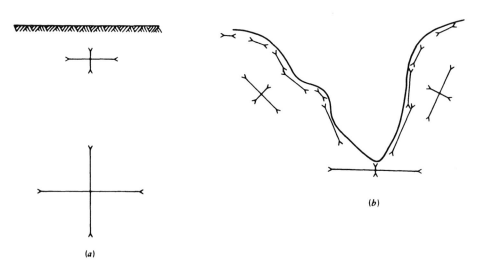

(a)

(b)

Figure 4.2 The influence of topography on initial stresses.

(Figure 4.2*b*). These stresses likewise approach zero where the rock slope is convex upward but grow larger where the slope is concave upward. Beneath the sharp notch of a V-shaped valley, the in situ stresses may be close to or at the strength of the rock.

Over any significant horizontal surface within the ground, the average vertical stress must equilibrate the downward force of the weight of overlying rock, hence the rule stated previously:

$$\bar{\sigma}_v = \gamma Z \tag{4.1}$$

where $\bar{\sigma}_v$ is the average total vertical stress at depth Z in rock with unit weight γ. This rule has been supported by numerous measurements (Figure 4.7*a*) and is one of the reliable formulas of stress in situ. However, it can be violated over limited horizontal distances owing to effects of geological structure. Figure 4.3, for example, shows how the vertical stress might vary along horizontal planes cutting through a succession of rigid and compliant beds folded into synclines and anticlines. Along line *AA'* the stress varies from perhaps 60% greater than γZ under the syncline to zero just beneath the anticline, the more rigid layer serving as a protective canopy and directing the flow of force down the limbs of the fold. A tunnel driven along line *BB'* could expect to pass from relatively understressed rock in the compliant shales to highly stressed rock as it crossed

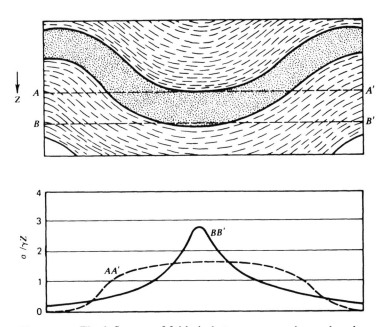

Figure 4.3 The influence of folds in heterogeneous, layered rock on vertical stresses.

into more rigid sandstone in passing under the trough of the syncline. If there is a low-strength sheared zone along the contact, produced by slip between the layers during folding, the vertical stress could be expected to jump in crossing the contact. Since geological structure can alter the vertical stresses and the direction of principal stresses, it is wise to investigate geological effects through analysis in important applications wherever geological heterogeneities can be expected to deflect the lines of force away from the vertical. Figure 4.4 shows the result of one such analysis, performed using the finite element method, in a region with heterogeneous geology superimposed on a sharply notched topography.

HORIZONTAL STRESS

In regard to the magnitude of the horizontal stresses, it is convenient to discuss the ratio of horizontal to vertical stresses. Let

$$K = \frac{\sigma_h}{\sigma_v} \tag{4.2}$$

In a region of recent sedimentation, such as the Mississippi Delta, the theory of elasticity can be invoked to predict that K will be equal to $\nu/(1 - \nu)$. This expression derives from the symmetry of one-dimensional loading of an elastic material over a continuous plane surface, which infers a condition of no horizontal strain; such a formula has no validity in a rock mass that has experienced cycles of loading and unloading. Consider an element of rock at depth Z_0 with initial value of $K = K_0$, which is then subjected to unloading by removal of ΔZ thickness of overburden (Figure 4.5). Due to unloading of $\gamma\Delta Z$ vertical stress, the horizontal stress is reduced by $\gamma\Delta Z\nu/(1 - \nu)$. Therefore, after erosion of a thickness of rock equal to ΔZ, the horizontal stress at depth $Z = Z_0 - \Delta Z$ will become equal to $K_0\gamma Z_0 - \gamma\Delta Z\nu/(1 - \nu)$, and

$$K(Z) = K_0 + \left[\left(K_0 - \frac{\nu}{1 - \nu}\right)\Delta Z\right] \cdot \frac{1}{Z} \tag{4.3}$$

Thus, erosion of overlying rock will tend to increase the value of K, the horizontal stress becoming greater than the vertical stress at depths less than a certain value.[1] The hyperbolic relationship for $K(Z)$ predicted by Equation 4.3 can be generated by other arguments. While the vertical stress is known to equal γZ, the horizontal stress could lie anywhere in the range of values between the two extremes $K_a\sigma_v$ and $K_p\sigma_v$ shown in Figure 4.6. K_a corresponds to conditions for normal faulting, Figure 4.6b, in which the vertical stress is the

---

[1] With the restriction $K \le K_p$ given by (4.5). Thermal effects have been ignored.

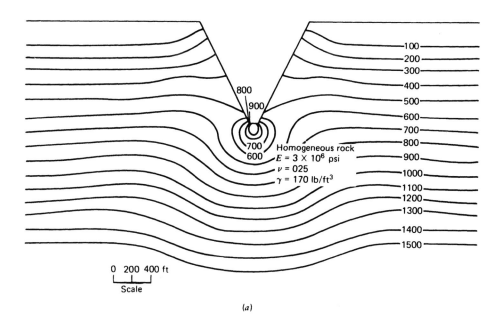

(a)

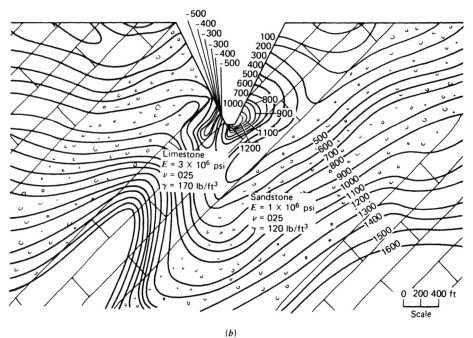

(b)

Figure 4.4 Comparison of maximum shear stresses beneath valleys in homogeneous (a) and heterogeneous (b) formations. Units of shear stress are hundreds of pounds per square foot.

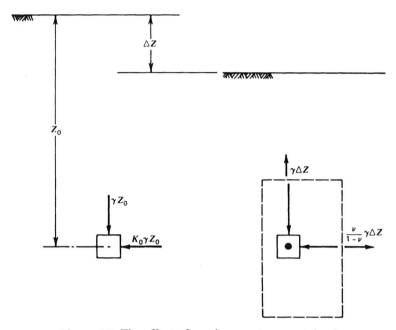

Figure 4.5 The effect of erosion on stresses at depth.

major principal stress and failure is by horizontal extension. Assuming Coulomb's law

$$K_a = \mathrm{ctn}^2\left(45 + \frac{\phi}{2}\right) - \left[\left(\frac{q_u}{\gamma}\right)\mathrm{ctn}^2\left(45 + \frac{\phi}{2}\right)\right]\cdot\frac{1}{Z} \qquad (4.4)$$

K_p corresponds to conditions for reverse faulting (Figure 4.6c), in which the vertical stress is the minor principal stress and failure is by horizontal compression, giving

$$K_p = \tan^2\left(45 + \frac{\phi}{2}\right) + \frac{q_u}{\gamma}\cdot\frac{1}{Z} \qquad (4.5)$$

Values of these extreme horizontal stresses are tabulated for an assumed set of rock properties in Table 4.1. If there is no existing fault, we observe that the range of possible values of K such that $K_a \leq K \leq K_p$ is quite vast. However, near a preexisting fault, q_u can be assumed equal to zero and the range of K is considerably reduced. Although tension is possible, it has rarely been measured and is to be considered an unusual situation.

Brown and Hoek (1978) examined a number of published values of in situ

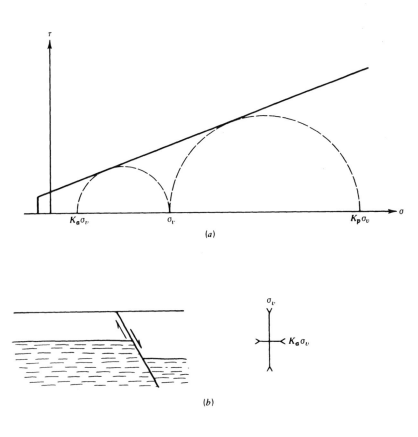

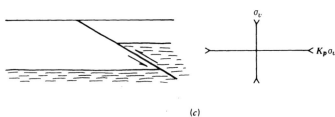

Figure 4.6 Stresses required to initiate normal and reverse faults.

stress (Figure 4.7*b*) and independently discerned a hyperbolic relation for the limits of $K(Z)$, as

$$0.3 + \frac{100}{Z} < \overline{K} < 0.5 + \frac{1500}{Z} \tag{4.6}$$

where Z is the depth in meters and $\overline{K}$ is the ratio of average horizontal stress to vertical stress. The range in extreme values of $\overline{K}$ given by this empirical crite-

Table 4.1 *Extreme Values for Possible Horizontal Stresses Corresponding to Conditions for Normal and Reverse Faulting* $\gamma = 25.9 \ kN/m^3$

Depth (m)	Vertical Stress σ_v (MPa)	Before Faulting Occurs; No Preexisting Fault Horizontal Stress σ_h				After Faulting Has Occurred and a Fault Exists Horizontal Stress σ_h			
		$q_u = 13.8$ MPa $\phi = 40°$		$q_u = 2$ MPa $\phi = 20°$		$q_u = 0$ $\phi = 40°$		$q_u = 0$ $\phi = 20°$	
		Normal Faulting (MPa)	Reverse Faulting (MPa)	Normal Faulting (MPa)	Reverse Faulting (MPa)	Normal Faulting (MPa)	Reverse Faulting (MPa)	Normal Faulting (MPa)	Reverse Faulting (MPa)
10	0.26	−2.94	14.99	−0.85	2.53	0.06	1.19	0.13	0.53
20	0.52	−2.88	16.18	−0.73	3.06	0.11	2.38	0.25	1.06
40	1.04	−2.77	18.56	−0.47	4.11	0.23	4.76	0.51	2.11
60	1.55	−2.66	20.95	−0.22	5.17	0.34	7.15	0.76	3.17
100	2.59	−2.43	25.72	0.29	7.28	0.56	11.91	1.27	5.28
150	3.89	−2.15	31.68	0.92	9.92	0.84	17.87	1.90	7.92
200	5.18	−1.87	37.64	1.56	12.57	1.13	23.82	2.54	10.57
400	10.36	−0.74	61.49	4.10	23.13	2.25	47.64	5.08	21.13
750	19.43	1.23	103.2	8.54	41.62	4.22	89.33	9.52	39.62
1000	25.90	2.64	133.0	11.72	54.83	5.63	119.1	12.70	52.83
2000	51.80	8.28	252.4	24.42	107.6	11.26	238.2	25.40	105.6

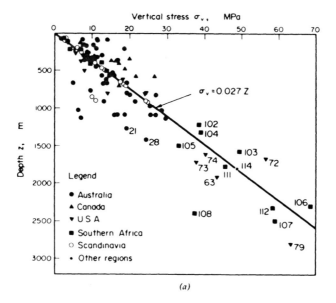

(a)

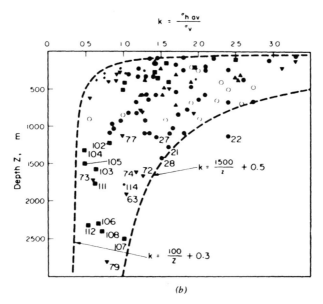

(b)

Figure 4.7 Results of stress measurements. (a) Vertical
stresses. (b) Average horizontal stresses [(a) and (b)
from Brown and Hoek, (1978).]

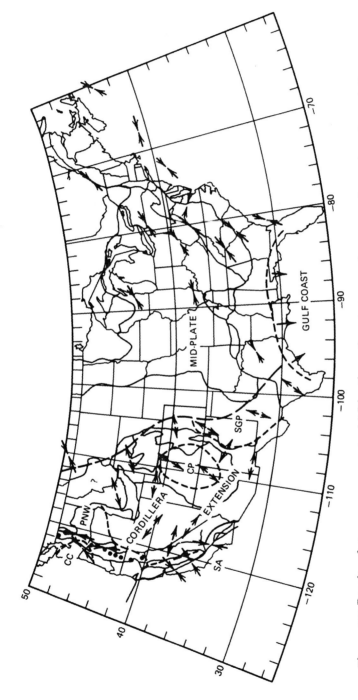

Figure 4.7 Results of stress measurements. (c) Directions of greatest stress in compression regions (inward directed arrows) and of least stress in extension regions (outward directed arrows). From Zoback and Zoback (1988) with permission.

rion is considerably less than the range K_a to K_p given by (4.4) and (4.5) when q_u is not equal to zero, due in part to the fact that average horizontal stress is being considered, whereas the previous criteria refer to maximum and minimum values of horizontal stress. In any event, all the equations for $K(Z)$ presented and the actually measured data are consistently found to be inverse with Z. Thus, even without measurements one can estimate, within broad limits, the variation of horizontal stress with depth. While the *magnitude* of the horizontal stress might be estimated only approximately, it is often possible to offer good estimates for the *directions* of the horizontal stresses.

HORIZONTAL STRESS DIRECTION

If the present state of stress is a remnant of that which caused visible geological structure, it will be possible to infer the directions of stresses from geological observations. Figure 4.8 shows the relationship between principal stress directions and different types of structures. The state of stress that causes a normal fault has σ_1 vertical, and σ_3 horizontal pointed perpendicularly to the fault trace as seen in the plan. In the case of reverse faulting, the stresses that caused the rupture have σ_3 vertical, while σ_1 is horizontal and directed perpendicular to the fault trace. Axial planes of folds also define the plane of greatest principal stress. Strike-slip faults are created by a state of stress in which σ_1 is horizontal and inclined about 30° with the fault trace, clockwise or counterclockwise as dictated by the sense of motion on the fault. These directions of horizontal stresses are not those of crustal blocks caught and squeezed between pairs of parallel faults; in such blocks, the primary stress state of the crust that is linked directly to the primary rupture surfaces will have superimposed on it the effects of the strain from accumulated fault motions, as discussed by Moody and Hill (1956).

Another line of observations comes from dikes and flank volcanoes formed around larger craters. Some dikes represent hydraulic fractures, in which case they lie perpendicular to σ_3. The perpendicular to a radius from a master crater to a flank volcano similarly identifies the direction of least horizontal stress.[2] Seismologists are able to indicate the directions of primary stresses from first motion analysis of earthquakes. If the directions of the vectors from the focus to different seismic stations are plotted on a stereographic projection of a unit reference hemisphere, it will be seen that two regions contain vectors to stations that received compressive first motion, while the other two regions contain vectors that received extensile first motion (Figure 4.8f). Two great circles are drawn to divide these fields and their point of intersection defines the direction of σ_2. The direction of σ_1 is 90° from the direction of σ_2 approxi-

[2] K. Nakamura (1977) Volcanoes as possible indicators of tectonic stress orientation—Principle and proposal. *J. Volcanol. Geothermal Res.* **2:** 1–16.

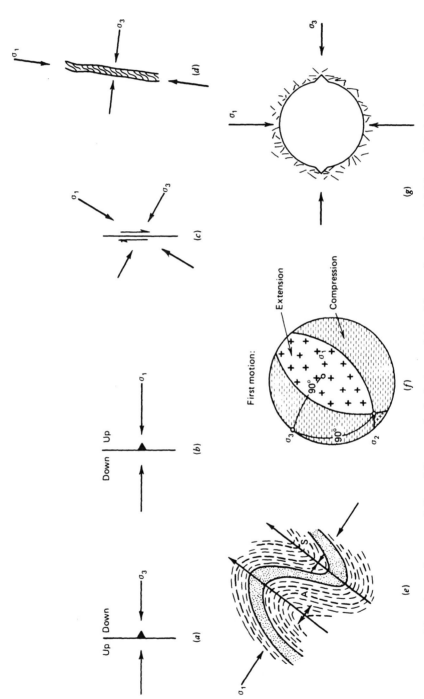

Figure 4.8 Directions of stresses inferred from geologic features. (*a*) to (*e*) are plan views. (*a*) Normal fault. (*b*) Reverse fault. (*c*) Strike slip fault. (*d*) Dike. (*e*) Folds. (*f*) Stereographic projection of first motion vectors from an earthquake. (*g*) Relation of stress directions to bore-hole breakouts.

mately along the great circle bisecting the angle between the dividing great circles in the extension first motion field. The direction of σ_3 is the perpendicular to the plane of σ_1 and σ_2. (Stereographic projection principles are presented in Appendix 5.)

Another approach to determining stress directions comes from the occurrence of rock breakage on the walls of wells and boreholes, which tends to create diametrically opposed zones of enlargement, termed "breakouts." These features can be seen in caliper logs, photographs, and televiewer logs of boreholes and have been found to be aligned from hole to hole in a region. Haimson and Herrick (1985) reported experimental results confirming that breakouts occur along the ends of a borehole diameter aligned with the least horizontal stress as depicted in Figure 4.8*g*.

Directions of horizontal stresses in the continental United States, inferred from a variety of techniques, are shown in Figure 4.7*c*, prepared by Zoback and Zoback (1988). This map also indicates the styles of deformation, that is, *extension* with the least principal stress horizontal or *contraction* with the greatest principal stress horizontal.

4.3 Techniques for Measurement of In-Situ Stresses

Stresses in situ can be measured in boreholes, on outcrops, and in the walls of underground galleries as well as back calculated from displacements measured underground. The available techniques summarized in Table 4.2 involve a variety of experimental approaches, with an even greater variety of measuring tools. Three of the best known and most used techniques are *hydraulic fracturing*, the *flat jack method*, and *overcoring*. As will be seen, they are complementary to each other, each offering different advantages and disadvantages. All stress measurement techniques perturb the rock to create a response that can then be measured and analyzed, making use of a theoretical model, to estimate part of the in situ stress tensor. In the hydraulic fracturing technique, the rock is cracked by pumping water into a borehole; the known tensile strength of the rock and the inferred concentration of stress at the well bore are processed to yield the initial stresses in the plane perpendicular to the borehole. In the flat jack test, the rock is partly unloaded by cutting a slot, and then reloaded; the in situ stress normal to the slot is related to the pressure required to null the displacement that occurs as a result of slot cutting. In the overcoring test, the rock is completely unloaded by drilling out a large core sample, while radial displacements or surface strains of the rock are monitored in a central, parallel borehole. Analysis using an unloaded thick-walled cylinder model yields stress in the plane perpendicular to the borehole. In each case stress is inferred, but

Table 4.2 *Methods for Measuring the Absolute State of Stress in Rocks*

Principle	Procedure	Reference
Complete strain relief	Overcore a radial deformation gage in a central borehole (U. S. Bureau of Mines method)	Merrill and Peterson (1961)
	Overcore a soft inclusion containing strain gages (LNEC and CSIRO methods)	Rocha et al. (1974), Worotnicki and Walton (1976)
	Overcore a borehole with strain gages on its walls (Leeman method)	Leeman (1971), Hiltscher et al. (1979)
	Drill around a rosette gage placed on a rock face	Olsen (1957)
	Overcore a rosette gage placed on the bottom of a drill hole (doorstopper method)	Leeman (1971)
	Overcore a soft photoelastic inclusion	Riley, Goodman, and Nolting 1977)
	Measure time dependent strains on faces of a rock after its removal from the ground	Emery (1962) Voight (1968)
Partial strain relief	Null displacements caused by cutting a tabular slot in a rock wall (flat jack method)	Bernède (1974) Rocha et al. (1966)
	Overcore a stiff photoelastic inclusion with down-hole polariscope (glass stress meter)	Roberts et al. (1964, 1965)
	Overcore a stiff inclusion to freeze stresses into it; measure frozen streses in the laboratory (cast inclusion method)	Riley, Goodman, and Nolting (1977)
	Overcore a stiff instrumented inclusion (stiff inclusion method)	Hast (1958) Nichols et al. (1968)
	Drill in the center of a rosette array on the surface of a rock face (undercoring method)	Duvall, in Hooker et al. (1974)
	Monitor radial displacements on deepening a borehole (borehole deepening method)	De la Cruz and Goodman (1970)

	⎧ Measure strain to fracture a borehole with a borehole jack (Jack fracturing technique)	De la Cruz (1978)
	Measure water pressures to create and extend a vertical fracture in a borehole (Hydraulic fracturing)	Fairhurst (1965) Haimson (1978)
Rock flow or fracture	⎨ Measure strains that accumulate in an elastic inclusion placed tightly in a viscoelastic rock	
	Core disking—observe whether or not it has ⎩ occurred	Obert and Stephenson (1965)
	⎧ Resistivity	
Correlation between rock properties and stress; other techniques	Rock noise (Kaiser effect)	Kanagawa, Hayashi, and Nakasa (1976)
	⎨ Wave velocity	
	X-ray lattice spacing measurements in quartz	Friedman (1972)
	Dislocation densities in ⎩ crystals	

displacements are actually measured. Precisions are seldom great and the results are usually considered satisfactory if they are internally consistent and yield values believed to be correct to within about 50 psi (0.3 MPa). The main problem of all stress measurement techniques is that the measurement must be conducted in a region that has been disturbed in the process of gaining access for the measurement; this paradox is handled by accounting for the effect of the disturbance in the analytical technique, as shown below.

HYDRAULIC FRACTURING

The hydraulic fracturing method makes it possible to estimate the stresses in the rock at considerable depth using boreholes. Water is pumped into a section of the borehole isolated by packers. As the water pressure increases, the initial compressive stresses on the walls of the borehole are reduced and at some points become tensile. When the stress reaches $-T_0$, a crack is formed; the down-hole water pressure at this point is p_{c1} (Figure 4.9a). If pumping is contin-

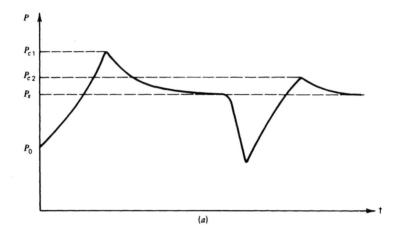

Figure 4.9 Hydraulic fracturing. (*a*) Pressure versus time data as water is pumped into the packed-off section. (*b*) Experiment in progress. (Photo by Tom Doe.)

ued, the crack will extend, and eventually the pressure down the hole will fall to a steady value p_s, sometimes called "the shut-in pressure."

To interpret the data from the hydraulic fracturing experiment in terms of initial stresses, we need to determine the orientation of the hydraulically induced fracture ("hydrofac"). The greatest amount of information coincides with the case of a vertical fracture, and this is the usual result when conducting tests below about 800 m. The orientation of a fracture could be observed by using down-hole photography or television; however, a crack that closes upon depressuring the hole to admit the camera would be difficult to see in the photograph. It is better to use an impression packer, such as one available from Lynes Company, which forces a soft rubber lining against the wall while internal pressure is maintained, recording the fracture as an impression on the rubber surface.

The analysis of the pressure test is simplified if it is assumed that penetration of the water into the pores of the rock has little or no effect on the stresses around the hole. Making such an assumption, it is possible to use the results of the known distribution of stress around a circular hole in a homogeneous, elastic, isotropic rock (the "Kirsch solution") to compute the initial stresses at the point of fracture. The tangential stress on the wall of the hole reaches the least magnitude at A and A' (Figure 4.10) where it is

$$\sigma_\theta = 3\sigma_{h,\min} - \sigma_{h,\max} \qquad (4.7)$$

When the water pressure in the borehole is p, a tensile stress is added at all points around the hole equal (algebraically) to $-p$. The conditions for a new, vertical tensile crack are that the tensile stress at point A should become equal to the tensile strength $-T_0$. Applying this to the hydraulic fracturing experiment yields as a condition for creation of a hydraulic fracture

$$3\sigma_{h,\min} - \sigma_{h,\max} - p_{c1} = -T_0 \qquad (4.8)$$

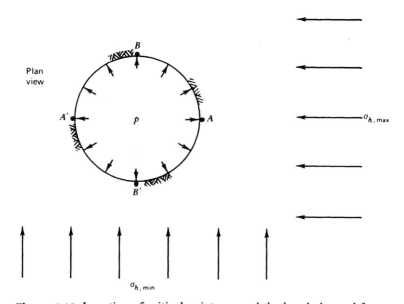

Figure 4.10 Location of critical points around the borehole used for hydraulic fracture.

Once formed, the crack will continue to propagate as long as the pressure is greater than the stress normal to the plane of the fracture. If the pressure of water in the crack were less than or greater than the normal stress on this crack, it would close or open accordingly. In rocks, cracks propagate in the plane perpendicular to σ_3. In the context of hydraulic fracturing with a vertical fracture, this means that the stress normal to the plane of the fracture is equal to the shut-in pressure p_s:

$$\sigma_{h,\min} = p_s \tag{4.9}$$

Equations 4.8 and 4.9 allow the major and minor normal stresses in the plane perpendicular to the borehole to be determined if the tensile strength of the rock is known. If the borehole pressure is dropped and once again raised above the value p_s, the hydraulic fracture will close and then reopen. Let the new peak pressure, smaller than p_{c1}, be called p_{c2}. Replacing T_0 and p_{c1} of Equation 4.8 with the values 0 and p_{c2}, respectively, and subtracting Equation 4.8 from the resulting equation yields a formula for the tensile strength of the rock around the borehole applicable to the conditions of the experiment:

$$T_0 = p_{c1} - p_{c2} \tag{4.10}$$

Assuming that the vertical stress equals γZ, and is a principal stress, the state of stress is now completely known, for the experiment yields the values and directions of the major and minor normal stress in the plane perpendicular to the borehole.

If the rock is pervious, water will enter cracks and pores creating an internal pressure gradient whereas the theory above presumed a sudden pressure drop across the borehole wall. The effect is to lower the value of p_{c1} and round the peak of Figure 4.9. Haimson (1978) shows how to modify the analysis to solve for the principal stresses in this case.

The hydraulic fracturing experiment does not yield the above results if the fracture is horizontal. Conditions for propagation of a horizontal fracture are met if the internal pressure becomes equal to the vertical stress plus the tensile strength. Assuming that the tensile strengths for propagation of horizontal and vertical fractures are the same, the vertical fracture could form only at depths below which the vertical stress obeys

$$\sigma_v \geq (3N - 1)\sigma_{h,\max} \tag{4.11}$$

where $N = \sigma_{h,\min}/\sigma_{h,\max}$. To permit an estimate of the minimum depth for vertical fracturing, it is useful to express Equation 4.11 in terms of $\overline{K}$, the ratio of mean horizontal stress to vertical stress. In these terms, a vertical fracture will form at a depth such that $\overline{K}$ is less than $(1 + N)/(6N - 2)$ where N is $\sigma_{h,\min}/\sigma_{h,\max}$ (with N restricted to values greater than $\frac{1}{3}$). The minimum depths for a vertical fracture, corresponding to the upper and lower limits of $\overline{K}(Z)$ given in Equation 4.6, are presented for various values of N in Table 4.3. When the

Table 4.3 *Minimum Depths for a Vertical Hydraulic Fracture*

$\sigma_{h,\min}/\sigma_{h,\max}$ (N)	Transition Value[a] of $\overline{K} = \overline{\sigma}_h/\sigma_v$ $(\overline{K}_T)$	Minimum Depth (meters) for a Vertical Hydrofrac Assuming	
		$Z = \left(\dfrac{100}{\overline{K} - 0.3}\right)$	$Z = \left(\dfrac{1500}{\overline{K} - 0.5}\right)$
≤0.33	∞	0	0
0.40	3.5	31	500
0.50	1.5	83	1500
0.60	1.0	143	3000
0.667	0.833	188	4505
0.70	0.773	211	5495
0.80	0.643	292	10,490
0.90	0.559	386	25,424
1.00	0.500	500	∞

[a] $\overline{K} = \dfrac{1 + N}{6N - 2}, N > \dfrac{1}{3}.$

value of N is small, or when the mean horizontal stress tends toward the lower values in the range of experience, vertical fractures can occur at shallow depths. This has in fact been experienced by the oil industry, which has produced more than a million hydrofracs for artificial stimulation of oil and gas wells.

THE FLAT JACK METHOD

Hydraulic fracturing can be performed only in a borehole. If one has access to a rock face, for example, the wall of an underground gallery, stress can be measured using a simple and dependable technique introduced by Tincelin in France in 1952. The method involves the use of flat hydraulic jacks, consisting of two plates of steel welded around their edges and a nipple for introducing oil into the intervening space. Through careful welding and the use of preshaping bends, or internal fillets, it is possible to achieve a pressure of 5000 psi or higher in such a jack without rupture. The first step is to install one or more sets of measuring points on the face of the rock. The separation of the points is typically 6 in., but must conform to the gage length of available extensometers. Then a deep slot is installed perpendicular to the rock face between the reference points (Figure 4.11b); this may be accomplished by drilling overlapping jackhammer holes, by using a template to guide the drill, or by diamond sawing (Rocha et al., 1966). As a result of cutting the slot, the pin separation will decrease from d_0 to a smaller value if the rock was under an initial compression normal to the plane of the slot (Figure 4.11c). The initial normal stresses could

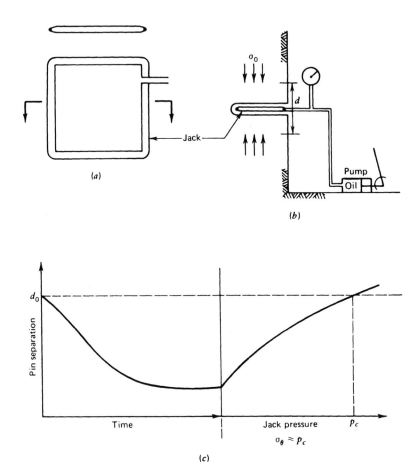

Figure 4.11 The flat jack test.

be calculated from the measured pin displacement if the elastic constants of the rock were known. However, a self-compensating method of stress determination is preferred making it unnecessary to determine the rock properties explicitly. The flat jack is inserted into the slot, cemented in place, and pressured. When the pins have been returned to d_0, their initial separation, the pressure in the jack (p_c) approximates the initial stress normal to the jack. In theory, the initial stress parallel to the slot and the geometric differences between the inside of the jack and the inside of the slot require a correction to this result (Alexander, 1960). However, the correction is often within the band of uncertainty anyway, and if a diamond sawed slot is used, it is negligibly small; thus p_c (the "cancellation pressure of the jack") is an acceptable estimate for the average stress normal to the jack.

In the flat jack test we have a large, rugged, and inexpensive method for determining one stress component of the stress tensor. The equipment can be fabricated on site and is virtually indestructible, an important consideration in any instrumentation or measurement program underground. A serious limitation of the method is that the measured stress lies in the region of disturbance of the gallery introduced for the purpose of taking the measurement. If the gallery is carefully executed, this disturbance might be calculated by conducting an independent stress concentration investigation, using numerical methods (e.g., the finite element method). In general, if the stresses normal to the plane of the jack are determined at three points around the section of the gallery, yielding values $\sigma_{\theta A}$, $\sigma_{\theta B}$, $\sigma_{\theta C}$ for the tangential stresses (stresses parallel to the surface of the opening) near the surface at these points, the initial stresses in the plane perpendicular to the gallery can be calculated by inverting the relationship:

$$\begin{Bmatrix} \sigma_{\theta,A} \\ \sigma_{\theta,B} \\ \sigma_{\theta,C} \end{Bmatrix} = \begin{pmatrix} a_{11} & a_{12} & a_{13} \\ a_{21} & a_{22} & a_{23} \\ a_{31} & a_{32} & a_{33} \end{pmatrix} \begin{Bmatrix} \sigma_x \\ \sigma_y \\ \tau_{xy} \end{Bmatrix} \tag{4.12}$$

where the coefficients a_{ij} are determined by the numerical study. For example, suppose flat jacks were placed at R and W, in the roof and side wall, respectively, of a perfectly circular underground opening; if the initial stresses were known to be horizontal and vertical, and if the tunnel radius were large compared to the width of the jacks, then Equation 4.12 would simplify to

$$\begin{Bmatrix} \sigma_{\theta,W} \\ \sigma_{\theta,R} \end{Bmatrix} = \begin{pmatrix} -1 & 3 \\ 3 & -1 \end{pmatrix} \begin{Bmatrix} \sigma_{\text{horiz}} \\ \sigma_{\text{vert}} \end{Bmatrix} \tag{4.13}$$

whereupon

$$\sigma_{\text{horiz}} = \tfrac{1}{8}\sigma_{\theta,W} + \tfrac{3}{8}\sigma_{\theta,R}$$

and $\tag{4.14}$

$$\sigma_{\text{vert}} = \tfrac{3}{8}\sigma_{\theta,W} + \tfrac{1}{8}\sigma_{\theta,R}$$

The stresses around an underground gallery vary inversely with the radius squared (see Equations 7.1). Therefore, if stresses are measured in a borehole at least one gallery diameter in depth, the results should correspond to the initial state of stress before driving the measurement gallery. This can be accomplished using the *overcoring test*.

OVERCORING

First one drills a small-diameter borehole and sets into it an instrument to respond to changes in diameter. One such instrument is the U. S. Bureau of Mines six-arm deformation gage (Figures 4.12a and 4.13a), a relatively rugged

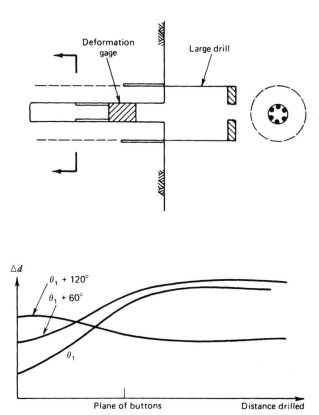

Figure 4.12 The overcoring method, using the Bureau of Mines gage.

tool that uses the bending of a cantilever equipped with strain gages to give output voltage proportional to displacement. There are three opposed pairs of carbon-carbide-tipped buttons, each pressing against a cantilever arm fixed to a base plate, tightened against the wall of the borehole by a spring. By choosing a button of appropriate size in each of the six positions, each of the cantilevers can be pre-bent to yield an initial output in the center of the linear region and the borehole diameter changes can be monitored along three diameters simultaneously, whether the borehole becomes smaller or larger. After the gage is inserted, the output wires are threaded through a hollow drill and out through the water swivel and a larger hole is cored concentrically over the first (Figure 4.13b). This produces a thick-walled cylinder of rock, detached from the rock mass and therefore free of stress. If the rock had been under an initial compression, the deformation gage will record an enlargement along two or all of the monitored directions in response to the "overcoring" (Figure 4.12b)—all radii

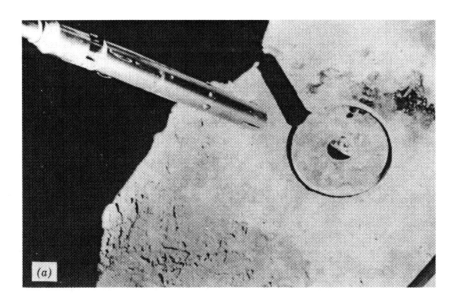

Figure 4.13 In-situ stress measurements by overcoring from a rock outcrop. (*a*) Six component borehole deformation gage and the overcored measuring hole. (*b*) experiment in progress. (Photos by Rick Nolting; Courtesy of TerraTek.)

expanding if the ratio of minor to major normal stress in the plane perpendicular to the borehole is greater than one-third. As a result of the experiment, the change in borehole diameter will be known along three diameters, 60° apart. Select an x axis conveniently in the plane perpendicular to the hole and let θ be the angle counterclockwise from $0x$ to a pair of buttons that yields results $\Delta d(\theta)$. Let the plane perpendicular to the borehole be the xz plane, with the borehole parallel to y. Then, the deformations measured are related to the initial stresses in the xyz coordinate system according to

$$\Delta d(\theta) = \sigma_x f_1 + \sigma_y f_2 + \sigma_z f_3 + \tau_{xz} f_4 \tag{4.15}$$

where $f_1 = d(1 + 2 \cos 2\theta) \dfrac{1 - \nu^2}{E} + \dfrac{d\nu^2}{E}$

$$f_2 = -\frac{d\nu}{E}$$

$$f_3 = d(1 - 2 \cos 2\theta) \frac{1 - \nu^2}{E} + \frac{d\nu^2}{E}$$

$$f_4 = d(4 \sin 2\theta) \frac{1 - \nu^2}{E}$$

In the above, E is Young's modulus, ν is Poisson's ratio, and d is the diameter of the borehole in which the measurement is conducted. Equation 4.15 excludes the two shear stress components τ_{xy} and τ_{zy} parallel to the borehole because these have no influence on the diameter of the borehole. Gray and Toews (1968) showed that only three linearly independent equations are obtainable from repeated diametral measurements in different orientations, so the general state of stress cannot be computed from diameter changes recorded in one borehole. However, a solution can be found if one of the stress components is known or can be assumed. If the measurement is conducted in a borehole perpendicular to a rock face and at shallow depth, σ_y might be taken as zero. If the value of σ_y were known, or assumed, on the other hand, the term $f_2\sigma_y$ could be taken to the left side of the equal sign in each of three equations representing measurements along different directions and the remaining three stress components could be determined. In this way, the state of stress in the plane perpendicular to the borehole could be computed as a function of σ_y alone. An alternative approach, discussed later, is to combine measurements from three or more nonperpendicular boreholes, adopting a single, universal coordinate system into which the unknown stresses from each borehole are transformed. The resulting set of equations will be redundant, and, furthermore, since it is impossible to occupy the same volume of rock in more than one measurement, the results will be scattered.

In the usual situation where measurements are conducted in one borehole parallel to y, and a value of σ_y is assumed for purposes of computation, diameter change measurements are conducted in directions θ_1, $\theta_1 + 60$, and $\theta_1 + 120$, yielding three equations in three unknowns:

$$\begin{Bmatrix} \Delta d(\theta_1) - f_2\sigma_y \\ \Delta d(\theta_1 + 60) - f_2\sigma_y \\ \Delta d(\theta_1 + 120) - f_2\sigma_y \end{Bmatrix} = \begin{pmatrix} f_{11} & f_{13} & f_{14} \\ f_{21} & f_{23} & f_{24} \\ f_{31} & f_{33} & f_{34} \end{pmatrix} \begin{Bmatrix} \sigma_x \\ \sigma_z \\ \tau_{xz} \end{Bmatrix} \qquad (4.16)$$

Inversion of Equations 4.16, after assuming a value for σ_y, yields the stress components in the plane perpendicular to the borehole.

The overcoring test thus can be used to measure the stresses at some distance from a rock face. There is a practical limit to how far one borehole can be drilled concentrically over another. With a template to collar the drillhole and homogeneous, nonfractured rock, it might be possible to proceed for as much as 30 m from a face; but normally the test has to be discontinued beyond about 5 m.

The Swedish State Power Board has perfected the mechanical aspects of overcoring and has succeeded in conducting Leeman-type triaxial measurements at depths of more than 500 m. These tests are performed by cementing strain gage rosettes to the walls of a 36-mm hole drilled exactly in the center of the bottom of a 76-mm-diameter borehole. Extending the larger borehole overcores the former and strains the rosettes (Hiltscher, Martna, and Strindell, 1979; Martna, Hiltscher, and Ingevald 1983).

The principal disadvantage of the U. S. Bureau of Mines overcoring test is the linear dependence of the stresses upon the elastic constants. The Bureau of Mines determines E and ν directly on the overcore by compressing it in a special large-diameter triaxial compression chamber, while the borehole deformation gage responds inside. Another approach, applicable in horizontal holes, is to assume a value for ν and use the value of E that makes the vertical component of stress, at some distance behind the wall, agree with the value of the unit weight of rock times depth below ground. Another approach altogether is to replace the deformation gage with a stiffer gage (e.g., glass or steel) forming a "stiff elastic inclusion." In such a case, the stresses inside the inclusion on overcoring are almost independent of the elastic modulus of the rock. However, the precision of measurement is reduced making the experiment more difficult. Another difficulty with the overcoring method is the requirement to use large drill cores (e.g., 6-in.-diameter). There is no theoretical demand that the outer diameter be any specific value, and, in fact, the stresses deduced from the experiment will be unaffected by choice of outer diameter. In practice, however, difficulty is experienced with rock breakage if the outer diameter is less than at least twice the inner diameter.

In the *doorstopper method* (Figure 4.14) strain gages are fixed to the center of the stub of rock at the bottom of the hole which is then isolated from the surrounding rock by continuing the original hole (Leeman, 1971). This permits

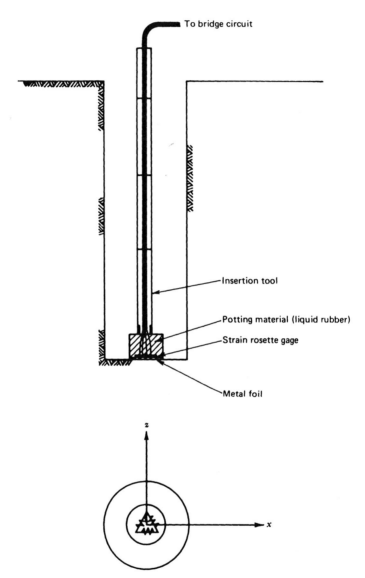

Figure 4.14 In situ stress measurement by the "doorstopper" technique.

the interpretation of stress at greater depth but the interpretation of the data is more precarious. The test is performed as follows. First, drill a borehole to the site of measurement. Then install a flat, noncoring drill bit to grind the bottom to a smooth flat surface. Clean the bottom surface and then cement onto it a piece of metal foil bearing a strain gage rosette on its upper surface. When the cement has hardened, thread the bridge wires through the drill and deepen the hole. This releases the stresses in the bottom, yielding strains ε_x, ε_z, γ_{xz} (with the y axis parallel to the borehole and the x, z axes along two perpendicular lines in the bottom, selected at will). Appendix 2 shows how to convert strain rosette readings to strain components ε_x, ε_z, γ_{xz}.

The changes in stress at the bottom of the hole ($\Delta\sigma_{x,B}$, $\Delta\sigma_{y,B}$, $\Delta\tau_{xy,B}$) can be calculated from the strain components by the stress-strain relationship for linear, elastic isotropic bodies:

$$\begin{Bmatrix} \Delta\sigma_{x,B} \\ \Delta\sigma_{z,B} \\ \Delta\tau_{xz,B} \end{Bmatrix} = \frac{E}{1-\nu^2} \begin{bmatrix} 1 & \nu & 0 \\ \nu & 1 & 0 \\ 0 & 0 & \dfrac{1-\nu}{2} \end{bmatrix} \begin{Bmatrix} \varepsilon_x \\ \varepsilon_z \\ \gamma_{xz} \end{Bmatrix} \tag{4.17}$$

The initial stresses in x, y, z coordinates are related to the stress changes on the bottom of the hole by

$$\begin{Bmatrix} \Delta\sigma_{x,B} \\ \Delta\sigma_{z,B} \\ \Delta\tau_{xz,B} \end{Bmatrix} = - \begin{pmatrix} a & c & b & 0 \\ b & c & a & 0 \\ 0 & 0 & 0 & d \end{pmatrix} \begin{Bmatrix} \sigma_x \\ \sigma_y \\ \sigma_z \\ \tau_{xz} \end{Bmatrix} \tag{4.18}$$

Constants a, b, c, and d have been evaluated by several independent workers. De la Cruz and Raleigh (1972) give the following values, based upon a finite element analysis:

$$\begin{aligned} a &= 1.30 \\ b &= (0.085 + 0.15\nu - \nu^2) \\ c &= (0.473 + 0.91\nu) \\ d &= (1.423 - 0.027\nu) \end{aligned} \tag{4.19}$$

As in the overcoring test, σ_y must be assumed or evaluated independently. Then

$$\begin{Bmatrix} \sigma_x \\ \sigma_z \\ \tau_{xz} \end{Bmatrix} = - \begin{pmatrix} a & b & 0 \\ b & a & 0 \\ 0 & 0 & d \end{pmatrix}^{-1} \begin{Bmatrix} \Delta\sigma_{x,B} + c\sigma_y \\ \Delta\sigma_{z,B} + c\sigma_y \\ \Delta\tau_{xz,B} \end{Bmatrix} \tag{4.20}$$

The "doorstopper" method can be pursued at the bottom of a shaft as well as in a drill hole.

Measurements Made Directly on the Rock Surface   If a machine-bored shaft or tunnel is available for rock mechanics work, stress measurements may be made directly on the wall if the rock is not highly fractured. There are at least two methods for doing this: overdrilling a strain gage *rosette* applied directly to the rock surface, and drilling a central hole amid a set of measuring points (*undercoring*).

Strain gage rosettes applied to the rock surface have been used in boreholes by Leeman (1971) with an ingenious device to transport, glue, and hold the rosettes at several points simultaneously. Upon overcoring the hole, these rosettes then report strain changes that can be transformed to yield the complete state of stress $(\sigma)_{xyz}$. In the present context, we can overcore strain gage rosettes cemented to points directly on the rock surface. Appendix 2 presents formulas for calculating the state of strain $(\varepsilon_x, \varepsilon_z, \gamma_{xz})$ from the readings of the component gages of the rosette when the rock to which they are attached is overcored. These strains can then be converted to stresses using (4.17).

Undercoring is a name applied by Duvall (in Hooker et al., 1974) to a procedure for measuring stresses on an exposed surface by monitoring radial displacements of points around a central borehole (Figure 4.15). Expressions for the radial and tangential displacements of a point located at polar coordinates r, θ from the central hole of a radius a are given in Equations 7.2 for plane strain; these expressions are changed to plane stress by substituting $v/(1 + v)$ in place of v as discussed in the derivation of Equations 7.1 and 7.2 (Appendix 4).

Equations 7.2 are developed for the condition where the major and minor principal stress directions in the measuring plane are known. For the stress measurement problem, these directions will not be known apriori so an arbi-

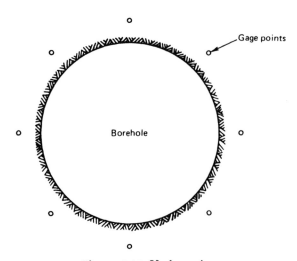

Figure 4.15 Undercoring.

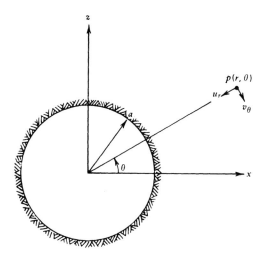

Figure 4.16 Coordinate system for the displacement equations.

trary choice of axes (x, z) is made (Figure 4.16). The stresses $\{\sigma\}_{xz}$ can then be determined from radial displacement measurements (u_r) at three positions (r, θ) using the following equation for each point in turn[3]:

$$u_r = \sigma_x f_1 + \sigma_z f_2 + \tau_{xz} f_3 \qquad (4.21)$$

where $f_1 = \dfrac{1}{2E} \dfrac{a^2}{r} [(1 + v) + H \cos 2\theta]$

$f_2 = \dfrac{1}{2E} \dfrac{a^2}{r} [(1 + v) - H \cos 2\theta]$

$f_3 = \dfrac{1}{E} \dfrac{a^2}{r} (H \sin 2\theta)$

$H = 4 - (1 + v) \dfrac{a^2}{r^2}$

With radial displacement, $u_{r,1}$ measured at r_1, θ_1, $u_{r,2}$ at r_2, θ_2, and $u_{r,3}$ at r_3, θ_3, Equation 4.21 yields

$$\begin{Bmatrix} u_{r,1} \\ u_{r,2} \\ u_{r,3} \end{Bmatrix} = \begin{bmatrix} f_{11} & f_{12} & f_{13} \\ f_{21} & f_{22} & f_{23} \\ f_{31} & f_{32} & f_{33} \end{bmatrix} \begin{Bmatrix} \sigma_x \\ \sigma_z \\ \tau_{xz} \end{Bmatrix} \qquad (4.22)$$

[3] We assume that the tangential displacement v_θ does not influence the measured radial displacement.

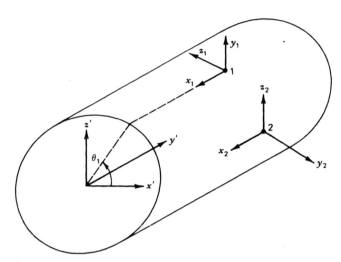

Figure 4.17 Coordinate systems for stress measurements on the walls of a tunnel.

which can be inverted to determine the stresses. This method cannot yield good percision unless the measuring points are close to the surface of the central hole, or the rock is deformable; otherwise, the values of u_r will be quite small. Duvall placed the measuring pins on a 10-in.-diameter circle and created a 6-in. central hole by reaming out an *EX* pilot hole. Vojtec Mencl used undercoring to measure stresses in the toe of a landslide in soft rock,[4] where measurable displacements were experienced despite relatively small stresses (0.6 MPa) because the value of E was quite low. A variant of undercoring using a central cylindrical expansion cell (dilatometer) to null the initial radial displacements of points around the central hole was used by Dean, Beatty, and Hogan at Broken Hill Mine, Australia.[5]

Virgin stresses (the initial stresses at the test site before it was excavated) can be calculated from the stress components measured on the rock walls if the stress concentrations at the measuring points are known. The problem at hand resembles that discussed previously in connection with the flat jack test. Since the shape of the bored gallery is circular with smooth walls, the required stress concentrations can be obtained from the classical Kirsch solution (whose derivation can be followed in Jaeger and Cook (1976)). For our purposes, the adopted coordinate system is shown in Figure 4.17; unprimed coordinates

---

[4] O. Zaruba and V. Mencl (1969) *Landslides and Their Control*, Elsevier, New York.

[5] Rock stress measurements using cylindrical jacks and flat jacks at North Broken Hill Ltd. from *Broken Hill Mine Monography No. 3* (1968), Australian Inst. Min. Metal. Melbourne, Australia (399 Little Collins St.).

$x_1 y_1 z_1$, $x_2 y_2 z_2$, etc., refer to local coordinate directions at each measuring site 1, 2, etc., with $y_1 y_2$ . . . always in the direction of the normal to the surface (radius of the tunnel or shaft) and $x_1 x_2$, etc., parallel to the axis of the measuring tunnel or shaft. The x', y', z' are global coordinates with y' parallel to the axis of the shaft or tunnel, and x' and z' any convenient orthogonal axes in its cross section.

The surface stress concentrations can then be obtained from the general Kirsch formulas given by Leeman (1971) (see Appendix 4), substituting $r = a$ to identify points on the wall at the site of the measurement: with the above coordinates, at each site

$$\sigma_r = \sigma_y = 0$$

$$\sigma_\theta = \sigma_z$$

$$\sigma_{\text{long}} = \sigma_x$$

$$\tau_{r\theta} = \tau_{yz} = 0$$

$$\tau_{\text{long},\theta} = \tau_{xz}$$

$$\tau_{\text{long},r} = \tau_{xy} = 0$$

$$\begin{Bmatrix} \sigma_z \\ \sigma_x \\ \tau_{xz} \end{Bmatrix} = \begin{pmatrix} d & 0 & e & 0 & 0 & f \\ g & 1 & h & 0 & 0 & i \\ 0 & 0 & 0 & n & p & 0 \end{pmatrix} \begin{Bmatrix} \sigma_{x'} \\ \sigma_{y'} \\ \sigma_{z'} \\ \tau_{x'y'} \\ \tau_{y'z'} \\ \tau_{z'x'} \end{Bmatrix} \tag{4.23}$$

where $d = 1 - 2 \cos 2\theta$ $h = 2\nu \cos 2\theta$

$e = 1 + 2 \cos 2\theta$ $i = -4\nu \sin 2\theta$

$f = -4 \sin 2\theta$ $n = -2 \sin \theta$

$g = -2\nu \cos 2\theta$ $p = 2 \cos \theta$

Two or more sites for surface stress measurement (e.g., (1) the roof $\theta_1 = 90°$, and (2) the wall $\theta_2 = 0$ (Figure 4.17) yield six equations whose solution determines the complete state of stress. Depending on the choice of sites, the coefficient matrix might be singular, necessitating a third location (with redundant data) to obtain a complete stress solution.

Principal Stresses   If the stresses are determined with reference to two arbitrarily chosen directions x and z in the plane of measurement, the values of normal stress will depend on the choice of axes. It is better to convert the results to the form of principal stresses and directions. (If the xz plane is not a principal plane, it is still possible to find, within it, directions in which the shear stress is zero; these are then called "secondary principal stresses.") Given σ_x, σ_z, and τ_{xz}, the principal stresses are found from

$$\sigma_{\text{major}} = \tfrac{1}{2}(\sigma_x + \sigma_z) + [\tau_{xz}^2 + \tfrac{1}{4}(\sigma_x - \sigma_z)^2]^{1/2}$$

and   (4.24)

$$\sigma_{\text{minor}} = \tfrac{1}{2}(\sigma_x + \sigma_z) - [\tau_{xz}^2 + \tfrac{1}{4}(\sigma_x - \sigma_z)^2]^{1/2}$$

The major principal stress, σ_1 acts in a direction θ, measured counterclockwise from $0x$, given by

$$\tan 2\theta = \frac{2\tau_{xz}}{\sigma_x - \sigma_z}$$ (4.25)

Since the arctan is multivalued, we must observe the following rules.[6] Let $\alpha = \tan^{-1}[2\tau_{xz}/(\sigma_x - \sigma_z)]$ with $-\pi/2 \leq \alpha \leq \pi/2$ then

$$2\theta = \alpha \qquad \text{if } \sigma_x > \sigma_z$$
$$2\theta = \alpha + \pi \qquad \text{if } \sigma_x < \sigma_z \text{ and } \tau_{xz} > 0$$
$$2\theta = \alpha - \pi \qquad \text{if } \sigma_x < \sigma_z \text{ and } \tau_{xz} < 0$$

Measurement of Stresses in Three Dimensions   Civil engineering and mining work rarely require that all stress components be known. If such knowledge is desired, methods exist to yield the complete state of stress from a single experiment (e.g., Leeman, 1971; Rocha et al., 1974). Also, data from techniques enumerated above can be combined to permit computation of the complete stress matrix. A procedure for doing this was already discussed for the case of strain measurements on the surface of a drilled shaft or tunnel. Data can also be combined for overcoring, doorstopper, and other approaches. In each case, the strategy is to transform the measured stress components to a global coordinate system to combine data from nonparallel directions at different measuring sites.

For example, consider overcoring measurements in several nonparallel boreholes. In hole A, adopt coordinate axes x_A, y_A, z_A with y_A parallel to the axis of the borehole; diametral displacements are measured in directions θ_{A1}, θ_{A2}, and $\theta_{A,3}$. Application of 4.15 to each direction yields

$$\begin{Bmatrix} \Delta d(\theta_{A,1}) \\ \Delta d(\theta_{A,2}) \\ \Delta d(\theta_{A,3}) \end{Bmatrix} = \begin{pmatrix} f_{11} & f_{12} & f_{13} & f_{14} \\ f_{21} & f_{22} & f_{23} & f_{24} \\ f_{31} & f_{32} & f_{33} & f_{34} \end{pmatrix} \begin{Bmatrix} \sigma_{x,A} \\ \sigma_{y,A} \\ \sigma_{z,A} \\ \tau_{xz,A} \end{Bmatrix}$$ (4.26)

---

[6] These rules were suggested to the writer by Professor Steven Crouch, University of Minnesota.

where the coefficients f_{ij} are defined for each θ for f_j of (4.15). Now transform the stresses in $x_A y_A z_A$ coordinates to some convenient set of axes x', y', z' (referred to henceforth as the "global axes"). This can be written

$$\begin{Bmatrix} \sigma_{x,A} \\ \sigma_{y,A} \\ \sigma_{z,A} \\ \tau_{xz,A} \end{Bmatrix} = \underset{(4 \times 6)}{(T_\sigma)} \underset{(6 \times 1)}{\{\sigma\}_{x'y'z'}} \tag{4.27}$$

in which (T_σ) is a 4×6 matrix corresponding to rows 1, 2, 3, and 5 of the coefficient matrix defined in Equation 23 of Appendix 1, and $\{\sigma\}_{x'y'z'}$ is the column of the six stress components of the same equation. Let (f_A) denote the 3×4 coefficient matrix in (4.26). Equations 4.26 and 4.27 can then be combined into

$$\underset{3 \times 1}{\{\Delta d\}_A} = \underset{3 \times 4}{(f_A)} \underset{4 \times 6}{(T_\sigma)_A} \underset{6 \times 1}{\{\sigma\}_{x'y'z'}} \tag{4.28}$$

Similarly for borehole B, nonparallel with A,

$$\underset{3 \times 1}{\{\Delta d\}_B} = \underset{3 \times 4}{(f_B)} \underset{4 \times 6}{(T_\sigma)_B} \underset{6 \times 1}{\{\sigma\}_{x'y'z'}} \tag{4.29}$$

Combining the six rows of (4.28) and (4.29) gives six equations with $\sigma_{x'y'z'}$ as the right-hand vector. Gray and Toews (1968), however, showed that the coefficient matrix thus derived is singular. Thus three nonparallel boreholes will be required to yield sufficient information to solve for $\{\sigma\}_{x'y'z'}$. One can reject surplus rows to achieve a solvable set of six equations. Even better, one can use a least-squares solution scheme. Panek (1966) and Gray and Toews (1975) showed how to handle the redundancy and minimize error associated with variation in the state of stress from one measuring site to another.

A similar procedure can be followed to combine the results from "door-stopper tests" in three nonparallel holes to determine the complete state of stress.

References

Alexander, L. G. (1960) Field and lab. test in rock mechanics, *Proceedings, Third Australia–New Zealand Conference on Soil Mechanics*, pp. 161–168.

Bernède, J. (1974) New Developments in the flat jack test (in French), *Proc. 3rd Cong. ISRM* (Denver), Vol. 2A, pp. 433–438.

Booker, E. W. and Ireland, H. O. (1965) Earth pressures at rest related to stress history, *Can. Geot. J.* 2: 1–15.

Brekke, T. L. (1970) A survey of large permanent underground openings in Norway, *Proceedings of Conference on Large Permanent Underground Openings*, pp. 15–28 (Universitets Forlaget, Oslo).

Brekke, T. L. and Selmer-Olsen, R. (1966) A survey of the main factors influencing the stability of underground construction in Norway, *Proc. 1st Cong. ISRM* (Lisbon), Vol. II, pp. 257–260.

Brown, E. T. and Hoek, E. (1978) Trends in relationships between measured in situ stresses and depth, *Int. J. Rock Mech. Min. Sci.* **15**: 211–215.

De la Cruz, R. V. (1978) Modified borehole jack method for elastic property determination in rocks, *Rock Mech.* **10**: 221–239.

De la Cruz, R. V. and Goodman, R. E. (1970) Theoretical basis of the borehole deepening method of absolute stress measurement, *Proceedings, 11th Symposium on Rock Mechanics* (AIME), pp. 353–376.

De la Cruz, R. V. and Raleigh, C. B. (1972) Absolute stress measurements at the Rangely Anticline, Northwestern Colorado, *Int. J. Rock Mech. Min. Sci.* **9**: 625–634.

Emery, C. L. (1962) The measurement of strains in mine rock, *Proceedings, International Symposium on Mining Research* (Pergamon), Vol. 2, pp. 541–554.

Fairhurst, C. (1965) Measurement of in-situ stresses with particular reference to hydraulic fracturing, *Rock Mech. Eng. Geol.* **2**: 129–147.

Friedman, M. (1972) X-ray analysis of residual elastic strain in quartzose rocks, *Proceedings, 10th Symposium on Rock Mechanics* (AIME), pp. 573–596.

Gray, W. M. and Toews, N. A. (1968) Analysis of accuracy in the determination of the ground stress tensor by means of borehole devices, *Proceedings, 9th Symposium on Rock Mechanics* (AIME), pp. 45–78.

Gray, W. M. and Toews, N. A. (1975) Analysis of variance applied to data obtained by means of a six element borehole deformation gauge for stress determination, *Proceedings, 15th Symposium on Rock Mechanics* (ASCE), pp. 323–356.

Haimson, B. C. (1976) Pre-excavation deep hole stress measurements for design of underground chambers—case histories, *Proceedings, 1976 Rapid Excavation and Tunneling Conference* (AIME), pp. 699–714.

Haimson, B. C. (1978) The hydrofracturing stress measurement technique-method and recent field results., *Int. J. Rock Mech. Sci.* **15**: 167–178.

Haimson, B. C. and Fairhurst, C. (1967) Initiation and extension of hydraulic fractures in rock, *Soc. Petr. Eng. J.* **7**: 310–318.

Haimson, B. C. and Herrick, C. G. (1985) In situ stress evaluation from borehole breakouts—Experimental studies, *Proceedings, 26th U.S. Symposium on Rock Mechanics* (Balkema), pp. 1207–1218.

Hast, N. (1958) The measurement of rock pressure in mines, *Sveriges Geol. Undersökning Arsbok* **52** (3).

Hiltscher, R., Martna, F. L., and Strindell, L. (1979) The measurement of triaxial stresses in deep boreholes and the use of rock stress measurements in the design and construction of rock openings, *Proceedings of the Fourth International Congress on Rock Mechanics*, Montreux, (ISRM) Vol. 2, 227–234.

Hooker, V. E., Aggson, J. R. Bickel, D. L., and Duvall, W. (1974) Improvement in the three component borehole deformation gage and overcoring technique, *U.S.B.M. Rep. Inv. 7894*; with Appendix by Duvall on the undercoring method.

Jaeger, J. C. and Cook, N. G. W. (1976) *Fundamentals of Rock Mechanics*, 2d ed., Chapman & Hall, London.

Kanagawa, T., Hayashi, M., and Nakasa, H. (1976) Estimation of spatial geostress in rock samples using the Kaiser effect of acoustic emission, *Proceedings, 3rd Acoustic Emission Symposium* (Tokyo; separately available from Central Research Inst. of Elec. Power Ind., Japan).

Leeman, E. R. (1971) The CSIR "Doorstopper" and triaxial rock stress measuring instruments, *Rock Mech.* **3:** 25–50.

Lindner, E. N. and Halpern, J. A. (1978) In situ stress in North America—a compilation, *Int. J. Rock Mech. Sci.* **15:** 183–203.

Martna, J., Hiltscher, R., and Ingevald, K. (1983), Geology and rock stresses in deep boreholes at Forsmark in Sweden, *Proceedings of the Fifth International Congress on Rock Mechanics*, Melbourne, (ISRM), Section F, pp. 111–116.

Merrill, R. H. and Peterson, J. R. (1961) Deformation of a borehole in rock, *U.S.B. Mines R.I. 5881*.

Moody, J. D. and Hill, M. J. (1956) Wrench fault tectonics, *Bull. Geol. Soc. Am.* **67:** 1207–1246.

Nichols, Jr., T. C., Abel, Jr., J. F., and Lee, F. T. (1968) A solid inclusion borehole probe to determine three dimensional stress changes at a point of a rock mass, *U. S. G. S. Bulletin 1258C*, pp. C1–C28.

Obert, L. and Duvall, W. (1967) *Rock Mechanics and the Design of Structures in Rock*, Wiley, New York.

Obert, L. and Stephenson, D. E. (1965) Stress conditions under which core discing occurs. *Trans. Soc. Min. Eng.* **232:** 227–234.

Olsen, O. J. (1957) Measurements of residual stress by the strain relief method, *Qtly. Colorado School of Mines*, Vol. 52, July, pp. 183–204.

Panek, L. A. (1966) Calculation of the average ground stress components from measurements of the diametral deformation of a drill hole, in *ASTM Spec. Tech. Publ. 402* (American Society of Testing and Materials). pp. 106–132.

Riley, P. B., Goodman, R. E., and Nolting, R. E. (1977) Stress measurement by overcoring cast photoelastic inclusions, *Proceedings, 18th Symposium on Rock Mechanics*, paper 4C4.

Roberts, A. et al. (1964, 1965) The photoelastic stress meter, *Int. J. Rock Mech. Sci.* **1:** 441–454; **2:** 93–103.

Rocha, M. (1971) A new method of integral sampling of rock masses, *Rock Mech.* **3:** 1.

Rocha, M., Baptista Lopes, J., and DaSilva, J. (1966) A new technique for applying the method of the flat jack in the determination of stresses inside rock masses, *Proc. 1st Cong. ISRM* (Lisbon), Vol. 2, pp. 57–65.

Rocha, M., Silverio, A., Pedro, J., and Delgado, J. (1974) A new development of the LNEC stress tensor gauge, *Proc. 3rd Cong. ISRM* (Denver), Vol. 2A, pp. 464–467.

Stauder, W. (1962) The focal mechanism of earthquakes, *Adv. Geophys.* **9:** 1–75.

Terzaghi, K. and Richart, R. E. (1952) Stresses in rock about cavities. *Geotechnique*, **3:** 57–90.

Tincelin, M. E. (1952) Measurement of pressure in the iron mines in the East: Methods (in French), *Supplement to Annales ITBTP*, October.

Voight, B. (1968) Determination of the virgin state of stress in the vicinity of a borehole from measurements of a partial anelastic strain tensor in drill holes, *Rock Mech. Eng. Geol.* **6:** 201–215.

Voight, B. and St. Pierre, B. H. P. (1974) Stress history and rock stress, *Proc. 3rd Cong. ISRM* (Denver), Vol. 2A, pp. 580–582.

Worotnicki, G. and Denham, D. (1976) The state of stress in the upper part of the earth's crust in Australia according to measurements in mines and tunnels and from seismic observations, *Proceedings, Symposium on Investigation of Stress in Rock (ISRM)* (Sydney, Australia), pp. 71–82.

Worotnicki, G. and Walton, R. J. (1976) Triaxial "hollow inclusion" gauges for determination of rock stresses in situ, *Proceedings, Symposium on Investigation of Stress in Rock* (Sydney), Supplement 1–8 (ISRM and Inst. of Engineers of Australia).

Zoback, M. D. and Healy, J. H. (1984) Friction, faulting, and "in situ" stress, *Ann. Geophys.* **2:** 689–698.

Zoback, M. D., Moos, D., and Mastin, L. (1985) Well bore breakouts and in situ stress, *J. Geophys. Res.* **90:** 5523–5530.

Zoback, M. L. and Zoback M. D. (1980) State of stress in the conterminous United States, *J. Geophys. Res.* **85:** 6113–6156.

Zoback, M. L. and Zoback, M. D. (1988), Tectonic stress field of the continental U.S., in *GSA Memoir* (in press), *Geophysical Framework of the Continental United States*, edited by L. Pakiser and W. Mooney.

Problems

1. Estimate the vertical and horizontal stresses at a depth of 500 m in a zone of normal faulting in Paleozoic sedimentary rocks. Use Figure 4.7b and assume the normal faulting is recent.

2. A vertical hydraulic fracture was initiated in a borehole at a depth of 3000 ft. Assume the ground is saturated continuously from the surface and that the pressure in the ground water is hydrostatic. The water pressure was first raised 710 psi above the original groundwater pressure and then it was not possible to raise it further. When pumping stopped, the water pressure fell to a value 110 psi above the original groundwater pressure. After a day, the pressure was raised again, but it could not be pumped to a value higher than 100 psi above the previous pressure (the "shut-in" pressure). Estimate the horizontal stresses at the site of measurement, the tensile strength of the rock, and the vertical pressure at the site.

3. A borehole is drilled and instrumented and then overcored, yielding the following values of the stress components in the plane perpendicular to the borehole:

$$\sigma_x = 250 \text{ psi}$$
$$\sigma_y = 400 \text{ psi}$$
$$\tau_{xy} = -100 \text{ psi}$$

The *x* axis was horizontal and to the right. Find the magnitudes and directions of the major and minor principal stresses in the plane perpendicular to the borehole.

4. A natural slope rises at 45° for 1000 m and then levels off. The rock has an unconfined compressive strength of 50 MPa. A tunnel is to be driven for an underground pressure pipe beginning with a portal at the base of the slope and continuing directly into the mountain. Based upon Norwegian experience, at what distance from the portal would you first expect to encounter rock pressure problems?

5. In a zone of active thrust faulting (low-angle reverse faulting), in rock with $\phi = 30°$, $q_u = 1000$ psi, and unit weight of 150 lb/ft^3, estimate the major and minor principal stresses at a depth of 3500 ft, assuming conditions for faulting. Compare your estimate with that of Figure 4.7.

6. A rock mass at a depth of 5000 m had a value of K (= ratio of horizontal to vertical stresses) equal to 0.8. If Poisson's ratio is 0.25, what should K become after erosion of 2000 m of rock?

7. What form would the data of the flat jack test assume if the initial stress normal to the plane of the jack were tensile? How could the data be worked to estimate the magnitude of the tensile stress?

8. Two flat jacks, 12 in. square, are placed in the wall and roof of an approximately circular test gallery 8 ft in diameter. Flat jack 1 is horizontal, and placed in the side wall. Flat jack 2 is vertical, with its edge parallel to the axis of the gallery. The cancellation pressures measured were 2500 psi with FJ 1 and 900 psi with FJ 2. Estimate the initial stresses (vertical and horizontal). List your assumptions.

9. The U. S. Bureau of Mines overcoring method is used to measure stresses in a borehole drilled perpendicularly to a tunnel wall. The site of the measurement (the plane of the measuring pins) is 5 ft deep in a test gallery 10 ft in diameter. The measuring borehole has a diameter of 1.25 in. The first pair of buttons is horizontal, pair 2 is oriented 60° counterclockwise from button pair 1; button pair 3 is 120° counterclockwise from pair 1. Deformations were measured as a result of overcoring as follows: pair 1 moved outward 3×10^{-3} in.; pair 2 moved outward 2×10^{-3} in.; and pair three moved outward 1×10^{-3} in. If $E = 2 \times 10^6$ psi and $\nu = 0.20$, determine the stress components in the plane perpendicular to the borehole, and the major and minor normal stresses in this plane and their directions. (Assume the initial stress parallel to the borehole is insignificant.)

10. Stress measurements in a horizontal rock outcrop using a series of vertical flat jacks all give a cancellation pressure of about 80 MPa. The rock is

granite with $E = 5 \times 10^4$ MPa and $\nu = 0.25$. If the rock started its life at a depth of 10 km with $\sigma_h = \sigma_y$ and was brought to the surface by erosion, what should be the value of the horizontal stress? ($\gamma = 0.027$ MN/m³.) If there is a discrepancy, explain why.

11. Making use of the effective stress principle (Chapter 3), derive a formula corresponding to Equation 4.8 expressing the effective pressure ($p_{c1} - p_w$) for crack initiation in hydraulic fracturing when the rock has a pore pressure p_w.

12. Bearpaw shale was loaded from 0 to 2000 psi vertically over a broad area so horizontal strain could be assumed equal to zero. Poisson's ratio was 0.40 during loading. Subsequently, the vertical load was reduced to 1000 psi. Poisson's ratio was 0.31 during unloading. (a) Estimate the horizontal pressure corresponding to the maximum and final vertical loads. (b) What natural events could bring about a similar stress history?

13. In a rock with $\nu = 0.3$ and $E = 3.0 \times 10^4$ MPa, "doorstopper" measurements yield the following strains in the arms of a 60° rosette gauge on the bottom of a vertical borehole 10 m deep: $\varepsilon_A = -20 \times 10^{-4}$ in the gage parallel to OX (east-west direction); $\varepsilon_B = -3.8 \times 10^{-4}$ in the gage oriented 60° counterclockwise from OX; and $\varepsilon_C = -5.0 \times 10^{-4}$ in the gage aligned 120° counterclockwise from OX. The hole is parallel to the y axis. Assuming σ_y is due to rock weight alone and $\gamma = 0.027$ MN/m³, compute the greatest and least normal stresses in the plane of the hole bottom (xz plane) and their directions.

Chapter 5

Planes of Weakness in Rocks

5.1 Introduction

Those who excavate into rock know it to be a material quite apart from what a mathematician might choose for tractable analysis. The engineer's rock is heterogeneous and quite often discontinuous. The latter is especially true if blasting has been excessively violent and uncontrolled, but even if the engineer were as careful as a sculptor, he or she would be confronted with naturally occurring planes of weakness traversing the rock mass and separating it into perfectly fitted blocks. Furthermore, the process of excavation will induce new fractures in the surrounding rock by virtue of stress readjustments (Figure 5.1a, b).

Small cracks that we have been referring to as *fissures* may be seen in many hand specimens. *Joints* (Figure 5.2) are usually present in rock outcrops. They appear as approximately parallel planar cracks separated by several centimeters up to as much as 10 m. One set of joints commonly forms parallel to bedding planes and there are usually at least two other sets in other directions. Igneous and metamorphic rocks may have regular jointing systems with three or more sets. Rocks that have been deformed by folding often contain roughly parallel seams of sheared and crushed rock produced by interlayer slip or minor fault development. These *shears* are usually spaced more widely than joints and are marked by several millimeters to as much as a meter thickness of soft or friable rock or soil. Shears parallel to bedding planes occur in unfolded strata near valley sides, due to interlayer slip as the rock mass relaxes horizontally. *Faults* that offset all other crossing structures may also occur in the rock of an engineering site. Thus there is a full range of planar weaknesses in rock masses with a statistical distribution of spacings and orientations at all scales. Figures 5.3a and b show histograms of discontinuities observed at two dam sites studied by B. Schneider (1967). The fractures were studied using a combination of aerial photographic interpretation, field observation, and microscopic study of

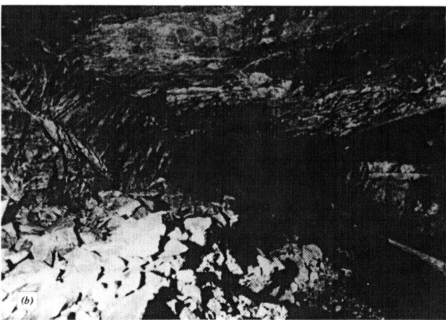

Figure 5.1 Fractures induced by mining a longwall face at great depth in quartzite, South Africa. In both photos, the new fractures terminate in the roof against a preexisting, flat-lying joint developed along a shale parting. (*a*) The slabs formed by the mining-induced fractures can be seen toppling towards the excavated space. (*b*) The new fractures terminate downward in a preexisting shear zone inclined towards the excavation. (Courtesy South African Chamber of Mines.)

Figure 5.2 Discontinuous rocks. (*a*) Rock-bolted sheet joints in Navajo Sandstone, Glen Canyon, Arizona. (*b*) Cross-bedded and sheet-jointed Navajo sandstone, Glen Canyon.

Figure 5.2 Discontinuous rocks. (*c*) Bedded sandstone, shale, and coaly sediments in the foundation of Bennett Dam, Peace River, B. C. (*d*) Grouting open joints between beds of quartzite, Akosombo Dam, Ghana.

Figure 5.2 Discontinuous rocks. (*e*) Stream controlled by primatic jointing. (*f*) Formation of an arch by deterioration of wall rock liberated by a persistent joint. [(*e*), (*f*) are in Devonian siltstone and shale, Enfield Glen, near Ithaca, N.Y.].

Figure 5.2 Discontinuous rocks. (g) A long fracture surface formed by linking of nonpersistent individual joints. (h) Short cross joints and one long discontinuity [as in (g)], Navajo sandstone, Zion Canyon, Utah.

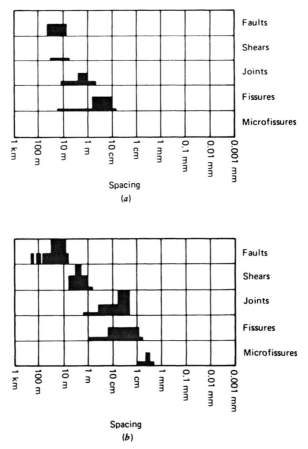

Figure 5.3 Relative distributions of discontinuities at (*a*) Ta Chien Dam Site, in quartzite and (*b*) Malpasset Dam Site, in schistose gneiss. [After B. Schneider (1967).]

stained thin sections of rock. At Ta-Chien dam site, Formosa, planes of weakness traversed the rock with spacings from 50 m down to 10 cm. At the Malpasset Dam site, the rock showed cracks of various types with spacings from more than 100 m to less than 5 mm. The Malpasset Dam failed because of the discontinuous character of the rock in the abutments.

Discontinuities that are spaced more than about 20 m apart can be shown individually in site sections and plans and considered individually in analysis. On the other hand, planes of weakness that are more closely spaced occur in large numbers and the only feasible way to appreciate their impact is often to appropriately modify the properties of the rock mass, for example, by reducing its modulus of elasticity. Figure 5.4 shows examples of single features (*S*) and

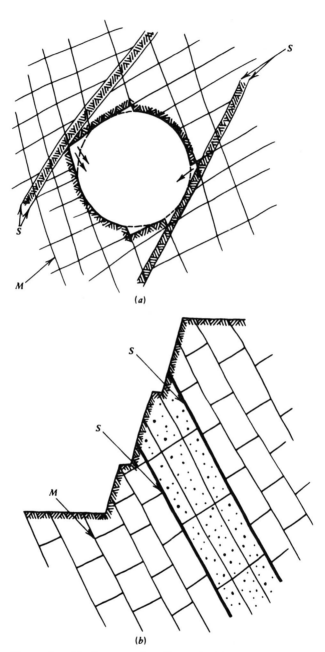

Figure 5.4 Singly occurring discontinuities (S) and multiple features (M) in the region of influence of excavations.

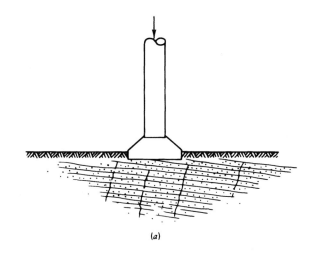

(a)

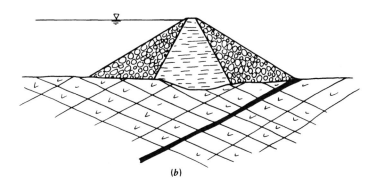

(b)

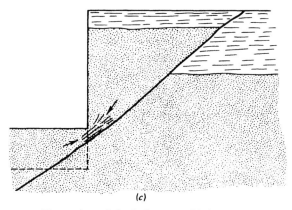

(c)

Figure 5.5 Examples of the influence of joints and other discontinuities on foundations and excavations.

multiple features (*M*) in the rock around a tunnel and a surface excavation. Notice how the locations of joints have affected the shape of the tunnel in Figure 5.4*a*. Normally the precise locations of the multiple features will not be shown on the drawings, a pattern like that of Figure 5.4*a* being diagrammatic only. Single features, however, can and should be plotted, for details of their occurrence can radically affect the quality and cost of the work.

The importance of planar weaknesses stems from the special properties that such features superimpose on rock. Basically, the rock mass becomes weaker, more deformable, and highly anisotropic because there is reduced shear strength and higher permeability parallel to discontinuities and increased compressibility as well as reduced tensile strength (essentially zero) perpendicular to them. These factors combine to create a variety of potential problems. Foundations on jointed rocks (Figure 5.5*a*) may settle significantly as the joints close under load even if the rock itself is very stiff. Dams underlain by discontinuous rock (Figure 5.5*b*) may initiate slip of rock blocks along one or more weak surfaces; more than one dam failure has been attributed to this mechanism (including Malpasset mentioned above—see Figure 9.15). Rock slopes may fail as rock blocks move on single or multiple weakness planes. For example, the rock slope in Figure 5.5*c* will most probably fail when the excavation is deepened to the position of the dashed lines. Figure 5.4*a* showed how blocks might fall from the roof of a tunnel due to intersecting joint planes. At a larger scale, whole chambers can collapse owing to unfortunate intersections of planar weaknesses. Another behavior pattern of jointed rock, in addition to rigid translation of blocks on planar surfaces, is bending of blocks under stress. Flexural cracking and rock falls may follow such bending in a roof in thinly bedded rock. Similarly, the rock cut of Figure 5.4*b* is subject to flexure and cracking of the inclined "cantilever beams" created by the steeply dipping joints and contacts.

Since joint planes introduce such strongly directional weaknesses, the most important joint attribute can be considered to be *orientation*. Fortunately, this can be determined relatively reliably.

5.2 Joint Orientations

It is rare to encounter a rock mass with truly random fracturing. In every instance where attitudes were measured for naturally occurring weakness planes, the author's experience has been that planar weaknesses cluster around one or more "preferred orientations." This is appreciated readily if the directions normal to the measured planes are plotted on a stereographic or equal area projection. (The principles of stereographic projection are presented in Appendix 5.) Either the upper hemisphere or the lower hemisphere normals may be plotted. The upper hemisphere normal will be preferred here because it has the same direction as that of the dip vector; that is, if the strike is north and the dip 30° east, the upward normal will rise 60° to the east. Figure 5.6*a* shows a series of normals that cluster about three preferred orientations roughly orthogonal to each other. In Figure 5.6*b*, there are two clusters but the scatter of values is very great in one case (set 1) while a second cluster (set 2) has been rotated and spread by folding so that its points are distributed about a segment of a great circle of the sphere. If the normals to planes are distributed evenly around a center, it is possible to pick a good representation for the preferred orientation of the distribution by guessing the location of the point of greatest concentration of normals. There are also methods for contouring the points (see, for example, Hoek and Bray (1977), cited in Chapter 8). Another approach to describing the preferred orientation is to sum the normals vectorially.

Each joint normal can be considered a unit vector and the orientation of the resultant of all the individuals of a cluster represents the preferred orientations (the "mean") of the set. The summation can be accomplished by accumulating the direction cosines (see Appendix 1). Let x be directed horizontally and north, y horizontally and west, and z vertically upward. If a normal to a joint plane rises at angle δ above horizontal in a direction β measured counterclockwise from north, the direction cosines of the normal to the joints are

$$l = \cos \delta \cos \beta$$
$$m = \cos \delta \sin \beta \qquad (5.1)$$

and

$$n = \sin \delta$$

If many joints are mapped in one set, the preferred, or mean orientation of the joint set is parallel with the line defined by direction cosines equal to the sums of all individual l's, m's and n's; dividing by the magnitude of this resultant vector gives the direction cosines (l_R, m_R, n_R) of the mean joint orientation:

$$l_R = \frac{\Sigma l_i}{|\overline{R}|} \qquad m_R = \frac{\Sigma m_i}{|\overline{R}|} \qquad n_R = \frac{\Sigma n_i}{|\overline{R}|} \qquad (5.2)$$

where

$$|\overline{R}| = [(\Sigma l_i)^2 + (\Sigma m_i)^2 + (\Sigma n_i)^2]^{1/2}$$

The angle of rise δ_R and the direction of rise β_R of the normal to the mean orientation are obtained with Equations 5.1 together with rules for the correct sign of the arc cosine:

$$\delta_R = \sin^{-1}(n_R) \qquad 0 \leq \delta_R \leq 90°$$
$$\beta_R = + \cos^{-1}\left(\frac{l_R}{\cos \delta_R}\right) \qquad \text{if } m_R \geq 0$$

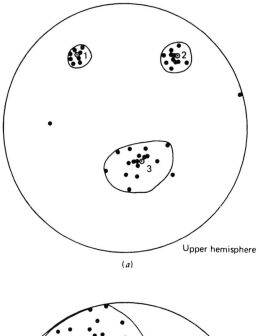

(a)

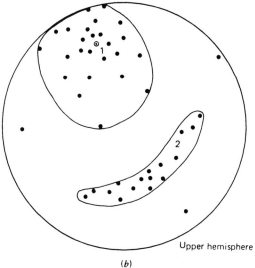

(b)

Figure 5.6 Distributions of normals to discontinuities, plotted on stereographic projections. (a) Two-well-defined sets and a third more disperse set. (b) One very disperse set and a second set distributed in a great circle girdle.

and   (5.3)

$$\beta_R = -\cos^{-1}\left(\frac{l_R}{\cos \delta_R}\right) \qquad \text{if } m_R < 0$$

(In the above, the number represented by the $\cos^{-1}$ term is presumed to lie between 0 and 180°.)

One can also estimate the scatter of normals about the mean orientation by comparing the length of the resultant with the number N of joints considered. If the joints were all parallel, the resultant would equal N, whereas if the joints were widely varying in orientation, the resultant would be considerably less than N. This is represented by a parameter K_F:

$$K_F = \frac{N}{N - |\overline{R}|}$$ (5.4)

K_F becomes very large as the dispersion of joint orientations becomes small.

According to the hemispherical normal distribution (Fisher, 1953), the probability P that a normal will make an angle of ψ degrees or less with the mean orientation is described implicitly by

$$\cos \psi = 1 + \frac{1}{K_F} \ln(1 - P)$$ (5.5)

Thus one can express the spread of values about the mean—the "dispersion"—corresponding to any degree of certainty. The standard deviation of the hemispherical normal distribution ($\overline{\psi}$) can also be used to express the dispersion of normals about the mean:

$$\overline{\psi} = \frac{1}{\sqrt{K_F}}$$ (5.6)

When calculating or estimating the orientation parameters δ_R, β_R, K_F, and $\overline{\psi}$ for each joint set, one must insure unbiased selection of individuals for analysis. Unfortunately, as pointed out by Terzaghi (1965), outcrops and drill holes introduce bias. Figure 5.7a shows that the joints that parallel an outcrop surface (i.e., whose normals are parallel to the normal to an outcrop) cannot be seen. If α_0 is the angle between a normal to a joint and the normal to the outcrop, bias can be overcome by weighting the calculations such that the single joint is replaced by a number of joints equal to $1/\sin \alpha_0$. Similarly, drill holes will not reveal joints whose normals are perpendicular to the axis of the hole (Figure 5.7b). Thus each joint individual oriented in a drill core should be weighted by treating it in the analysis of orientations as if it were $1/\cos \alpha_H$ joints where α_H is the angle between the normal to the joint and the axis of the hole.

Generally, orientations of joints cannot be determined from drill hole data because the core rotates an unknown amount as it is returned to the surface. Methods for orienting core were reviewed by Goodman (1976).

The joint orientation parameters discussed here are fundamental properties of the rock mass. In general, each joint set will also have a characteristic physical description and a corresponding set of physical properties including, most importantly, the parameters necessary to represent joint strength.

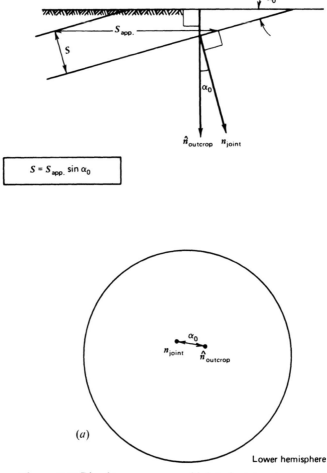

Figure 5.7 Bias in occurrence of joints in (*a*) outcrops.

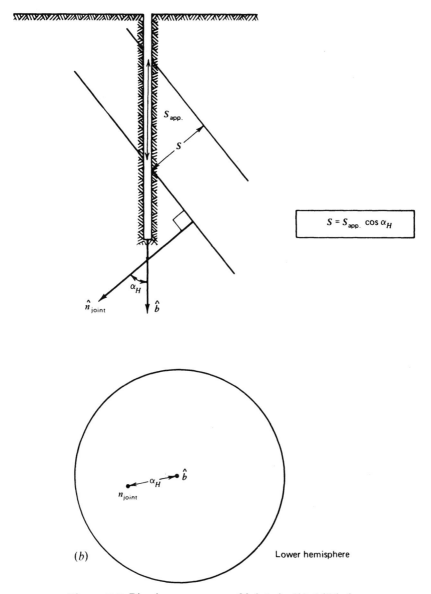

$$S = S_{app.} \cos \alpha_H$$

Figure 5.7 Bias in occurrence of joints in (*b*) drill holes.

5.3 *Joint Testing*

When a rock mass is excavated, some joints will close up while others will open and some blocks will slide against others along joint surfaces. The properties governing joint deformability and strength enable the magnitudes and directions of these movements to be calculated in practical problems. To obtain the required descriptive properties, the engineer has two choices: (1) to use experience and judgment to select reasonable values for the joint properties based on careful descriptions of joint characteristics as observed by geologists or geotechnical engineers in outcrops and in core samples, or (2) to attempt to measure the properties directly in field or laboratory tests. The latter is preferable but it is not often possible to obtain good samples for this purpose.

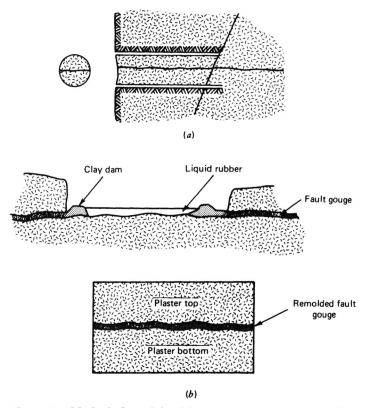

Figure 5.8 Methods for gaining joint samples. (*a*) Oriented drilling. (*b*) Molding and casting.

Figure 5.8 Methods for gaining joint samples. (*c*) Models of joints obtained as in (*b*): the left rectangle is a mold of a ripple-marked bedding surface supplied by H. Schneider; the middle rectangle is a cast of the same surface in plaster prepared for direct shear testing; the right squares are the top and bottom of a portion of the same surface after shear testing in a smaller direct shear machine (note the wear along the slopes of the ripples).

Samples for laboratory testing can be acquired by drilling a large-diameter core parallel to a joint plane that outcrops, as depicted in Figure 5.8*a*. Alternatively, a section of the joint surface could be molded with liquid rubber in a field and facsimiles of the upper and lower blocks later cast in plaster, cement, or sulfur in the laboratory (Figures 5.8*b,c*). The model joints will correctly represent the roughness of the surface and clay or mineral filling material collected in the field can be spread on the model surfaces to simulate the actual field condition. Good results can be obtained this way if the tests are scaled by the ratio of normal stress to compressive strength. For example, to study the shear of a joint under 500 psi normal pressure in limestone having a compressive strength of 16,000 psi, a model joint of sulfur (concrete cylinder capping compound "Cylcap") having a compressive strength of 8000 psi should be tested at a normal pressure of 250 psi.

Both triaxial and direct shear test methods may be adapted for testing specimens with joints. In the *direct shear test* (Figure 5.9*a*), the joint surface is oriented parallel to the direction of applied shear load and the two halves of the sample are fixed inside a shear box, using Cylcap, concrete, plaster, or epoxy.

To avoid a moment and rotation of one block relative to the other, the shear load may be inclined slightly, as in Figure 5.9b; but this prevents shear testing at very low normal loads. Rotation during shear can occur in the shear apparatus shown in Figure 5.9c, where the reactions to the normal and shear forces are supplied by cables. A test in which rotations can occur during shearing tends to underestimate the shear strength compared to one in which rotations are prevented. Both conditions of loading exist in nature.

The stress conditions inside the shear box are represented by the Mohr circle in Figure 5.9d. The normal stress σ_y and shear stress τ_{xy} on the failure plane define point A'. The normal stress σ_x parallel to the joint is unknown and

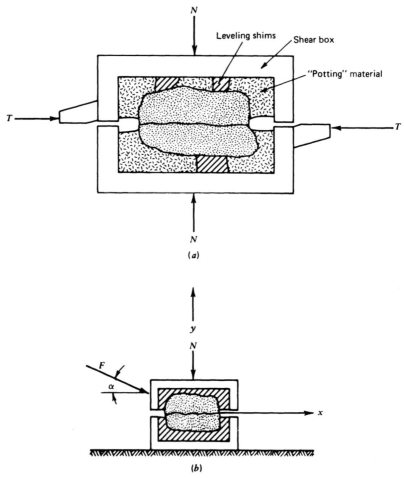

Figure 5.9 Direct shear testing. (a) The arrangement of the specimen in a shear box. (b) A system for testing with inclined shear force to avoid moments.

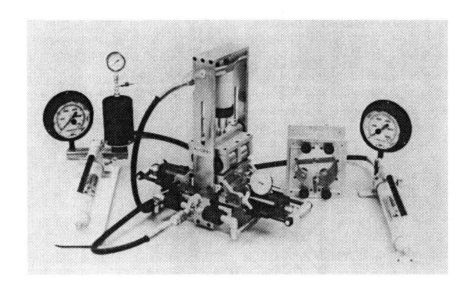

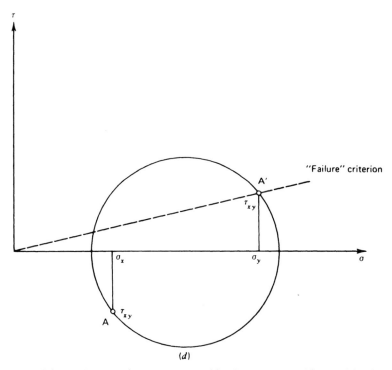

(d)

Figure 5.9 Direct shear testing. (c) A portable shear-test machine and load-maintaining system based upon the ISRM-suggested method which accommodates samples up to 115 by 115 mm. Courtesy of Roctest Inc., Plattsburg, N.Y. (d) Approximation to the state of stress in the shear box; x and y are parallel and perpendicular to the joint surface.

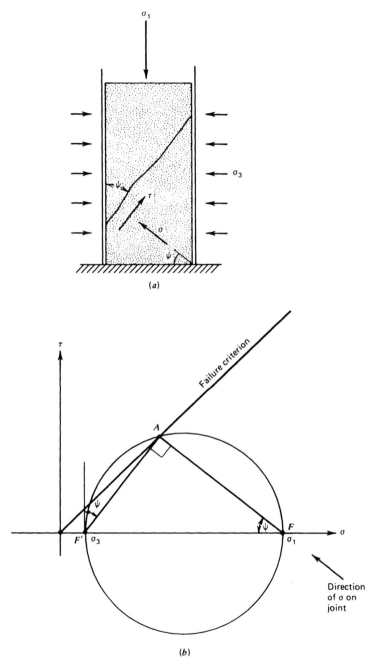

Figure 5.10 Triaxial testing with jointed specimens. (*a*) Arrangement of joint. (*b*) Stress conditions.

Figure 5.10 Triaxial testing with jointed specimens. (*c*) A natural joint in graphite schist oriented for a multistage triaxial test, viewed after the test. The joint was initially closed and was opened by the first load stage.

can range from zero to a large proportion of σ_y depending on the system used for fixing the specimen inside the shear box. Since the shear stress perpendicular to the joint plane must be equal to τ_{xy}, the stress is represented by the Mohr circle with diameter AA' (Figure 5.9*d*). This can lie partly in the tensile stress field if σ_x is small. Accordingly, direct shear is not a good way to test intact specimens of rock or soil. However, for sliding of joints, the direct shear test is advantageous because normal and shear displacements during shearing can both be measured easily and because shearing can proceed for long distances such that wear is developed and the strength falls to its minimum value; the latter is termed the *residual strength*. The direct shear test can also be conducted in the field by excavating a joint block in place. A series of direct shear tests conducted at different confining pressures generate a series of points A_1, A_2, etc., through which a joint shear strength curve or line will be drawn.

Another approach for joint testing is the *triaxial compression test* discussed in Chapter 3. If the joint is oriented at an angle ψ equal to 25 to 40° with the direction of axial load, as shown in Figure 5.10*a*, the joint will slip before the rock has had a chance to fracture and the joint strength criterion can be established. The strength criterion is *not* an envelope to a series of Mohr circles as in the case of intact rock, because failure is not free to develop in the most critical orientation but is constrained to occur along the joint at angle ψ. Figure 5.10*b*

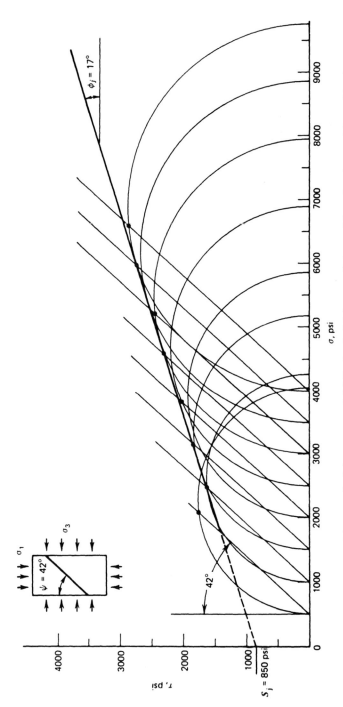

Figure 5.11 Results of a multistage triaxial test on an initially healed joint parallel to schistosity, in graphite schist (same specimen as in Figure 5.10c).

shows a simple construction to locate the relevant point on the Mohr circle. Using Bray's procedure (Appendix 1) for constructing the Mohr circle for the triaxial test with $\sigma_x = p$ and $\sigma_y = \sigma_{axial}$, the focus is located at the right side of the circle at point F (Figure 5.10). The normal and shear stress on the joint plane corresponding to the stresses at peak load are then found by drawing a line from F making an angle of ψ with the horizontal and piercing the Mohr circle at point A (or alternatively making an angle of ψ with vertical from F' at the left side of the circle). A series of triaxial tests conducted at different normal pressures will yield a series of points A_1, A_2, etc., through which a joint failure criterion can be drawn.

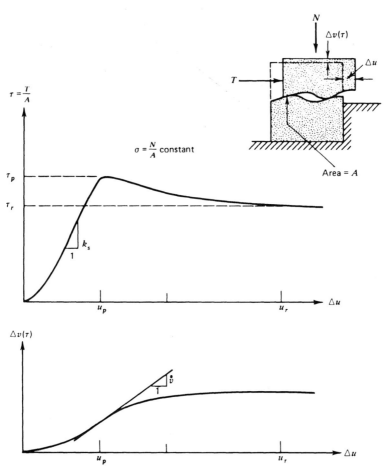

Figure 5.12 Tangential and normal displacements during direct shear of a rough joint.

If the triaxial test is started at a low confining pressure and the confining pressure is quickly raised after the peak axial stress is reached, it will be possible to generate a series of Mohr circles for slip from a single test. Such a procedure is called "a multistage test." Figure 5.11 shows a Mohr circle family generated from a multistage test for the joint in graphite schist shown in Figure 5.10c. (A multistage test can also be programmed in direct shear.) If real rock joints rather than models are tested, identical specimens cannot be supplied for a series of previously untested rough joints; but a multistage test could be used to investigate the strength criterion. Since wear will develop at higher normal pressures, a multistage test will not give the same results as a series of identical virgin specimens tested at different normal pressures; however, some compromise is required if actual rock is to be tested.

Figure 5.12 shows the type of data obtained from the direct shear test. The *shear displacement* across the joint, Δu, is the difference in displacements of the upper and lower blocks measured parallel to the joint plane. If the joint is rough, a mean joint plane is recognized, passing through hills and over valleys of the joint surface. Because of these undulations, the joint will tend to thicken or "dilate" during shearing. The *dilatancy* Δv is the difference between the normal displacements of the upper and lower block as a result of shear displacement. Opening (thickening) is reckoned as positive dilatancy. As the shear stress builds, a period of adjustment with slight dilatancy is followed by rapid increase in the rate of dilatancy; the dilatancy rate is greatest as the peak shear stress (the "shear strength") is attained. Thereafter, the shear stress falls continuously and the joint dilates continuously until attaining the residual displacement, some millimeters or even centimeters beyond the peak. In the field, with very rough joint surfaces, residual displacement may be achieved only after as much as a meter of movement. The dilatancy and, to a considerable degree, the joint strength are controlled by the joint roughness.

5.4 Joint Roughness

Imagine a joint surface with identical asperities rising at an angle i from the mean joint plane (Figure 5.13). Let the friction angle of a smooth joint be ϕ_μ. At the moment of peak shear stress, the resultant force on the joint, $\mathbf{R}$, is then oriented at an angle ϕ_μ with the normal to the surface on which motion is about to occur; since this surface is inclined i degrees with the joint plane, the joint friction angle is $\phi_\mu + i$ when referred to the direction of the mean joint plane. The accuracy and utility of this simple concept was demonstrated by Patton (1966).

Values of ϕ_μ have been reported by many authors. A reasonable value appears to be 30° with most values in the range 21 to 40°. Byerlee (1978) found

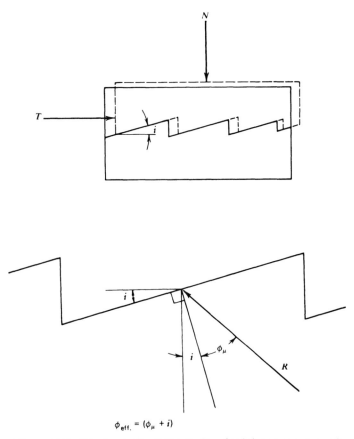

Figure 5.13 The basis for Patton's law for joint shear strength.

$\phi_\mu = 40°$ fit a variety of rocks with saw cut or ground surface up to σ equals 2 kbar (with $\tau_p = 0.5 + 0.6\,\sigma$ kbar applicable at higher pressure—up to 16 kbar). Frequently, ϕ_μ can be much lower when mica, talc, chlorite, or other sheet silicate minerals form the sliding surface or when clay gouge is present. There is little opportunity for drainage of water from the pores of saturated clay locked in between joint walls and values of ϕ_μ as low as 6° have been reported for montmorillonite clay-filled shears. The roughness angle i can be any value from 0 to 40° or more at low pressures as discussed later.

If the normal pressure is relatively large, it will be easier to shear the joint through the teeth along its surface than to lift over them. Mobilizing some rock strength by failure through the teeth generates a shear strength intercept S_J and a new friction angle ϕ_r related to sliding on surfaces broken through the rock and thus approximated by the residual friction angle for intact rock specimens. (The residual friction angle is the slope of the linear envelope to a series of

Mohr circles through residual stress values for a series of triaxial compression tests with intact rock specimens.) Figure 5.14 shows the bilinear failure criterion for joints representing the merging of Patton's law and the condition for shearing through asperites:

$$\tau_p = \sigma \tan(\phi_\mu + i) \qquad \text{for } \sigma \text{ small}$$

and   (5.7)

$$\tau_p = S_j + \sigma \tan \phi_r \qquad \text{for } \sigma \text{ large}$$

For many purposes, it is sufficient to replace ϕ_r in the second equation by ϕ_μ since these values are close. Actual data show a transition from the initial slope at $\phi_\mu + i$ to the final slope at ϕ_r. Theories of joint strength effecting this transition smoothly were presented by Ladanyi and Archambault (1970), Jaeger (1971), and Barton (1973) and were reviewed by Goodman (1976).

Roughness controls not only the peak shear strength at low normal pressures but the shape of the shear stress versus shear deformation curve and the rate of dilatancy. This is shown in Figures 5.15 and 5.16 based on work by Rengers (1970) and Schneider (1976). Suppose the joint surface were accurately profiled as in Figure 5.15a. Any two points a distance nS apart along the profile will define a line inclined i degrees with the mean plane through the joint surface. If the measuring points are moved all along the surface and the measuring base length is varied, a series of angles will be measured; these angles have been plotted against base length in Figure 5.15b, and envelopes to all the points have been drawn. It is seen that the maximum angles presented by a rough surface approach zero as the measuring distance becomes appreciably

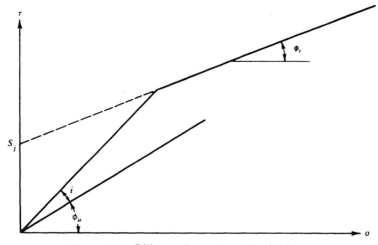

Figure 5.14 Bilinear shear strength criterion.

(a) Profile:

(b)

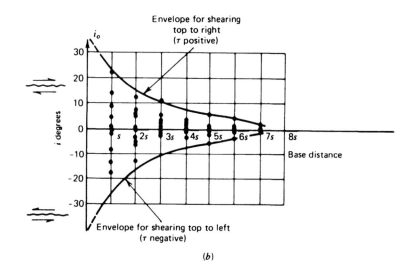

(c)

(d)

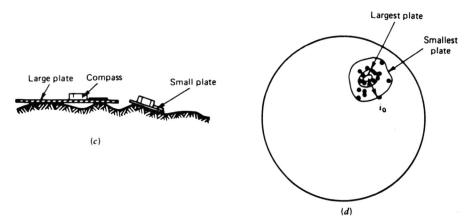

Figure 5.15 Rengers' analysis of roughness. (a) A rough surface. (b) Envelopes to roughness angles as a function of base length. (c) and (d) Approximation to Rengers' roughness angles by Fecker and Rengers' method.

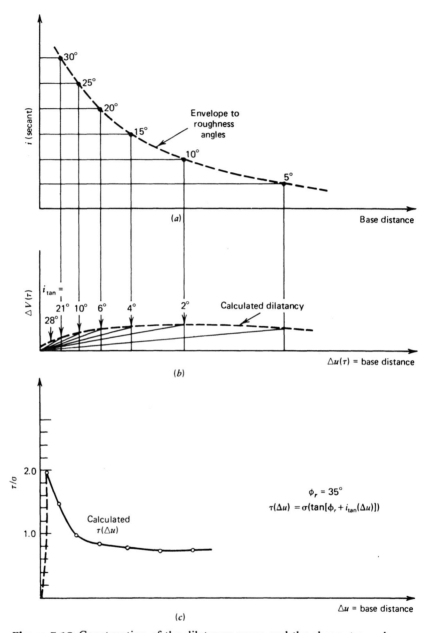

Figure 5.16 Construction of the dilatancy curve and the shear stress-shear deformation curve from the Rengers' envelope according to the method of H. Schneider (1976).

larger than the wavelengths of the roughness and waviness of the surface. The envelopes drawn in Figure 5.15b present a series of i values corresponding to shear displacements numerically equal to each base distance nS. The upper envelope corresponds to shearing of the upper block to the right while the lower envelope corresponds to shearing of the upper block to the left. Consider shearing to the right. In Figure 5.16a, the appropriate envelope has been replotted. Then, directly below it (Figure 5.16b), the dilatancy curve has been constructed by drawing a series of secants described by the appropriate i value for each base distance. Next we construct the shear stress versus shear displacement curve (Figure 5.16c) from the dilatancy curve assuming that each shear stress value $\tau(\Delta u)$ can be calculated from the normal stress acting on a surface whose effective friction angle is the sum of the current i value and the residual friction angle (Schneider, 1976); that is,

$$\tau(\Delta u) = \sigma \tan[\phi_r + i_{tan}(\Delta u)] \tag{5.8}$$

where $i_{tan}(\Delta u)$ is the inclination of the tangent to the dilatancy curve at any value of Δu.

We have seen that the roughness profile of a weakness "plane" is fundamentally valuable. It will be useful then to have more than one way of determining it. Figure 5.15c shows another method demonstrated by Fecker and Rengers (1971). If the orientations of flat plates placed down on a single rough surface are compared, it will be seen that they are scattered about a mean value. For a plate of a given size, the maximum angle from the mean of the series of measurements, in other words the i value, can be obtained by plotting the normals on a stereographic projection, drawing an envelope to all the points, and measuring the angle between the envelope and the mean orientation (Figure 5.15d). Alternatively, this can be done mathematically, emulating the procedure discussed in Section 5.2 where a series of nonparallel planes were averaged. The difference here is that one plane only is being measured, the different points representing the different orientations measured at different places on its surface.

5.5 Interrelationships among Displacements and Strengths

When a block containing a joint plane is subjected to a shear stress parallel to the joint, it can undergo both shear Δu and normal displacement Δv. If compressed normal to the joint, it will tend to shorten by joint closure and if pulled apart normal to the joint, the block will eventually separate into two blocks as the joint opens. All these phenomena are coupled together as shown in Figure 5.17. The upper figure (5.17a) shows the compression behavior of the joint; it is

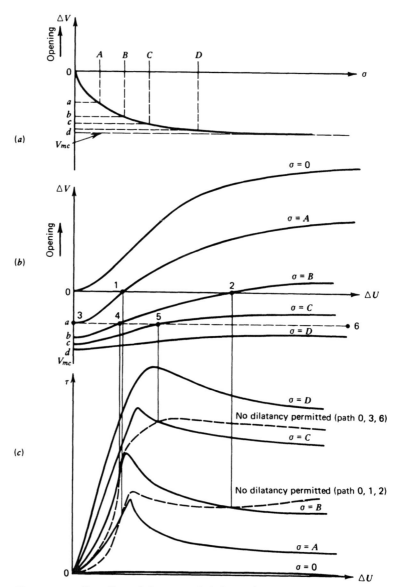

Figure 5.17 Coupling of the normal deformation, shear deformation, and dilatancy laws for rough joints and analysis of path dependency.

highly nonlinear and becomes asymptotic to a maximum closure (V_{mc}) related to the initial thickness or aperture of the joint.

Suppose a virgin specimen were sheared without normal pressure. Dilatancy would occur as shown in the upper curve of Figure 5.17*b* while the shear stress would never rise above zero since there is no frictional resistance in this case (lowermost curve of Figure 5.17*c*). If the specimen had been compressed initially to a value *A*, *B*, *C*, or *D*, the dilatancy and shear stress versus shear displacement relationships would have been as depicted by the families of curves in 5.17*b* and *c*. As the normal pressure grows, the dilatancy is gradually reduced because a greater proportion of the asperities becomes damaged during shearing. All the curves of dilatancy and shear assume the normal stress was maintained constant at the indicated initial value throughout the shearing process. This would not have been true if the normal displacement had been restricted, as, for example, when a block moves into a tunnel between parallel joint surfaces. However, the shear stress versus shear displacement function for such a condition can be determined from the data presented. Suppose, for example, the joint normal stress were zero to start but no dilatancy was allowed during shearing. Then, the joint would acquire normal stress *A* when it had sheared to point 1, with shear resistance appropriate to the point on the shear stress/displacement curve corresponding to normal stress *A*. Thus, as shear progressed, the shear stress would rise with displacement along the dashed locus 0, 1, 2. In similar fashion, one could construct the shear stress/shear displacement curve for a joint initially compressed to normal stress *A* and then sheared without dilatancy (path 0, 3, 6). In both cases, note that considerable additional shear strength was acquired by the restriction of normal displacement and the behavior became plastic rather than brittle; that is, there was little or no drop in strength after a peak stress had been reached. This helps explain why rock bolt reinforcement has been successful in stabilizing rock slopes and excavations.

Mathematical relationships have already been presented for the variation of peak stress versus normal pressure (Equation 5.7). Additional formulas have been demonstrated linking the decline of peak dilatancy with normal stress (Ladanyi and Archambault, 1970) and for the compression of joints (Goodman, 1976).

5.6 Effect of Water Pressure

Joints obey the effective stress principle discussed in Chapter 3. The water pressure in a joint directly counteracts the strengthening effect of the normal stress applied to the joint. To calculate the water pressure required to cause a fault or joint to slip, one needs to determine the amount by which the Mohr circle corresponding to the current state of stress has to travel to the left to bring the normal and shear stress on the fault or joint plane to the limiting condition represented by the criterion of failure (Figure 5.18). This calculation is slightly more complicated than for the case of unjointed rock, because now, in addition to initial stresses and strength parameters, the orientation of the joint plane (ψ with the direction of σ_1) needs to be considered. If the initial stresses are σ_3 and σ_1, the water pressure that will produce fault slip is

$$p_w = \frac{S_j}{\tan \phi_j} + \sigma_3 + (\sigma_1 - \sigma_3) \left(\sin^2 \psi - \frac{\sin \psi \cos \psi}{\tan \phi_j} \right) \tag{5.9}$$

p_w is the minimum of values calculated from (5.9) using (a) $S_j = 0$ and $\phi_j = \phi + i$ and (b) $S_j \neq 0$ and $\phi_j = \phi_r$.

This simple application of the effective stress principle has been shown to explain satisfactorily the occurrence of earthquakes due to water injection in a deep waste disposal well near Denver, Colorado (Healy et al., 1968) and in the Rangely oil field, western Colorado (Raleigh et al., 1971). It can be used to consider the likelihood of triggering earthquakes by construction of reservoirs

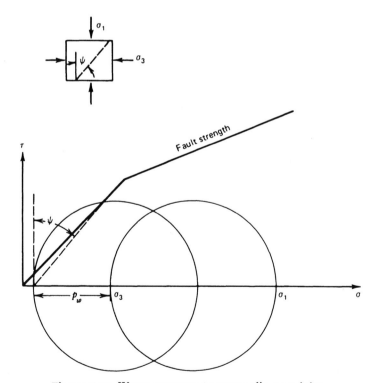

Figure 5.18 Water pressure to cause slip on a joint.

near active faults. However, the initial stresses in the crust as well as the frictional properties of the fault will have to be known.

References

Barton N. (1973) Review of a new shear strength criterion for rock joints, *Eng. Geol.* 7: 287–332.

Barton, N. (1976) The shear strength of rock and rock joints, *Int. J. Rock Mech. Min. Sci.* 13: 255–279.

Barton, N. et al. (1978) Suggested methods for the quantitative description of discontinuities in rock masses, for ISRM Commission on Standardization of Lab and Field Tests, *Int. J. Rock Mech. Min. Sci.* 15: 319–368.

Barton, N. R. and Choubey, V. (1977) The shear strength of rock joints in theory and practice, *Rock Mech.* 10: 1–54.

Bray, J. W. (1967) A study of jointed and fractured rock, I. Fracture patterns and their failure characteristics, *Rock Mech. Eng. Geol.* 5: 117–136.

Byerlee, J. (1978) Friction of rocks, *Pure and Applied Geophysics,* American Geophysical Union.

Fecker, E. and Rengers, N. (1971) Measurement of large scale roughness of rock planes by means of profilograph and geological compass, *Proceedings, International Symposium on Rock Fracture,* Nancy (ISRM), paper 1–18.

Fisher, R. A. (1953) Dispersion on a sphere, *Proc. R. Soc. London, Ser. A* 217: 295.

Goodman, R. E. (1970) The deformability of joints, *ASTM Spec. Tech. Publ. 477,* pp. 174–196.

Goodman, R. E. (1976) reference given in Chapter 1.

Healy, J. H., Rubey, W. W., Griggs, D. T., and Raleigh, C. B. (1968) The Denver earthquake, *Science* 161: 1301–1310.

Jaeger, J. C. (1971) Friction of rocks and the stability of rock slopes—Rankine Lecture, *Geotechnique* 21: 97–134.

Ladanyi, B. and Archambault, G. (1970) Simulation of the shear behavior of a jointed rock mass, *Proceedings, 11th Symposium on Rock Mechanics* (AIME), pp. 105–125.

Patton, F. D. (1966) Multiple modes of shear failure in rock, *Proc. 1st Cong. ISRM* (Lisbon), Vol. 1, pp. 509–513.

Priest, S. D. and Hudson, J. A. (1976) Discontinuity spacings in rock, *Int. J Rock Mech. Min. Sci.* 13: 135–148.

Raleigh, C. B., Healy, J. H., Bredehoeft, J. D., and Bohn, J. P. (1971) Earthquake control at Rangely, Colorado, *Trans AGU* 52: 344.

Rengers, N. (1970) Influence of surface roughness on the friction properties of rock planes, *Proc. 2nd Cong. ISRM* (Belgrade), Vol. 1, pp. 229–234.

Schneider, B. (1967) Reference given in Chapter 6.

Schneider, H. J. (1976) Comment in *Proceedings of International Symposium on Numerical Methods in Soil Mechanics and Rock mechanics,* G. Borm and H. Meissner, Eds., pp. 220–223 (Inst. für Bodenmechanik und Felsmechanik of Karlsruhe University, D-7500, Karlsruhe 1, Germany).

Terzaghi, R. (1965) Sources of error in joint surveys, *Geotechnique* 15: 287.

Problems

1. Determine by mathematical calculation the mean orientation and the Fisher distribution parameter k_f for each of the joint sets represented by the following data collected in the field:

Joint or Other Plane	Strike (°)	Dip (°)	Joint or Other Plane	Strike (°)	Dip (°)
1	S40 E	35 NE	16	S38 W	62 NW
2	S42 E	35 NE	17	S36 W	63 NW
3	S40 E	39 NE	18	S38 E	41 NE
4	S30 W	60 NW	19	S25 E	38 NE
5	S35 W	61 NW	20	S30 W	58 NW
6	S41 E	34 NE	21	N30 E	30 SE
7	S32 W	59 NW	22	N35 E	32 SE
8	S35 W	62 NW	23	N22 E	28 SE
9	S38 E	37 NE	24	N45 E	60 NW
10	S40 E	37 NE	25	N55 E	58 NW
11	S33 W	61 NW	26	N50 E	59 NW
12	S33 W	64 NW	27	N30 W	90
13	S40 E	37 NE	28	N40 W	88 NE
14	S41 E	36 NE	29	N40 W	1 NE
15	S40 W	62 NW	30	N30 E	24 SE

2. Plot the normals to the joint planes of Problem 1 on an upper hemisphere stereographic projection and compare the calculated preferred orientations with what seem to be the points where the greatest density of normals occur.

3. A multistage triaxial test with a sawed joint oriented 45° with the axis of the core yielded the following data. Determine ϕ_μ.

Confining Pressure (p) (MPa)	Maximum Axial Stress (MPa)
0.10	0.54
0.30	1.63
0.50	2.72
1.00	5.45

4. A reverse fault in the rock of Problem 3 has a dilatancy angle of 5° and is inclined 20° with the horizontal. What is the maximum horizontal stress that could be sustained at a depth of 2000 m in this rock?

5. Trace the roughness profile of Figure 5.15a on a sheet of paper; then cut along it carefully with scissors to produce a model of a direct shear specimen. Slide the top to the right past the bottom, without rotation and without "crushing," and draw the path of any point on the top block. Compare this path to the constructed dilatancy curve of Figure 5.16b. Mark the locations of potential crushing at different shear displacements.

6. A normal fault that is partly cemented with calcite mineralization dips 65° from horizontal. The fault slipped when the water pressure reached 10 MPa at a depth of 600 m. If S_j = 1 MPa and ϕ_j = 35°, what was the horizontal stress before the fault slipped?

7. S_j = 0 and ϕ_j = 28.2° for a sawed joint oriented 50° from vertical in a saturated triaxial compression specimen. The confining pressure is 1.5 MPa and the axial stress σ_1 = 4.5 MPa with zero joint water pressure. What water pressure will cause the joint to slip if σ_1 and σ_3 are held constant?

8. The following data were taken in a direct shear test conducted in the field along a rock joint, with area 0.50 m². The weight of the block above the joint is 10 kN.

T, Shear Force (kN)	0	1.0	2.0	3.0	5.0	6.5	6.0	5.5	5.4	5.3
u, Shear Displacement (mm)	0	0.5	1.0	1.5	3.0	5.2	7.5	9.5	11.5	≥12

Assuming that joint cohesion is zero, and that ϕ_μ = ϕ_{resid}, determine the peak and residual friction angles, the shear stiffness (MPa/m), and the dilatancy angle at peak and post peak displacements.

9. A jointed shear test specimen is drilled at angle α with the normal to the shear plane and a model rock bolt is installed and tensioned to force F_B (see figure). Then a pair of shear forces T are applied until the joint slips.
   (a) What is the bolt tension F_B just sufficient to prevent slip under shear force T.
   (b) What is the value of α that minimizes the value of F_B required to prevent joint slip?
   (c) How are the answers to be changed if the joint tends to dilate during shear, with dilatancy angle i, and the bolt has stiffness k_b?

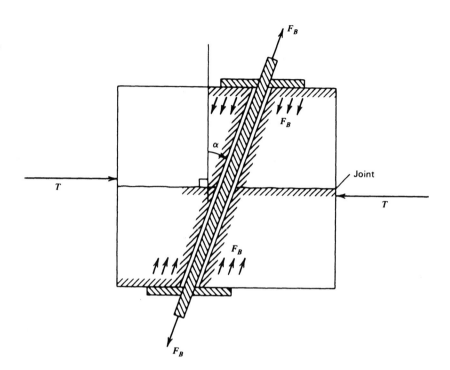

10. John Bray (1967) derived the following expression for the limiting effective stresses for joint slip:

$$K_f \equiv \frac{\sigma_3'}{\sigma_1'} = \frac{\tan |\psi|}{\tan(|\psi| + \phi_j)}$$

where ψ is the angle between the *direction* of σ_1 and the joint plane. (The derivation for this useful formula is given in Appendix 4 in the derivations to equations 7.11 to 7.16.) Draw a polar plot of the ratio σ_3'/σ_1' for limiting conditions as a function of $\psi(-\pi/2 \leq \psi \leq \pi/2)$ for values of (a) $\phi_j = 20°$ and (b) $\phi_j = 30°$. Label the regions on these diagrams corresponding to *slip* and *safe* principal stress ratios.

11. Use the expression given in Problem 10 to re-solve Problem 5-7. (*Hint:* Substitute $\sigma_1 - p_w$ and $\sigma_3 - p_w$ in place of σ_1', σ_3')

12. (a) Sedimentation is increasing the thickness of overburden (z) and the vertical stress (σ_v) in a rock mass. Assume the rock strength is given by $S_i = 1$ MPa and $\phi = 30°$. With $\nu = 0.2$, and $\gamma = 0.025$ MPa/m, draw the limiting Mohr circle that causes shear failure in the rock and determine the corresponding values of z, σ_v, and σ_h.

(b) Now assume that shear fractures have formed, in the orientation determined by the shear failure in (a). The new shear joints have $\phi_j = 20°$. Draw the new Mohr circle after the failure and determine the new value of σ_h (no change in σ_v has occurred).

(c) Assume additional sedimentation increases the value of σ_v to 1.5 times its value in (b). What are the corresponding values of z and σ_h. Draw the corresponding Mohr circle.

(d) Now erosion begins, reducing σ_v. Assuming the corresponding reduction in σ_h is given by $\Delta\sigma_h = [\nu/(1 - \nu)]\,\Delta\sigma_v$, draw a series of Mohr circles and determine the value of z when $\sigma_h = \sigma_v$.

(e) With further erosion, the shear joints formed in (a) are no longer relevant to the stress circles since the major stress is now horizontal. New joints form when the Mohr circle contacts the rock strength envelope. Draw this circle and determine the corresponding values of z, σ_h, and σ_v.

(f) Assume the Mohr circle is now limited by the new joints. Find the appropriate new value for σ_h (no change in σ_v).

(g) Draw graphs showing the variations of σ_v and σ_h with z increasing to the max found in (c) and then decreasing to zero.

13. The average fracture frequency (λ) across a rock core is the total number of natural fractures divided by the total length drilled.

(a) Suppose there is only one set of joints and that λ is the fracture frequency measured in a direction normal to them. Derive an expression giving λ in a direction θ with respect to the normal.

(b) There are two orthogonal sets of fractures with λ values, respectively λ_1 and λ_2. Derive an expression for λ measured in a direction θ from λ_1.

(c) Given $\lambda_1 = 5.0$ and $\lambda_2 = 2.0$ fractures per meter. Find the values of θ and λ such that the fracture frequency is a maximum. What is the average fracture *spacing* in this direction?

14. Barton (1973) proposed an empirical criterion of peak shear strength for joints:

$$\tau = \sigma_n \tan[JCR \log (JCS/\sigma_n) + \varphi_b]$$

where JCS is the compressive strength of the wall rock, and JCR is the joint roughness coefficient. (In this expression, the argument of tan is understood to be expressed in degrees.) Compare this equation with Equation 5.8.

Chapter 6

Deformability of Rocks

6.1 Introduction

Deformability means the capacity of rock to strain under applied loads or in response to unloads on excavation. The strains in rock concern engineering even when there is little risk of rock failure because locally large rock displacements can raise stresses within structures. For example (Figure 6.1), a dam seated astride varying rock types having dissimilar deformability properties will develop shear and diagonal tension stresses by virtue of the unequal deflections of the foundation. The dam can be structured to handle these deflection tendencies if the properties of the rock are known and if the variation of properties within the foundation are determined. Furthermore, in mass concrete structures like gravity dams, the deformability of the rock also relates to thermal stresses in the concrete, which are calculated by the product of a thermal expansion coefficient, a temperature change, and a deformability term.

There are many situations in which rock displacements should be calculated. To design pressure tunnels, one should know the expansion of the lining under operating pressure, as well as the amount of recovery when the pressure is lowered. The same is true of arch dams pressing against their abutments. Tall buildings on rock may transmit sufficient load to their foundations that rock settlement becomes significant for design. For long-span, prestressed roof structures and bridges anchorages in rock, structures pushing against rock, or gravity blocks seated on rock, knowledge of the rock displacements and rotations is basic to design details. And for any excavation that is monitored, the expectable displacements should be calculated to provide a framework with which to interpret the measurements.

ELASTIC AND NONELASTIC BEHAVIOR

It is not sufficient to characterize rock deformability by elastic constants alone, for many rocks are nonelastic. *Elasticity* refers to the property of reversibility

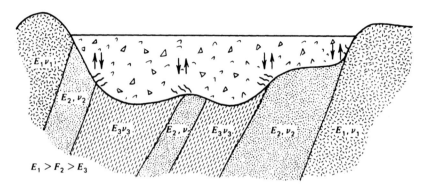

Figure 6.1 Shear stress developed in a concrete dam due to variable deformability in the foundation rock.

of deformation in response to load. Many fresh, hard rocks are elastic when considered as laboratory specimens. But on the field scale, where the rock can be expected to contain fissures, fractures, bedding planes, contacts, and zones of altered rock and clays with plastic properties, most rocks do not exhibit perfect elasticity. The extent of irrecoverability of strain in response to load cycles may be as important for the design as the slope of the load/deformation curve. Figure 6.2 offers an example. As the reservoir behind the arch dam rises, the rock under the arch responds along curve 1. The concave upward curvature of this load/deflection path is typical for fractured rocks on first ("virgin") loading because the fractures close and stiffen at low loads. When the reservoir is lowered for whatever reason, the rock unloads along path 2, with a permanent deflection. The dam will try to follow the loading, but since it

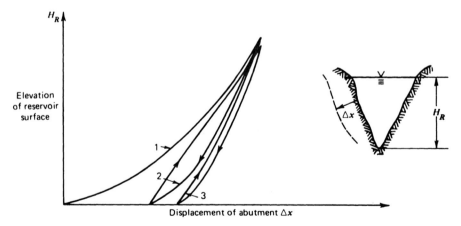

Figure 6.2 Permanent foundation deformation caused by cycles of reservoir filling and emptying.

is often more elastic than the rock, it will move away from the rock on unloading. This could open joints in the rock or concrete or simply lower the compressive stress flowing through the structure. Repeated cycles of loading and unloading in response to cyclic operation of the reservoir would produce the series of loops ("hysteresis") depicted in Figure 6.2. Some sites have been considered unacceptable for concrete dams because of large hysteresis even though the *modulus of elasticity* of the rock itself was considered reasonable. A criterion in this regard is presented later in conjunction with the plate-bearing test.

6.2 Elastic Constants

The deformations of linearly elastic *isotropic* solids can be calculated for known increments of stress if only two material constants are specified. In Chapter 4, these were taken as Young's modulus E (the "modulus of elasticity") and Poisson's ratio ν. Hooke's law, generalized to three dimensions, is

$$\begin{Bmatrix} \varepsilon_x \\ \varepsilon_y \\ \varepsilon_z \\ \gamma_{xy} \\ \gamma_{yz} \\ \gamma_{zx} \end{Bmatrix} = \begin{bmatrix} \dfrac{1}{E} & -\dfrac{\nu}{E} & -\dfrac{\nu}{E} & 0 & 0 & 0 \\[2mm] -\dfrac{\nu}{E} & \dfrac{1}{E} & -\dfrac{\nu}{E} & 0 & 0 & 0 \\[2mm] -\dfrac{\nu}{E} & -\dfrac{\nu}{E} & \dfrac{1}{E} & 0 & 0 & 0 \\[2mm] 0 & 0 & 0 & \dfrac{2(1+\nu)}{E} & 0 & 0 \\[2mm] 0 & 0 & 0 & 0 & \dfrac{2(1+\nu)}{E} & 0 \\[2mm] 0 & 0 & 0 & 0 & 0 & \dfrac{2(1+\nu)}{E} \end{bmatrix} \begin{Bmatrix} \sigma_x \\ \sigma_y \\ \sigma_z \\ \tau_{xy} \\ \tau_{yz} \\ \tau_{zx} \end{Bmatrix} \quad (6.1)$$

The quantities E and ν can be determined directly from tests where known stress is applied and strains are measured. Where strains are applied and stress changes are measured, it is more natural to use Lame's constant λ and the shear modulus G as the two elastic constants; these are defined by

$$\begin{Bmatrix} \sigma_x \\ \sigma_y \\ \sigma_z \\ \tau_{xy} \\ \tau_{yz} \\ \tau_{zx} \end{Bmatrix} = \begin{bmatrix} \lambda + 2G & \lambda & \lambda & 0 & 0 & 0 \\ \lambda & \lambda + 2G & \lambda & 0 & 0 & 0 \\ \lambda & \lambda & \lambda + 2G & 0 & 0 & 0 \\ 0 & 0 & 0 & G & 0 & 0 \\ 0 & 0 & 0 & 0 & G & 0 \\ 0 & 0 & 0 & 0 & 0 & G \end{bmatrix} \begin{Bmatrix} \varepsilon_x \\ \varepsilon_y \\ \varepsilon_z \\ \gamma_{xy} \\ \gamma_{yz} \\ \gamma_{zx} \end{Bmatrix} \quad (6.2)$$

The relationships between the two sets of constants are

$$G = \frac{E}{2(1 + \nu)}$$ (6.3)

and

$$\lambda = \frac{E\nu}{(1 + \nu)(1 - 2\nu)}$$ (6.4)

One other constant is quite useful, the *bulk modulus K*, which expresses the relationship between hydrostatic pressure p and volumetric strain $\Delta V/V$. Let

$$p = K \frac{\Delta V}{V}$$ (6.5)

then

$$K = \frac{E}{3(1 - 2\nu)}$$ (6.6)

(The *compressibility* is the reciprocal of K.)

Many rock masses are *anisotropic*, that is, directional in their behavior, due to regular bedding or jointing or oriented fabric or microstructure that makes the rock itself anisotropic. It is rarely feasible to deal with extreme anisotropy but *orthotropic* symmetry can be entered into computations without mathematical burden. In the latter case, there are three mutually perpendicular directions of symmetry, referred to as *principal symmetry directions*. If the rock has three perpendicular sets of joints, for example, it should behave orthotropically. If x, y, and z are chosen parallel to the orthotropic symmetry directions, Hooke's law now requires constants:

$$\begin{Bmatrix} \varepsilon_x \\ \varepsilon_y \\ \varepsilon_z \\ \gamma_{xy} \\ \gamma_{yz} \\ \gamma_{zx} \end{Bmatrix} = \begin{bmatrix} \frac{1}{E_x} & -\frac{\nu_{yx}}{E_y} & -\frac{\nu_{zx}}{E_z} & 0 & 0 & 0 \\ -\frac{\nu_{yx}}{E_y} & \frac{1}{E_y} & -\frac{\nu_{zy}}{E_z} & 0 & 0 & 0 \\ -\frac{\nu_{zx}}{E_z} & -\frac{\nu_{zy}}{E_z} & \frac{1}{E_z} & 0 & 0 & 0 \\ 0 & 0 & 0 & \frac{1}{G_{xy}} & 0 & 0 \\ 0 & 0 & 0 & 0 & \frac{1}{G_{yz}} & 0 \\ 0 & 0 & 0 & 0 & 0 & \frac{1}{G_{zx}} \end{bmatrix} \begin{Bmatrix} \sigma_x \\ \sigma_y \\ \sigma_z \\ \tau_{xy} \\ \tau_{yz} \\ \tau_{zx} \end{Bmatrix}$$ (6.7)

The Poisson's ratio ν_{ij} determines the normal strain in symmetry direction j when stress is added in symmetry direction i. In the orthotropic rock mass

$$\frac{\nu_{ij}}{E_i} = \frac{\nu_{ji}}{E_j} \qquad (6.8)$$

The nine independent constants are reduced to five if the rock is isotropic within a plane (*transversely isotropic*). This arises when two types of rock are regularly interlayered. It also comes about when flat minerals like mica, talc, chlorite, graphite, or serpentine are arrayed in parallel orientation, or when long minerals (e.g., amphiboles) are oriented with their long axes randomly pointed within parallel planes. The number of elastic constants is reduced to four when transversely isotropic symmetry arises from one set of regular joints, for example, parallel to bedding, in otherwise isotropic rock (see Problem 6.11).

To describe transversely isotropic elasticity, let s and t be any two perpendicular directions in the plane perpendicular to the axis of symmetry (e.g., in the plane of bedding) and let n be the direction parallel to the axis of symmetry (e.g., in the direction normal to the bedding planes). Then $E_s = E_t$ and $\nu_{ts} = \nu_{st}$ and Equation 6.7 becomes

$$
\begin{Bmatrix} \varepsilon_n \\ \varepsilon_s \\ \varepsilon_t \\ \gamma_{ns} \\ \gamma_{nt} \\ \gamma_{st} \end{Bmatrix} =
\begin{bmatrix}
\dfrac{1}{E_n} & -\dfrac{\nu_{sn}}{E_s} & -\dfrac{\nu_{sn}}{E_s} & 0 & 0 & 0 \\[2mm]
-\dfrac{\nu_{sn}}{E_s} & \dfrac{1}{E_s} & -\dfrac{\nu_{st}}{E_s} & 0 & 0 & 0 \\[2mm]
-\dfrac{\nu_{sn}}{E_s} & \dfrac{\nu_{st}}{E_s} & \dfrac{1}{E_s} & 0 & 0 & 0 \\[2mm]
0 & 0 & 0 & \dfrac{1}{G_{ns}} & 0 & 0 \\[2mm]
0 & 0 & 0 & 0 & \dfrac{1}{G_{ns}} & 0 \\[2mm]
0 & 0 & 0 & 0 & 0 & \dfrac{2(1 + \nu_{st})}{E_s}
\end{bmatrix}
\begin{Bmatrix} \sigma_n \\ \sigma_s \\ \sigma_t \\ \tau_{ns} \\ \tau_{nt} \\ \tau_{st} \end{Bmatrix} \qquad (6.9)
$$

In routine engineering work, it is usual to assume the rock to be isotropic, but with schists this may be inappropriate. Measurements can be made on samples cored in different directions to determine additional deformability constants. When anisotropy derives from regular structures, the orthotropic constants can be calculated, as shown later.

6.3 Measurement of Deformability Properties by Static Tests

Stress-strain relationships can be observed in static and dynamic tests conducted in the laboratory or in the field. Deformability properties can then be obtained from the data, assuming that some idealized model describes the rock behavior in the test configuration. Deformability properties can also be back calculated from instrumental data on the movements of a structure or excavation, if the initial and final states of stress are known, using methods of Chapter 4 inversely.

The most widely used testing procedures for deformability measurements are laboratory compression and bending tests, wave velocity measurements in the lab or field, field loading tests using flat jacks or plate bearing apparatus, and borehole expansion tests.

LABORATORY COMPRESSION TESTS

An unconfined compression test on a core of rock, with carefully smoothed ends and length to diameter ratio of 2, yields a stress-strain curve like that of Figure 6.3*a*. The axial strain can be measured with strain gages mounted on the specimen or with an extensometer attached parallel to the length of the specimen; the lateral strain can be measured using strain gages around the circumference, or an extensometer across the diameter. The ratio of lateral to axial strain magnitudes determines Poisson's ratio ν. With hard rocks, it is usually not acceptable to determine axial strain from measured shortening of the testing space ("crosshead motion") because relatively large displacement occurs at the ends where the rock contacts the platens of the testing machine.

Figure 6.3*b* shows the difficulty of defining exactly what is meant by E. It is not simply the slope of the virgin loading curve, for this embraces nonrecoverable as well as elastic deformation. The unloading curve, or reloading curve after a cycle of load and unload, are better measures of E. This definition allows E to be determined even after the peak load when the rock has become fractured (Figure 6.3*b*).

Deere (1968) presented a classification graph for intact rock specimens based upon the ratio of elastic modulus to unconfined compressive strength, together with the absolute value of the latter. For most rocks, the ratio E/q_u lies in the range from 200 to 500 but extreme values range as widely as 100 to about 1200. In general, the "modulus ratio" E/q_u is higher for crystalline rocks than for clastic rocks, with sandstones higher than shales. Table 6.1 gives the measured ratio of "modulus of deformation" to unconfined compressive strength, and corresponding values of Poisson's ratio, for the set of rocks previously considered in Table 3.1. The substitution of *modulus of deformation* in place of *modulus of elasticity* indicates that the deformability property embraces both recoverable and nonrecoverable deformation. In general, whenever the modulus value is calculated directly from the slope of the rising portion of a virgin loading curve, the determined property should be reported as a modulus of deformation rather than as a modulus of elasticity. Unfortunately, this is not universal practice at present.

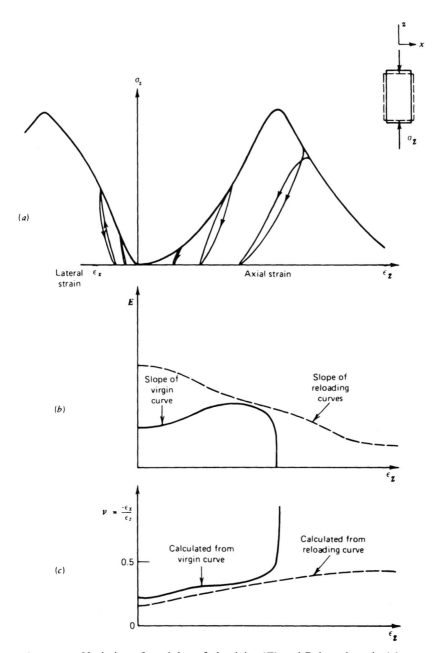

Figure 6.3 Variation of modulus of elasticity (E) and Poisson's ratio (ν) with axial strain in a compression test.

Table 6.1 *Modulus Ratio E/q$_u$ and
Poisson's Ratio ν for the Rock
Specimens of Table 3.1*[a]

Description	E/q_u	ν
Berea sandstone	261	0.38
Navajo sandstone	183	0.46
Tensleep sandstone	264	0.11
Hackensack siltstone	214	0.22
Monticello Dam greywacke	253	0.08
Solenhofen limestone	260	0.29
Bedford limestone	559	0.29
Tavernalle limestone	570	0.30
Oneota dolomite	505	0.34
Lockport dolomite	565	0.34
Flaming Gorge shale	157	0.25
Micaceous shale	148	0.29
Dworshak Dam gneiss	331	0.34
Quartz mica schist	375	0.31
Baraboo quartzite	276	0.11
Taconic marble	773	0.40
Cherokee marble	834	0.25
Nevada Test Site granite	523	0.22
Pikes Peak granite	312	0.18
Cedar City tonalite	189	0.17
Palisades diabase	339	0.28
Nevada Test Site basalt	236	0.32
John Day Basalt	236	0.29
Nevada Test Site tuff	323	0.29

[a] *E* reported here includes both recoverable and nonrecoverable deformation, mixed in unknown proportions.

The negative slope of the tail of the complete stress-strain curve is not a stress-strain curve in the conventional sense but is a yield function; in particular, it is the envelope of yield points from all reloading curves. Figure 6.3c shows the value of ν calculated from lateral deformation of a compression specimen on its virgin loading curve. The ratio of lateral to axial strain begins at a value near 0.2 and increases gradually until near the peak load when it begins to accelerate, even surpassing the theoretical maximum value of ν for isotropic materials—0.5. (Equation 6.6 shows that *K* approaches infinity as ν tends toward 0.5.) The rock cannot be described as elastic as it moves on the yield surface after the peak since it is cracked and large lateral deformations occur with movement of rock wedges. However, on unloading and reloading, lateral strains occur with $\nu < 0.5$. Again, we can conclude that the elastic constants should be defined with respect to the reloading curve.

The full description of rock deformability should include not only the elastic coefficients E and ν, but the permanent deformation associated with any applied stress level. Figure 6.4 shows how to determine the *modulus of permanent deformation M*, defined as the ratio of a stress to the permanent deformation observed on releasing that stress to zero. M is determined by running a series of load cycles during a compression test. We may compute similarly the Poisson's ratio ν_p corresponding to permanent lateral deformation increments.

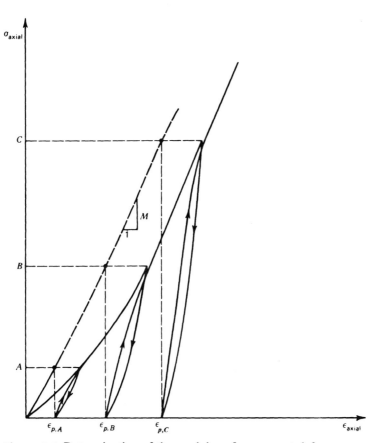

Figure 6.4 Determination of the modulus of permanent deformation M from load cycling data in a compression test.

PLATE-BEARING TEST

The deformability of rock may be measured in the field by loading a rock surface and monitoring the resulting deformation. This is easily arranged in an underground gallery as shown in Figure 6.5*a*. The site must be selected carefully to exclude loose, highly fractured rock that might be unrepresentative of the average rock condition. A relatively flat rock surface is sculptured and leveled with mortar to receive circular bearing plates 50 cm to 1 m in diameter.

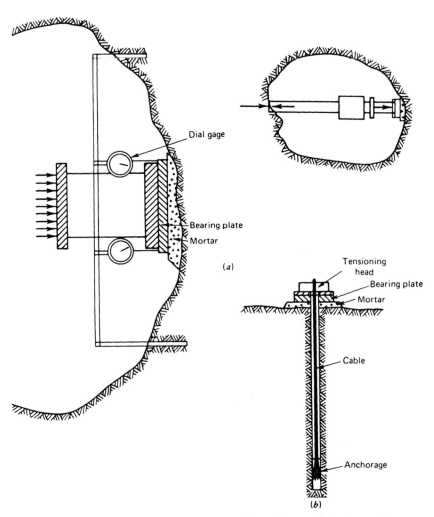

Figure 6.5 Plate-bearing test setup. (*a*) In a gallery. (*b*)At the surface.

The depth of the rock volume affected is proportional to the diameter of the loaded area, so it is desirable to choose a large bearing plate; but it proves difficult to apply loads greater than about 200 tons so it may be necessary to reduce the plate size to achieve desired contact pressure levels. The load can be applied by hydraulic cylinders or screw jacks reacting against the opposite wall of the gallery. Flat jacks, in series to allow sufficient "travel," have also been adapted for this function. Displacement must be measured at several points on the bearing plate to correct for rotations and plate bending. Displacements are usually monitored by mounting dial gages on a rigid inertial reference bar passing over the plate. It is also possible to use a benchmark set at depth in a borehole centered on the plate (Figure 6.5a). Plate-bearing tests can be run at the surface by jacking against a cable anchored at depth in a borehole drilled through the center of the bearing plate (Zienkiewicz and Stagg, 1966).

The data obtained from the plate-bearing test are the radius of the plate a, the plate pressure p (contact force per unit plate area), and the mean displacement $\bar{\omega}$ corrected for rotation. The following equation can be derived from Timoshenko and Goodier (1951) if we assume that the rock is a homogeneous infinite half space of elastic isotropic material:

$$\bar{\omega} = \frac{Cp(1 - \nu^2)a}{E} \qquad (6.10)$$

Assuming a value for ν permits calculation of E.

C is a constant depending on the boundary conditions. If the plate is perfectly rigid, $C = \pi/2$, whereas if the plate is flexible, $C = 1.70$; we see that there is little difference in the calculated E corresponding to extremes in boundary conditions as long as mean displacement of the plate is measured. However, determining the mean displacement of a flexible, bending bearing plate corresponding to a constant pressure boundary condition would require more dial gages than there is generally room to accommodate. Unless the rock is very hard, it will be simpler to attempt to achieve constant plate displacement conditions by using thick steel plates and a stiff arrangement.

The plate-bearing test is conducted in a gallery more often than on a semi-infinite medium, but Equation 6.10 is still used to compute E. A more important influence on the results is departure from nonideal rock conditions. This can be appreciated by using deep bench marks for displacement measurement. It should be noted that almost any departure from conditions assumed will tend to increase the measured displacements, so the plate-bearing test tends to underestimate the modulus of elasticity. Tests conducted vertically in galleries will usually give even lower values of E because joints in the roof rock tend to open under gravity.

Permanent and elastic deformations in plate bearing results can be separated if the load is cycled during the test. The elastic modulus should be calculated from the slope of the reloading portion of a load cycle:

$$E = Ca(1 - v^2) \frac{p}{\bar{\omega}_{elas}} \qquad (6.11)$$

where $\bar{\omega}_{elas}$ is the mean plate displacement on reapplying plate pressure from nearly zero to p. The modulus of permanent deformation M can be calculated from

$$M = Ca(1 - v^2) \frac{p}{\bar{\omega} - \bar{\omega}_{elas}} \qquad (6.12)$$

Using plates of 14- to 50-cm radius, with plate pressure up to 200 bars, Schneider (1967) tested foundations of a number of dam sites in various rock types. He observed that permanent deformation greater than 0.01 mm/bar characterized sites that had been considered to be unacceptable for foundations of concrete dams. This translates to a value of $M = 7700$ MPa (1.1×10^6 psi) for $a = 50$ cm, $v = 0.3$.

BOREHOLE AND GALLERY TESTS

Rock deformability may also be measured statically in boreholes. The *dilatometer test* (Figure 6.6a) is a borehole expansion experiment conducted with a rubber sleeve. The expansion of the borehole is measured by the oil or gas flow into the sleeve as the pressure is raised, or by potentiometers or linear variable differential transformers built inside the sleeve. The *gallery test* is a similar experiment conducted inside a bulkheaded section of a tunnel. The cost of gallery tests has tended to minimize their application in recent years. The *borehole jack* is similar to the dilatometer except that the loads are applied unidirectionally across a diameter. Interpretation is similar but the formula requires attention to the more difficult boundary conditions (Goodman et al., 1972; Heuze and Salem, 1979). For the dilatometer or gallery test, in which the pressure p is applied uniformly over the borehole or gallery surface of radius a, the modulus of elasticity can be calculated from the measured radial deformation Δu by

$$E = (1 + v)\Delta p \frac{a}{\Delta u} \qquad (6.13)$$

One problem with borehole deformability tests is that they affect a relatively small volume of rock and therefore contain an incomplete sample of the fracture system. Some would argue that the system is indeterminate and therefore that the test is useless. However, the borehole tests have the unique advantage of giving an indication of the range of properties of the rock remote from the surface at an early stage of investigation. Based upon the results of such a program of tests, it is possible to appreciate potential site difficulties, and it should be possible to subdivide the volume of rock in a foundation into

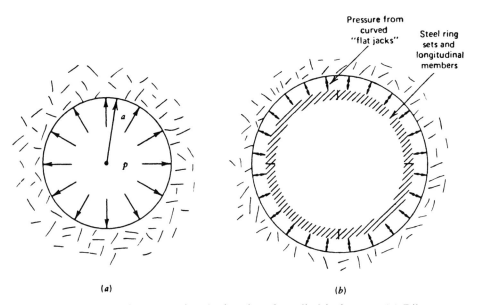

Figure 6.6 Schemes for pressuring the interior of a cylindrical space. (*a*) Dilatometer or gallery test, with fluid pressure inside the test region. (*b*) Radial jacking, or TIWAG test, with pressure supplied by jacks reacting against interior sets.

approximately homogeneous subregions. Further testing can be carried out to characterize each of these subregions. Field tests of a larger scale require galleries and are more expensive. They too present difficulties in interpretation since usually no field test will be as large as the rock volume affected by an actual structure.

RADIAL JACKING TESTS

Among the largest in situ tests used to measure deformability of rock are *radial jacking tests* (Figure 6.6*b*), an adaption of the "TIWAG" test used in Europe. Loads are applied to the circumference of a tunnel by a series of jacks reacting against circular steel ring members. The test allows the direction of load to be varied according to the plan for pressuring the jacks. Such tests were conducted by the Bureau of Reclamation at the site of the Auburn Dam—a site where seams of talc schist raised questions about deformability and stability of the abutments. The tests were expensive but were defensible in terms of the total cost of this enormous project. However, lab tests, borehole tests, and plate-bearing tests were also run at the Auburn Dam site and through these a good understanding of the variation and distribution of deformability values throughout the dam site was gained.

FLAT JACK TESTS

The flat jack test, previously discussed in connection with stress measurements, yields deformability properties as a by-product. A large volume of rock can be loaded to pressure up to 70 MPa or higher using stainless steel flat jacks with special welding details.

The area of typical flat jacks is of the order of 600 cm² and much larger jacks have been used; thus very large loads are applied to the rock. Recall that the pressuring stage of the flat jack test provides data on the variation of pin separation $2\Delta y$ with applied jack pressure p (Figure 6.7). If load cycles are programmed, the reloading relationships will permit calculation of E using a relationship derived by Jaeger and Cook (1976):

$$E = \frac{p(2c)}{2\Delta y} \left[(1 - \nu) \left(\sqrt{1 + \frac{y^2}{c^2}} - \frac{y}{c} \right) + \frac{1 + \nu}{\sqrt{1 + \frac{y^2}{c^2}}} \right] \qquad (6.14)$$

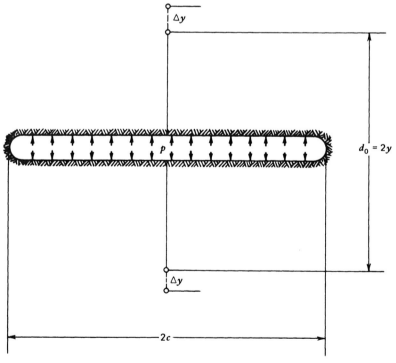

Figure 6.7 The flat jack test, with a slot obtained by drilling overlapping holes.

where y is the distance from the jack center to each of a pair of measuring pins, and $2c$ is the length of the jack. (The modulus of permanent deformation M can also be obtained as discussed for the plate bearing test.)

6.4 Dynamic Measurements

The velocity of stress waves may be measured in laboratory rock specimens and in the field. The laboratory *pulse velocity test* is run using sections of cylindrical core with smooth, parallel ends to which piezoelectric crystals are cemented (Figure 6.8*a*). A high-frequency electrical pulse transmitted to one crystal creates a stress wave that is received by the second crystal and reconverted to an electrical signal. A delay line allows the received wave form to be

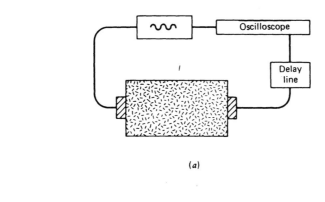

(a)

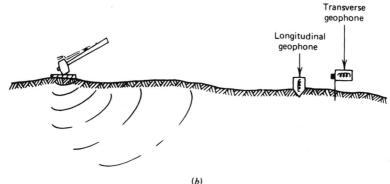

(b)

Figure 6.8 Scheme for dynamic modulus determination. (a) Pulse velocity method in the laboratory. (b) Sound velocity measurements in the field.

aligned to the sending wave form on an oscilloscope and the required delay to achieve this measures the travel time for the pulse through the specimen. The use of longitudinal and shear cut crystals permits both longitudinal waves and transverse waves to be observed so that both longitudinal wave velocity V_l and transverse wave velocity V_t can be determined. If the rock were an ideal elastic, isotropic solid of small diameter compared to the length, then E and G could be calculated from

$$E = V_l^2 \rho \tag{6.15}$$

and

$$G = V_t^2 \rho \tag{6.16}$$

where ρ is the mass density of the rock. Recalling that $G = E/[2(1 + \nu)]$ (Equation 6.3)

$$\nu = \frac{1}{2} \left(\frac{V_l^2}{V_t^2} \right) - 1 \tag{6.17}$$

In the field, wave velocity can be measured by swinging a sledgehammer against an outcrop and observing the travel time (milliseconds, typically) to a geophone standing on the rock at a distance of up to about 50 m. Portable seismographs available from several commercial sources are suited to such measurements. Another method is to record the time for a shock to travel between points in drill holes spaced 50 to 100 m apart. Both downhole hammers and explosive sources are used for such measurements. If the "signature" of the wave arriving at the geophone is displayed, both compressional wave velocities V_p and shear wave velocities V_s can be determined. Then, assuming the rock to be homogeneous, isotropic, and elastic,

$$V_p = \sqrt{\frac{\lambda + 2G}{\rho}} \tag{6.18}$$

and

$$V_s = \sqrt{G/\rho} \tag{6.19}$$

Recalling Equations 6.3 and 6.4

$$\nu = \frac{(V_p^2/V_s^2) - 2}{2[(V_p^2/V_s^2) - 1]} \tag{6.20}$$

and

$$E = 2(1 + \nu)\rho V_s^2 \tag{6.21}$$

or

$$E = \frac{(1 - 2v)(1 + v)}{(1 - v)} \rho V_p^2 \qquad (6.22)$$

The stress loadings sent through the rock by these methods are small and transient. Most rock masses and even rock specimens depart significantly from the ideal materials hypothesized with respect to Equations 6.15 to 6.22. Consequently, elastic properties calculated from these equations are often considerably larger than elastic properties calculated from static loading tests like plate bearing. This is particularly true in the case of fractured rocks. To distinguish elastic properties measured by static methods from those obtained dynamically, the subscripts s and d will be introduced (E_s, v_s and E_d, v_d for static and dynamic constants, respectively).

6.5 *Fractured Rocks*

Plate-bearing tests in fractured rocks reported by Schneider (1967) typically yielded a load deformation curve of the form shown in Figure 6.9, with a yield

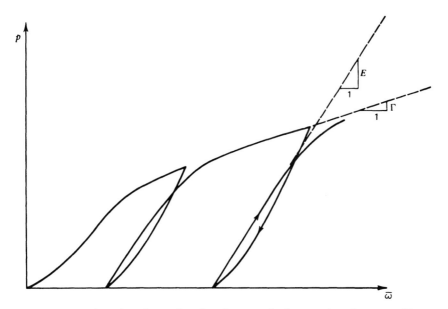

Figure 6.9 Typical data from plate-bearing tests in fractured rock reported by Schneider (1967); p is the average pressure on the plate and $\bar{\omega}$ is the average displacement of the plate.

point effect. The slope of the envelope of load cycles, the "yield function," is termed Γ. Schneider found that in highly fractured rock with open cracks, the ratio E/Γ was as high as 45. He proposed the following classification of results (Table 6.2):

Table 6.2

Class	E/Γ
Compact rock	<2
Moderately open	$2-10$
Very open	>10

If the rock is regularly crossed by a single set of joints, it is possible to calculate elastic constants for an "equivalent" continuous material representative of the rock mass. We assume the rock itself is isotropic and linearly elastic, with constants E and ν (Figure 6.10). The joints are assumed to be regularly spaced a distance S. Let k_s—the shear stiffness—be the slope of the shear stress-shear displacement curve until slip (Figure 5.12). We adopt axes n, t normal and parallel to the joints and therefore in the principal symmetry directions of the rock mass. When shear stress τ_{nt} is applied, each rock block undergoes a displacement equal to $(\tau_{nt}/G)S$ and each joint slips a distance (τ_{nt}/k_s) (Figure 6.10b). The shear deformation of a continuous material will be equivalent to that of the jointed rock mass if it has shear modulus G_{nt} such that $(\tau_{nt}/G_{nt})S$ is the sum of rock and joint displacements given above. Therefore,

$$\frac{1}{G_{nt}} = \frac{1}{G} + \frac{1}{k_s S} \tag{6.23}$$

Similarly, we assign the joint a "normal stiffness" k_n equal to the slope of the joint compression curve σ versus Δv (Figure 5.17a). Since the compression curve is highly nonlinear, k_n depends on the normal stress. The equivalent continuous material has modulus of elasticity E_n such that $(\sigma_n/E_n)S$ is the sum of rock deformation $(\sigma/E)S$ and joint deformation (σ/k_n) (Figure 6.10a). Therefore,

$$\frac{1}{E_n} = \frac{1}{E} + \frac{1}{K_n S} \tag{6.24}$$

The Poisson's ratio giving strain in the n direction caused by a normal stress in the t direction is simply ν:

$$\nu_{tn} = \nu \tag{6.25}$$

The modulus of elasticity in the t direction is simply E:

$$E_t = E \tag{6.26}$$

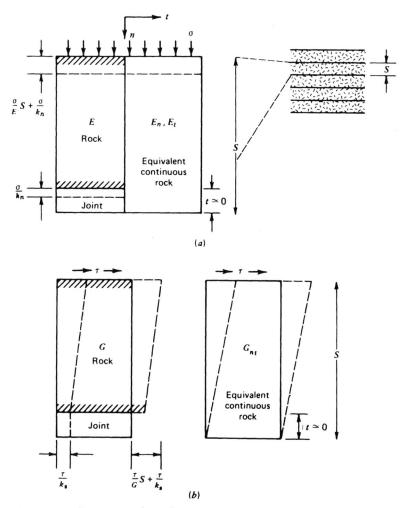

Figure 6.10 Representation of a regularly jointed rock by an "equivalent" transversely isotropic material.

Finally, symmetry of the stress-strain relationship requires $\nu_{tn}/E_t = \nu_{nt}/E_n$ giving

$$\nu_{nt} = \frac{E_n}{E} \nu \tag{6.27}$$

Equations 6.23 to 6.27 permit calculation of all five constants of the equivalent transversely isotropic medium representing a regularly jointed rock mass.

If the rock is highly fractured in several directions, Equation 6.24 can be used to find a "reduced modulus" representing the rock mass. The procedure is as follows. For each test specimen or test site, determine a characteristic average spacing between the joints of each set. From the measured value of the modulus of elasticity and assigning a value E to intact rock, calculate a value for k_n using Equation 6.24. Then, input this value of k_n in calculations with any specified fracture spacing. The rock mass modulus can in this way be related to degree of fracturing (Raphael and Goodman, 1979) or to RQD (discussed in Chapter 2) (Kulhawy 1978).

Bieniawski (1978) showed that a rock mass modulus could be assigned approximately if the rock were rated by the geomechanics classification system discussed in Chapter 2. Figure 6.11 shows his values of in situ modulus of deformation, determined by various large-scale field tests at a number of sites, plotted against the rock mass rating (RMR). For rocks rating higher than 55, the data points are fit approximately by

$$E = 2\,\mathrm{RMR} - 100$$

For softer rocks (10 < RMR < 50), Serafim and Pereira (1983) gave the following correlation between rock mass modulus of elasticity and RMR:

$$E = 10^{(\mathrm{RMR}-10)/40}$$

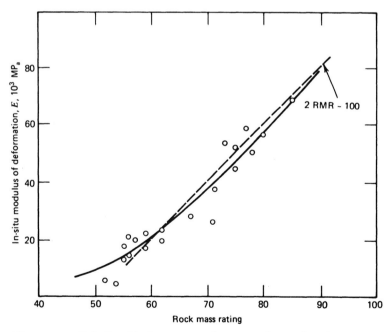

Figure 6.11 Relationship between rock mass rating and rock mass deformability. (After Bieniawski, 1978.)

The term *modulus of deformation* signifies that the value of E is calculated from the data of the loading portion of the load/deformation curve using both elastic and permanent deformation. The units of E in the above are GPA ($=10^3$ MPa). The data points embrace mudstone, sandstone, diabase, slate, phyllite, and quartzite.

Dynamic moduli E_d in fractured rocks tend to be considerably higher than rock mass moduli measured by static load tests E_s or computed as above. Schneider (1967) determined values of the ratio E_d/E_s up to 13 in fractured hard rocks. He observed that high frequencies are selectively attenuated in fractured rock. This was also shown by King et al. (1975) (Figure 6.12). One would

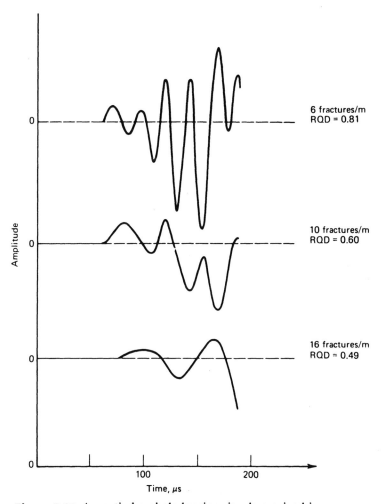

Figure 6.12 Acoustic borehole logging signals received in pegmatite fractured in various amounts. (After King et al., 1975.)

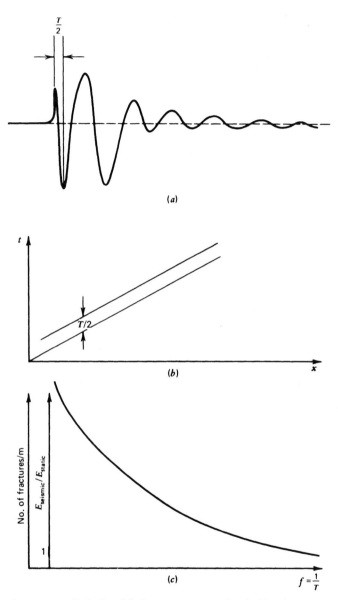

Figure 6.13 Relationship between transmitted vibration
frequency and degree of fracturing observed by Schneider (1967). (a) Typical wave group traveling through
the rock as a result of a hammer blow. (b) Measurement
of period using a first return seismograph by separation
of time-distance graphs caused by changing the polarity
of the instrument. (c) Inverse relationship between frequency and degree of fracturing.

expect then that measurements of frequency or wave length received at a standard distance from a constant type of seismic source would tend to correlate with E_d/E_s. Figure 6.13 confirms such a relationship for dam sites studied by Schneider using a hammer blow source with an engineering seismograph (MD1). The instrument yielded only the time for the arrival of the first wave having a signal above an adjustable threshold. Switching the polarity on the seismograph and repeating the experiment will cause a change in arrival time of approximately half the period, as shown in Figure 6.13*a, b*. Thus the frequency, velocity, and wavelength can be determined using a hammer source and a simple seismograph. The dynamic elastic constants can then be related to the static elastic constants through site calibration studies.

Alternatively, site studies can establish a direct relationship between in situ static modulus of elasticity and shear wave frequency. For example, Figure 6.14 shows the relationship between in situ static modulus of deformation and shear wave frequency using a hammer seismograph with standardized tech-

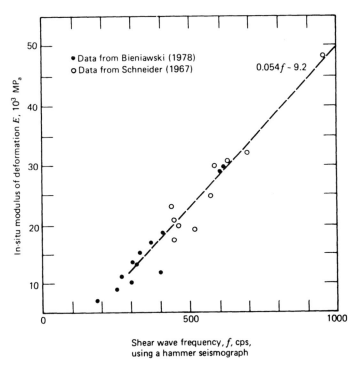

Figure 6.14 Relationship between transmitted vibration frequency and rock mass deformability. (After Bieniawski, 1978.)

nique (Bieniawski, 1978). Both Schneider's results and Bieniawski's results are fit by the same straight line:

$$E = 0.054f - 9.2$$

where E is measured in gigapascals (GPa),[1] f is the shear wave frequency from the hammer blow received at distances up to 30 m on a rock surface, and f is the frequency in hertz (cycles per second).

6.6 The Influence of Time on Rock Deformation

Thus far we have omitted all reference to *time* as a parameter of rock deformations. Since no effect can be truly instantaneous, time must be implicit in all the equations connecting stress and strain. In many cases, rock deformations can be calculated satisfactorily ignoring the influence of time, but sometimes they cannot.

Stress or displacements can change with time when the loads or pressures on the rock change, as, for example, due to flow of water; the geometry of the loaded or excavated region changes, as, for example, by further excavation; the deformability properties of the rock change, as, for example, by weathering or hydration; or the rock responds slowly to changes in stress or strain. All but the last factor can be accommodated by appropriate superposition of stress increments in a series of elastic analyses. However, the last reason for time dependency which we might term *viscous* behavior, requires further discussion.

VISCOUS BEHAVIOR AND CREEP

We can view *solids* as bodies that retain their shape indefinitely, while *liquids* assume the shape of their containers. An apparently solid material that distorts slowly and continuously in response to shearing stresses is then at least partly a viscous liquid. *Dynamic viscosity* η, depicted by a "dashpot" (Figure 6.15b), expresses the proportionality between shear stress τ and shear strain rate $\dot{\gamma}$:

$$\tau = \eta\dot{\gamma} \tag{6.28}$$

Since strain is dimensionless, the dimensions of η are $FL^{-2}T$, for example, psi/min or MPa/s.[2] Most rocks exhibit both "instantaneous" and delayed deformation when loaded and are therefore spoken of as *viscoelastic*. As with

---

[1] One gigapascal (1 GPa) = 1000 MPa.

[2] The common units "poise" (P) will not be used here; 1 P = 0.1 Pa/s = 1.450×10^{-5} psi/s.

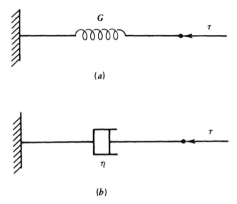

Figure 6.15 Elements of linear visco-elastic models. (*a*) Linear spring. (*b*) Linear dashpot shock absorber).

elasticity, real deformation data can display various non-linearities while the majority of theory concerns *linear viscoelasticity*.

Laboratory data are mostly in the form of strain-time curves from *creep tests*. In such tests, an increment of stress is applied quickly and then held constant while the gradually increasing strain is recorded. An alternative experiment, termed a *relaxation test*, monitors the decline in stress when strain is held constant. Figure 6.16 shows the general form of the creep curve for rock. Immediate "strain" is followed by *primary creep* in which strain occurs at a decreasing rate with time. In some rocks, the primary creep curve approaches a steady rate of creep, termed *secondary creep*. In specimens stressed near peak strength, secondary creep may turn upward in *tertiary creep*, in which strain rate increases with time, resulting in failure (creep rupture).

We can call on two types of mechanisms to explain creep in rocks—mass flow and cracking. Some rocks (e.g., rock salt, tar sands, and compaction shales) will creep at relatively low deviator stress, even with unfissured, intact specimens. In the case of salt and potash the process of creep involves movement of dislocations and intracrystalline gliding, while creep in uncemented clay rocks involves migration of water and movements of clay platelets ("consolidation"). Bituminous rocks like tar sand are inherently viscous, especially at higher temperatures. Hard rocks like granite and limestone can also exhibit creep at deviatoric stresses sufficient to cause new crack growth (e.g., when s_1 exceeds about one half q_u in an unconfined compression specimen). An increment of applied stress will provoke a change in the network of cracks through lengthening of old cracks and initiation of new ones. Such a process is nonlinear because the rock is changed by each new load increment; to calculate stresses and deformations in nonlinear viscoelastic materials, the properties will have to be determined and used as functions of stress. There are probably

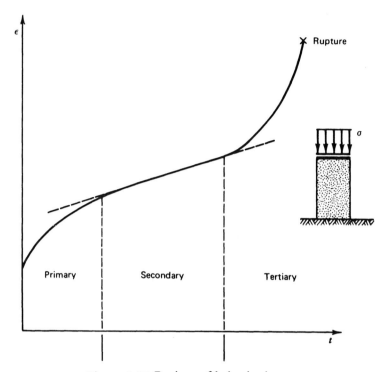

Figure 6.16 Regions of behavior in creep.

no ideally linear viscoelastic rocks, not even salt. However, a theory of linear viscoelasticity can still be used incrementally to approach time-dependent problems in much the same way that the theory of linear elasticity is used to calculate stress and strain for time-independent problems.

LINEAR VISCOELASTIC MODELS

It is possible to fit creep curves empirically using exponential or power functions. If creep data are forced to fit models composed of springs and dashpots, however, the results are more readily usable; therefore, this will be our procedure.

The theory of linear elasticity of an isotropic body is based on two constants, as we have seen previously. One of these (K) may be associated with purely volumetric deformation under hydrostatic loads. Then a second constant alone (G) must account for all distortion. The question we now face is, how many additional constants will be required to represent time-dependent deformation?

Figure 6.17 shows five possible models with one, two, or three additional constants. The series arrangement in Figure 6.17a is termed a *Maxwell body* or

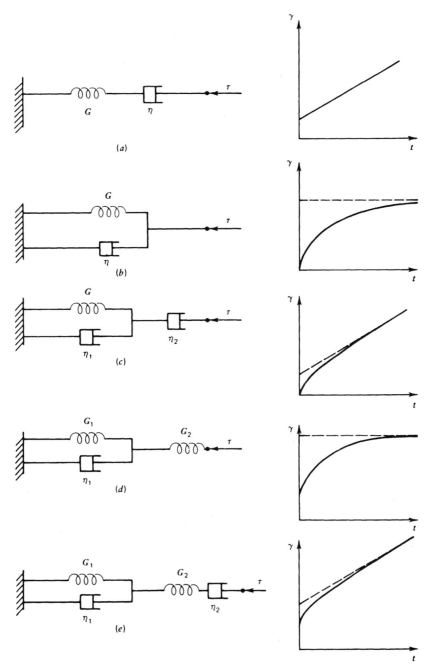

Figure 6.17 Simple linear viscoelastic models and their response to the creep test. (*a*) Two-constant liquid (Maxwell body). (*b*) Two-constant solid (Kelvin body). (*c*) Three-constant liquid (generalized Maxwell body). (*d*) Three-constant solid. (*e*) Four-constant liquid (Burgers body).

a two-constant liquid. It flows continuously at a constant rate when a shear stress is applied suddenly and held constant. Figure 6.17b is a *Kelvin* or *Voight body* or a two-constant solid. A suddenly applied constant shear stress causes shear strain at an exponentially decreasing rate, approaching zero as *t* grows without bound.

Figures 6.17c and *d* show a three-constant liquid and a three-constant solid, respectively. The former, termed a *generalized Maxwell body,* initially has an exponential rate of shear strain, decaying to become asymptotic to a constant rate of shear strain. The latter, termed a *generalized Kelvin body,* shows an initial "instantaneous" strain followed by shear strain at an exponentially decreasing rate, eventually tapering off completely.

Finally, Figure 6.17e shows a four-constant liquid termed the *Burgers body,* composed of a Maxwell and a Kelvin body in series. Its response to a suddenly applied and sustained shear stress is a combination of all the elements we have seen in the previous models—initial "instantaneous" shear strain followed by shear strain at an exponentially decreasing rate, becoming asymptotic to a line representing a constant rate of shear strain. In view of the form of the general creep curve, Figure 6.16, this is the simplest model that can be used to trace strain up to the onset of tertiary creep. More complicated models can be invoked by adding additional springs and dashpots, but the Burgers body will suffice and is preferable for many practical purposes. An informative comparison of various spring-dashpot models and empirical formulations for creep data was reported by Afrouz and Harvey (1974) for sedimentary rocks. Of the spring-dashpot models, the Burgers representation was consistently the best.

DETERMINING VISCOELASTIC CONSTANTS FROM LABORATORY TESTS

The simplest procedure for evaluating viscoelastic constants is through unconfined compression of cylindrical rock specimens over prolonged periods. This requires constant stress and constant temperature and humidity over the whole test duration, which may be hours, weeks, or longer. Load may be applied by dead weights acting through levers bearing directly on the specimen or through an oil pressure. Servocontrolled hydraulic pressure systems and compressed springs are also used. Careful testing procedure insures correction of load for changes in the cross sectional area of the specimen (see, for example, Rutter, 1972) and measurement of strain without long-term drift.

The axial strain with time $\varepsilon_1(t)$ in a Burgers body subjected to constant axial stress σ_1, is

$$\varepsilon_1(t) = \frac{2\sigma_1}{9K} + \frac{\sigma_1}{3G_2} + \frac{\sigma_1}{3G_1} - \frac{\sigma_1}{3G_1} e^{-(G_1 t/\eta_1)} + \frac{\sigma_1}{3\eta_1} t \qquad (6.29)$$

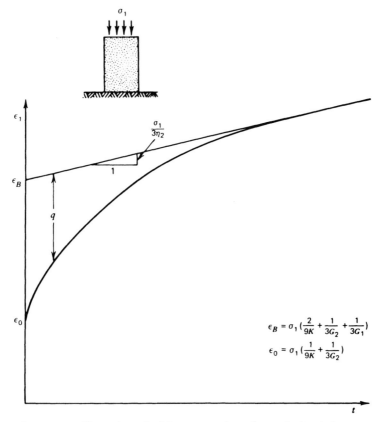

Figure 6.18 Creep in uniaxial compression of a rock that behaves as a Burgers body under deviatoric stress but as an elastic body under hydrostatic compression.

where $K = E/[3(1 - 2\nu)]$ is the bulk modulus, assumed to be independent of time[3] and η_1, η_2, G_1, and G_2 are properties of the rock to be evaluated as follows.

Figure 6.18 is a graph of ε_1 versus t corresponding to Equation 6.29. At $t = 0$, there is an intercept $\varepsilon_0 = \sigma_1(2/9K + 1/3G_2)$ while strain at large t falls along the line with intercept $\varepsilon_B = \sigma_1(2/9K + 1/3G_2 + 1/3G_1)$ and slope $\sigma_1/3\eta_2$. Load cannot be applied instantaneously and it may be preferable in practice to find

---

[3] The assumption that volumetric strain is non-time-dependent is not satisfied generally, but it simplifies problems without usually introducing large errors. It is feasible to define additional constants to represent time dependency of volumetric strain, but this will not be done here.

the intercept ε_0 by regression. Let q equal the positive distance between the creep curve and the line asymptotic to the secondary creep curve (Figure 6.15). Then

$$\log q = \log \left(\frac{\sigma_1}{3G_1}\right) - \frac{G_1}{2.3\eta_1} t \tag{6.30}$$

A semilog plot $\log_{10} q$ versus t has intercept $\sigma_1/3G_1$ and slope $-G_1/2.3\eta_1$ determining G_1 and η_1.

If lateral strains ε_3 are monitored as well as axial strains ε_1, the volumetric strain is determined by $\Delta V/V = \varepsilon_1 + 2\varepsilon_3$ (see Equation 3.6) while the mean stress is $\sigma_1/3$. Therefore K can be calculated by

$$K = \frac{\sigma_1}{3(\varepsilon_1 + 2\varepsilon_3)} \tag{6.31}$$

and G_2 can be calculated from

$$\frac{\sigma_1}{3G_2} = \varepsilon_B - \sigma_1 \left(\frac{1}{3G_1} + \frac{2}{9K}\right) \tag{6.32}$$

As an example, we will find the Burgers constants for Indiana limestone using the data from creep tests by Hardy et al. (1970) presented in Figure 6.19. The rock is homogeneous limestone with a mean grain size of 14 mm and porosity of 17.2%. The unconfined compressive strength q_u was 9000–11,000 psi (dry). Cylinders of limestone, 1.12 in. in diameter and 2.25 in. long, were loaded by dead weights on levers, in increments so that creep curves were obtained at different axial stresses. Time dependency was absent in tests with axial stress less than 40% of q_u, and secondary creep was unimportant when σ_1 was less than about 60% of q_u. Table 6.3 summarizes the pertinent data for one specimen. Straight lines were drawn asymptotic to each creep curve in Figure 6.19, giving slopes $\Delta\sigma_1/3\eta_2$ and intercepts ε_B. Regression analysis on a pocket calculator determined the constants $\Delta\sigma_1/3G_1$ and G_1/η_1. (Semilog plotting could have been used instead.) The determined values of K, G_1, G_2, η_1, and η_2 are given in Table 6.4. Note that G_1 and the viscosity terms are large for the first two increments when there was no time dependency and become progressively smaller as the axial stress augments. G_2 and K are almost independent of stress. This is nonlinear viscoelasticity of the type derived from the initiation and growth of cracks. These deformability constants have real physical meaning: G_2 is the *elastic shear modulus*; G_1 controls the *amount of delayed elasticity*; η_1 determines the *rate of delayed elasticity*; and η_2 describes the *rate of viscous flow*.

Table 6.3 *Incremental Creep of Indiana Limestone*[a]

Increment	Initial σ_1[b] (psi)	Step $\Delta\sigma_1$ (psi)	Slope of Asymptote $\Delta\sigma_1/3\eta_2$	Intercept of Asymptote ε_B[c]	Initial Axial Strain ε_0[c]	Initial Lateral Strain ε_3[c]	$q(t) = (\Delta\sigma_1/3G_1)e^{-(G_1t/\eta_1)}$	
							$\Delta\sigma_1/3G_1$	G_1/η_1
1 + 2	0	3693	0	685	685	−175	Not time dependent	
3	3693	2030	0	436	407	−128	16.7×10^{-6}	0.32
4	5723	699	0.105 $\mu\varepsilon$/min	139	125	−33	9.7×10^{-6}	0.28
5	6392	782	0.16 $\mu\varepsilon$/min	179	150	−39	16.2×10^{-6}	0.28
6	7174	781	0.41 $\mu\varepsilon$/min	183	147	−41	19.1×10^{-6}	0.295
7	7955	782	0.42 $\mu\varepsilon$/min	203	142	−42	33.5×10^{-6}	0.27

[a] Data from Hardy et al. (1970).
[b] $q_u = 9565$ psi in test lasting 40 min.
[c] All values of strain given are $\mu\varepsilon$ (10^{-6} in./in.) that were measured from the respective increment of load.

209

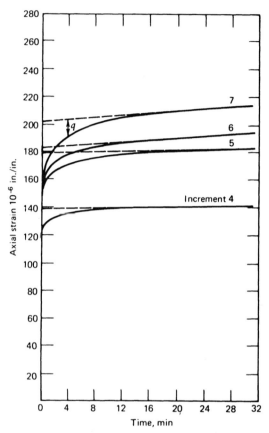

Figure 6.19 Creep of Indiana limestone in unconfined compression. (Data from Hardy et al., 1970.)

Table 6.4 *Burgers Body Constants Fit to Data of Table 6.3*

Increment	Percent q_u after Loading	K (10^6 psi)	G_1 (10^6 psi)	G_2 (10^6 psi)	η_1 (10^6 psi/min)	η_2 (10^6 psi/min)
1 and 2	39	3.7	Large	2.7	∞	∞
3	60	4.5	28.9	2.1	84	∞
4	67	3.8	23.0	2.5	71.8	2120
5	78	3.6	16.1	2.3	57.5	1630
6	83	4.0	13.6	2.2	46.1	640
7	91	4.5	7.8	2.0	28.9	620

DETERMINING VISCOELASTIC CONSTANTS
FROM FIELD TESTS

Any field test in which the load can be sustained for days or weeks can be used to evaluate viscoelastic constants of rock masses. Corrections might be necessary for changes in environmental conditions for tests conducted on the surface (e.g., the plate-bearing test using cables). But in boreholes and in underground galleries, temperature and humidity are often unvarying.

The *dilatometer test* is convenient for creep tests. Unlike the laboratory compression test on rock cylinders, there is no change in the mean stress when a dilatometer is pressured and expanded against the borehole. Therefore, there is no term in K influencing the time history of radial displacement. The outward radial displacement at the wall of the borehole ($r = a$) in a Burgers material is

$$u_r(t) = \frac{pa}{2G_2} + \frac{pa}{2G_1} - \frac{pa}{2G_1} e^{-(G_1 t/\eta_1)} + \frac{pa}{2\eta_2} t \qquad (6.33)$$

in which p is the internal pressure in the dilatometer.

The displacement with time follows a curve, Figure 6.20, like that for the compression test and the analysis of the data is as previously discussed, except that the intercepts and slopes have different values. At $t = 0$, the radial displacement is

$$u_0 = \frac{pa}{2G_2} \qquad (6.33a)$$

The asymptote to the displacement-time curve has intercept

$$u_B = pa \left(\frac{1}{2G_2} + \frac{1}{2G_1} \right) \qquad (6.33b)$$

and slope $pa/2\eta_2$. Again letting q equal the positive vertical distance between the asymptote and the displacement-time curve at any time,

$$\log q = \log \frac{pa}{2G_1} - \frac{G_1}{2.3\eta_1} t$$

Thus a series of sustained pressure increments in dilatometer tests will allow the constants G_1, G_2, η_1, and η_2 to be determined.

The plate-bearing test can also be conducted to yield the viscoelastic constants. Now, however, there are terms with the bulk modulus K because the mean pressure as well as the deviatoric stress changes when the pressure is applied. For a constant pressure p applied suddenly to a flexible bearing plate of

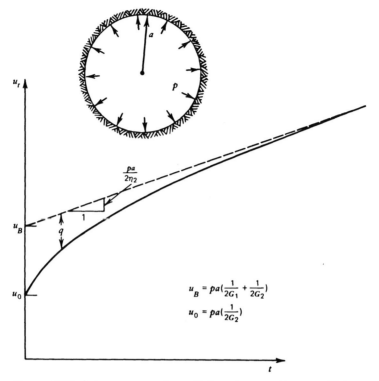

Figure 6.20 Creep response to a dilatometer or gallery test if the rock behaves as a Burgers body under deviatoric stress.

circular shape and radius a, the mean displacement $\bar{\omega}$ varies with time according to

$$
\bar{\omega} = \frac{1.70pa}{4} \left(\frac{1}{G_2} + \frac{1}{K} + \frac{t}{\eta_2} + \frac{1}{G_1}(1 - e^{-(G_1 t/\eta_1)}) \right.
$$
$$
\left. + \frac{3}{(G_1 + 3K)}(1 - e^{-(G_1 + 3K)t/\eta_1}) - \frac{G_2}{K(3K + G_2)} e^{-3KG_2 t/[\eta_2(3K + G_2)]} \right)
$$
(6.34)

As a simplification, it may be acceptable to assume that the rock is incompressible ($K = \infty$, $v = \frac{1}{2}$). In this instance, (6.34) reduces to

$$
\bar{\omega} = \frac{1.70pa}{4} \left(\frac{1}{G_2} + \frac{t}{\eta_2} + \frac{1}{G_1}(1 - e^{-(G_1 t/\eta_1)}) \right)
$$
(6.34a)

The initial displacement is then $\omega_0 = (1.70pa/4)(1/G_2)$ and after the delayed elasticity has occurred, the settlement of the plate tends to the line

$$\overline{\omega}_B = \frac{1.70pa}{4} \left(\frac{1}{G_2} + \frac{1}{G_1} + \frac{t}{\eta_2} \right) \tag{6.34b}$$

An analysis of the field test results imitating that performed for the creep test will therefore yield values for G_1, G_2, η_1, and η_2. (If v is less than 0.5, all but η_2 will be in error.)

For a test with a rigid bearing plate, it is tempting to replace the factor 1.70 by $\pi/2$, in analogy to the elastic case. However, this would not be strictly correct in the viscoelastic case.

Data from long-term plate-bearing tests in schists and sandstones are presented by Kubetsky and Eristov (1970).

TERTIARY CREEP

Secondary creep at stresses approaching peak values will end with tertiary creep and rupture, possibly violent (Figure 6.16). Recalling Figure 3.13, the key parameter identifying the time when tertiary creep begins is the accumulated strain, for when the stress and cumulative strain define a point on the right side of the complete stress-strain curve, rupture will occur. Data by John (1974) for norite, a basic igneous intrusive rock, demonstrate this principle. Creep tests were run from different starting stresses in unconfined compression. Figure 6.21a presents plots of axial stress versus log time. The horizontal lines, showing the paths of creep tests, terminate in points along a negatively inclined locus. Strain varies with time (although nonuniformly over the family of creep tests), so Figure 6.21 can be mapped into stress-strain space.

EFFECT OF STRESS RATE

Figure 6.21a also shows that the strength of norite falls when the stress rate is slowed. A specimen loaded to peak stress at 2.1 MPa/s (over about 100 hr) developed about two-thirds of the strength of a specimen loaded to peak stress at 1.8×10^4 MPa/s (over about a twentieth of a second). Figure 6.21b, also from John, shows stress-strain curves for unconfined compression of norite at different stress rates, demonstrating reduction in stiffness when the stress rate is lowered. Note that the elastic modulus (the slope of the axial stress versus axial strain curve) is unaffected by stress rate until a certain stress has been reached and thereafter there is pronounced curvature and apparent yielding before rupture (rupture points are not shown). These observations can be explained by viscoelasticity theory.

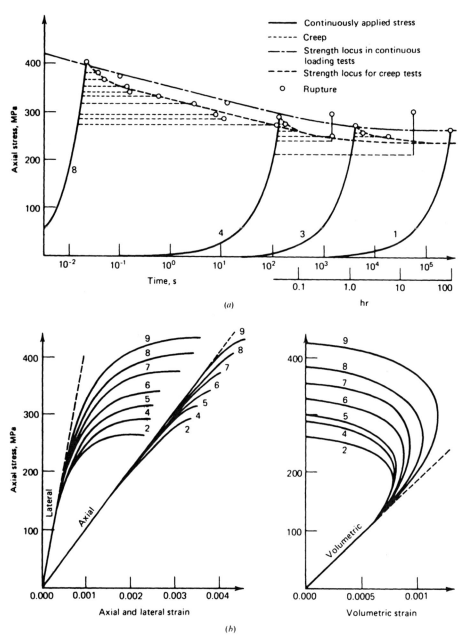

(a)

(b)

Figure 6.21 Results of dynamic tests on norite by M. John (1974). (*a*) Stress history on various loading paths. (*b*) Deformation versus axial stress at varying stress rates. Stress rates are keyed as follows (MPa/s): (1) = 8.4×10^{-4}, (2) = 4.1×10^{-3}, (3) = 6.4×10^{-2}, (4) = 2.1, (5) = 2.5×10^{1}, (6) = 2.2×10^{2}, (7) = 3.9×10^{3}, (8) = 1.8×10^{4} and (9) = 2.8×10^{5}.

When a constantly increasing stress σ_1 at a stress rate $\dot{\sigma}_1$ is applied to a rock behaving as a Burgers body in distortion and an elastic material in hydrostatic compression, the axial strain is

$$\varepsilon_1 = \sigma_1 \left(\frac{1}{3G_2} + \frac{1}{3G_1} + \frac{2}{9K} \right) - \frac{\dot{\sigma}_1 \eta_1}{3G_1^2}(1 - e^{-(G_1\sigma_1/\eta_1\dot{\sigma}_1)}) + \frac{\sigma_1^2}{6\eta_2\dot{\sigma}_1} \quad (6.35)$$

The stress-strain relation E is then dependent on $\dot{\sigma}_1$. For example, consider Indiana limestone whose Burgers body constants and bulk modulus were listed in Table 6.4. Applying Equation 6.35 to each increment of stress in turn, with the constants η_1, η_2, G_1, G_2, and K approximately selected from Table 6.4, yields the values of E presented in Table 6.5. In Figure 6.22, these have been plotted incrementally defining stress-strain paths OA, OB, OC, and OD for the four values of stress rate arbitrarily selected for the example. Let us assume that the complete stress-strain curve has a uniquely defined right side defined

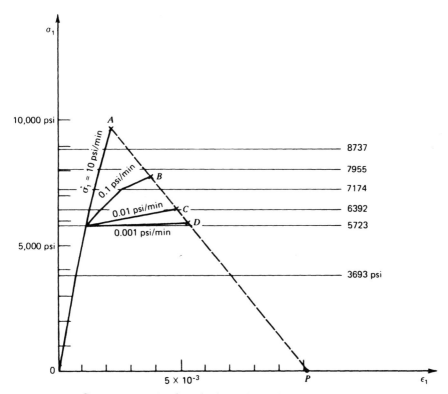

Figure 6.22 Stress versus strain calculated for Indiana limestone in unconfined compression at various stress rates.

Table 6.5 Calculated Young's Modulus for Different Stress Rates for Indiana Limestone with Properties Based on Creep Data by Hardy et al. (1970)—see Table 6.4[a]

Initial Axial Stress σ_1 (psi)	Stress Increm. $\Delta\sigma_1$ (psi)	$\dot{\sigma}_1 = 0.001$ psi/min		$\dot{\sigma}_1 = 0.01$ psi/min		$\dot{\sigma}_1 = 0.1$ psi/min		$\dot{\sigma}_1 = 10$ psi/min	
		$\Delta\varepsilon_1$ (10^{-6})	E (10^6 psi)	$\Delta\varepsilon_1$ (10^{-6})	E (10^6 psi)	$\Delta\varepsilon_1$ (10^{-6})	E (10^6 psi)	$\Delta\varepsilon_1$ (10^{-6})	E (10^6 psi)
0	3693	678	5.45	678	5.48	678	5.45	678	5.45
3693	2030	418	4.86	418	4.86	418	4.86	418	4.86
5723	669	35,320	0.0189	3,650	0.183	483	1.38	135	4.98
6392	782	62,700	0.0125	6,400	0.122	794	0.99	174	4.49
7174	781	159,000	0.0049	16,060	0.0486	1759	0.444	186	4.19
7955	782	165,000	0.0048	16,620	0.0470	1829	0.428	200	3.91

[a] Specimen ruptured at 9565 psi at stress rate $\dot{\sigma} = 240$ psi/min.

by line *AP* (for zero confining pressure). The stress-strain curves must reach peak values where they meet this line, at points *A, B, C,* and *D*. In this idealized example, we can appreciate how the rate of loading can alter both deformability and strength of a rock.

References

Afrouz, A. and Harvey, J. M. (1974) Rheology of rocks within the soft to medium strength range, *Int. J. Rock Mech. Min. Sci.* **2:** 281–290.

Benson, R. P., Murphy, D. K., and McCreath, D. R. (1969) Modulus testing of rock at the Churchill Falls Underground Powerhouse, Labrador, *ASTM Spec. Tech. Rept. 477,* 89–116.

Bieniawski, Z. T. (1978) Determining rock mass deformability—Experience from case histories, *Int. J. Rock Mech. Min. Sci.* **15:** 237–248.

Deere, D. V. (1968) Geological considerations, in Stagg, K. G. and Zienkiewicz, O. C. (Eds.), *Rock Mechanics in Engineering Practice,* Wiley, New York.

Flügge, W. (1975) *Viscoelasticity,* 2d ed., Springer, Berlin.

Goodman, R. E. and Duncan, J. M. (1971) The role of structure and solid mechanics in the design of surface and underground excavation in rock, *Proceedings, Conference on Structure, Solid Mechanics and Engineering Design,* Part 2, paper 105, p. 1379, Wiley, New York.

Goodman, R. E., Van, T. K., and Heuze, F. E. (1972) The measurement of rock deformability in boreholes, *Proceedings, 10th Symposium on Rock Mechanics* (AIME). pp. 523–555.

Hardy, H. R., Jr., Kim, R. Y., Stefanko, R., and Wang, Y. J. (1970) Creep and microseismic activity in geologic materials, *Proceedings, 11th Symposium on Rock Mechanics* (AIME), pp. 377–414.

Heuze, F. E. and Salem, A. (1979), Rock deformability measured in-situ—Problems and solutions, *Proceedings, International Symposium on Field Measurements in Rock Mechanics* (Balkema, Rotterdam), Vol. 1, pp. 375–388.

Jaeger, J. C. and Cook, N. G. W. (1976) *Fundamentals of Rock Mechanics,* 2d ed., Chapman & Hall, London.

John, M. (1974) Time dependence of fracture processes of rock materials (in German), *Proc. 3rd Cong. ISRM* (Denver), Vol. 2A, pp. 330–335.

King, M. S., Pobran, V. S., and McConnell, B. V. (1975) Acoustic borehole logging system, *Proceedings, 9th Canadian Rock Mechanics Symposium* (Montreal).

Kubetsky, V. L. and Eristov, V. S. (1970) In-situ investigations of creep in rock for the design of pressure tunnel linings, *Proceedings, Conference on In-situ Investigations in Soils and Rocks* (British Geot. Soc.), pp. 83–91.

Kulhawy, F. H. (1975) Stress-deformation properties of rock and rock discontinuities, *Eng. Geol.* **9:** 327–350.

Kulhawy, F. H. (1978) Geomechanical model for rock foundation settlement, *J. Geotech. Eng. Div.* (ASCE) **104** (GT2): 211–228.

Lane, R. G. T. and Knill, J. L. (1974) Engineering properties of yielding rock, *Proc. 3rd Cong. ISRM* (Denver), Vol. 2A, pp. 336–341.

Raphael, J. M. and Goodman, R. E. (1979) Strength and deformability of highly fractured rock, *J. Geotech. Eng. Div.* (ASCE) **105** (GT11): 1285–1300.

Rutter, E. H. (1972) On the creep testing of rocks at constant stress and constant force, *Int. J. Rock Mech. Min. Sci.* **9**:191–195.

Schneider, B. (1967) Moyens nouveaux de reconaissance des massifs rocheux, *Supp. to Annales de L'Inst. Tech. de Batiment et des Travaux Publics*, Vol. 20, No. 235–236, pp. 1055–1093.

Serafim, J. L. and Pereira, J. P. (1983) Considerations of the geomechanics classification of Bieniawski, *Proceedings, International Symposium on Engineering Geology and Underground Construction* (L.N.E.C., Lisbon, Portugal) Vol. 1, Section II, pp. 33–42.

Timoshenko, S. and Goodier, J. N. (1951) *Theory of Elasticity*, 2d ed., McGraw-Hill, New York.

Van Heerden, W. L. (1976) In-situ rock mass property tests, *Proceedings of Symposium on Exploration for Rock Engineering*, Johannesburg, Vol. 1, pp. 147–158.

Wawersik, W. R. (1974) Time dependent behavior of rock in compression, *Proc. 3rd Cong. ISRM* (Denver), Vol. 2A, pp. 357–363.

Zienkiewicz, O. C. and Stagg, K. G. (1967) The cable method of in-situ rock testing, *Int. J. Rock Mech. Min. Sci.* **4**: 273–300.

Problems

1. Show that the stress-strain relationship connecting deviatoric strain e_{ij} and deviatoric stress τ_{ij} consists of six uncoupled identical statements:

$$\tau_{ij} = 2Ge_{ij}$$
$$i, j = 1, 3$$

("Deviatoric strain" is discussed in Appendix 2.)

2. Suppose a triaxial compression test is conducted by simultaneous change in σ_1 and p; derive expressions for E and v in terms of the axial and lateral strains and the stresses σ_1 and p.

3. Describe a procedure for triaxial testing that raises the deviatoric stress while the nondeviatoric stress remains constant.

4. The following forces and displacements were measured in an unconfined compression test of a cylindrical claystone specimen 5.0 cm in diameter and 10.0 cm long.

Axial Force (N)	Axial Shortening (mm)	Lateral Extension (mm)	Axial Force (N)	Axial Shortening (mm)	Lateral Extension (mm)
0	0	0	0	0.080	0.016
600	0.030		2,500	0.140	
1000	0.050		5,000	0.220	
1500	0.070		6,000	0.260	
2000	0.090		7,000	0.300	
2500	0.110	0.018	7,500	0.330	0.056
0	0.040	0.009	0	0.120	0.025
2500	0.110		7,500	0.330	
3000	0.130		9,000	0.400	
4000	0.170		10,000	0.440	0.075
5000	0.220	0.037	0	0.160	0.035

Compute E and ν corresponding to elastic deformation and their counterparts M and ν_p for permanent deformation from the above data.

5. A triaxial compression test is performed as follows: (a) An all-around pressure is first applied to the jacketed rock specimen. Nondiviatoric stress $\bar{\sigma}$ is plotted against nondeviatoric strain ε and the slope $D_1 = \Delta\bar{\sigma}/\Delta\bar{\varepsilon}$ is determined. (b) Then deviatoric stress is increased while nondeviatoric stress is held constant and the axial deviatoric stress $\sigma_{1,\text{dev}}$ is plotted against the axial deviatoric strain $\varepsilon_{1,\text{dev}}$. The slope $D_2 = \Delta\sigma_{1,\text{dev}}/\Delta\varepsilon_{1,\text{dev}}$ is determined from the graph. Derive formulas expressing E and ν in terms of D_1 and D_2.

6. (a) Derive a relationship between E, the modulus of elasticity computed from the reloading curve of stress and strain; M, the modulus of permanent deformation; and E_{total}, the modulus of deformation computed from the slope of the loading curve of stress and strain. (b) Show how M varies with axial strain throughout the complete stress-strain curve.

7. In a full seismic wave experiment, the compressional and shear wave velocities were measured as $V_p = 4500$ m/s, $V_s = 2500$ m/s. Assuming the density of the rock is 0.027 MN/m^3, calculate E and ν.

8. What physical phenomena could explain a plate-bearing pressure versus displacement curve like that of Figure 6.9?

9. A rock mass is cut by one set of joints with spacing $S = 0.40$ m. (a) If the joint normal and shear deformations are assumed to be equal to that of the rock itself, express k_s and k_n in terms of E and ν. (b) Assuming $E = 10^4$ MPa and $\nu = 0.33$, calculate all the terms of the strain-stress relationship for an equivalent transversely isotropic medium, (corresponding to Equation 6.9).

10. Modify Equations 6.23 and 6.24 accordingly for a rock mass with three mutually perpendicular sets of joints.

11. Show that for rock cut by one set of joints with spacing S, the normal strains and normal stresses referred to n, s, t coordinates are related by

$$\begin{pmatrix} \varepsilon_n \\ \varepsilon_s \\ \varepsilon_t \end{pmatrix} = \frac{1}{E} \begin{bmatrix} p & -\nu & -\nu \\ -\nu & 1 & -\nu \\ -\nu & -\nu & 1 \end{bmatrix} \begin{pmatrix} \sigma_n \\ \sigma_s \\ \sigma_t \end{pmatrix}$$

where $p = 1 + E/k_nS$ and where E and ν are Young's modulus and Poisson's ratio of the rock, k_n is the normal stiffness of the joints, and n is the direction perpendicular to the joint planes.

12. A jacketed rock cube, with edge length 50 cm, is subject to an all around pressure p. The pressure versus volumetric strain curve recorded is given in the figure. Assume the rock contains three mutually perpendicular joint sets all spaced 5 cm apart. Calculate the normal stiffness of the joints k_n at each of the normal pressures corresponding to the start of unloading paths (2.4, 4.8, and 10.3 MPa).

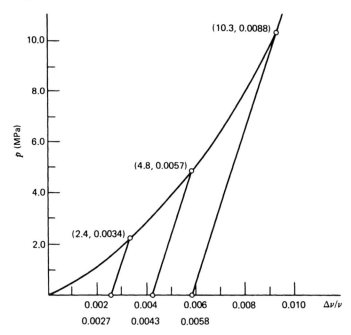

13. Let ν_p, ν_t, and ν be respectively the Poisson's ratios for plastic, total, and elastic strain; that is, for strain applied in the x direction, $\nu = -\varepsilon_y/\varepsilon_x$, etc. Derive a formula expressing ν_t as a function of E, M, ν, and ν_p.

Applications of Rock Mechanics in Engineering for Underground Openings

7.1 Introduction

Engineering underground space has many facets, some of which are unrelated to rock conditions. However, rock mechanics has direct bearing on many of the critical aspects of engineering work, for example, planning the location, dimensions, shapes, and orientations of chambers, selecting supports, arranging for construction access, engineering blasting, and designing instrumentation. Rock mechanics provides information of utmost relevance: measurement of initial stress; monitoring stresses developed in the peripheries of openings; measurement of material properties; analysis of stresses, deformations, temperatures, and water flow in support of design; and interpretation of instrumentation readings, especially displacements.

The uses of underground space are many, varying from simple openings in the dry, to large complexes of openings in three dimensions, filled with hot or cold fluids of varying viscosity and pressure. Tunnels built for highways and railroads may be short sections beneath valley sides, or very long structures underneath major mountain ranges. Ventilation requirements for highway tunnels tend to make these very large (e.g., 15 m wide). Water supply and sewage tunnels are generally smaller, but they may be very long and frequently operate under internal pressure. For hydroelectric power production, pressure headrace tunnels lead water to surface or underground penstocks and thence into surface or underground power stations. Large water pressures are supported by rock alone in some cases. The main machine hall chambers are rooms with spans of the order of 25 m, while access tunnels and other openings may also be quite large. These chambers are feasible only if the rock is essentially self-supporting (Figure 7.1).

Pumped storage projects may require rock tunnels, underground power-houses and other openings as well (Figure 7.2). Energy storage is now also demanding underground space—for storage of oil (Figure 1.4), and eventually for hot air or hot water used in peak demand energy conversion schemes of various kinds. Liquefied natural gas (LNG) may be stored in rock caverns more widely if the problems of thermal cracking and loss of product can be over-come. Nuclear wastes are to be stored in specially mined repositories in rock salt, chosen for its high heat conductivity and general continuity without frac-tures. In the field of mining, there are two types of underground openings: those that are intended to be stable while the ore is removed, and those that are intentionally collapsed to produce broken rock that is drawn off as the ground caves. For defense, deep cavities are required to protect installations from shock. Finally, industry has need for underground space for product storage, offices, and even public facilities like swimming pools.

With such a vast range of underground usage, many kinds of rock mechan-ics considerations need to be addressed. However, there are certain features common to all underground works. They are usually inaccessible until actual construction. Occasionally, when an existing installation is being expanded, the engineer will have access to the site at the initiation of the job; more usually, however, he or she will have to begin deliberations from information acquired in drill holes, shafts, and galleries. All underground workings are constructed in rock that is *initially stressed* and all openings cause changes in the initial stress when they are constructed. Most underground workings are made below the *water* table. And all openings are constructed in an environ-ment of even *temperature* equal to the mean surface temperature plus the product of geothermal gradient and depth. The gradient of temperature varies from 0.5°C/100 m to as much as 5°/100 m.

When working with rock mechanics below ground, there are certain condi-tions that should be appreciated. The underground environment is very often hostile for instruments due to water, blasting, and truck traffic. Working space is often cramped, poorly lit, and wet. As a result, experiments and concepts for instrumentation underground should be as simple as possible and the equip-ment must be rugged. Overly sophisticated testing technique or data handling, and overly precise measurements are to be avoided. However, almost any data taken underground near the site of the job will be more useful than data ob-tained remotely or from boreholes. As far as possible, then, major experiments and measurements should be deferred until the opportunity to work in the

Figure 7.1 Photos of Churchill Falls underground powerhouse during construction; courtesy of D. R. McCreath, Acres Consulting Services, Niagara Falls, Canada. (a) The machine hall, 297 m long, 25 m wide, and 47 m high; it was excavated at a depth of about 300 m in gneiss. (b) The surge chamber, with draft tube entries on the right. The keyhole shape was determined by finite element analysis to reduce the extent of zones of tensile stress in the rock. The opening is approximately 275 m long, 19.5 m wide at the maximum section, 12 m wide at the base, and 45 m high. [See Benson et al. (1971).]

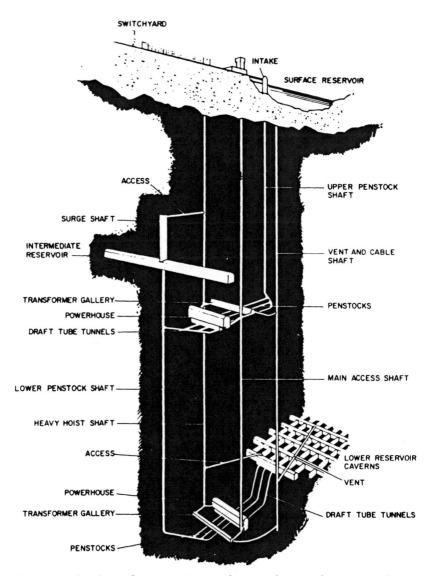

Figure 7.2 A scheme for a two-stage underground pumped storage project. (Reproduced from McCreath and Willett (1973) with permission.)

underground site itself. This may require that certain planning decisions be postponed until access is gained. For example, an underground powerhouse constructed at a depth of 800 m in Colombia was fixed in location but not in orientation, pending the completion of the access tunnel, which was built in a separate, early construction contract. Stress measurements, deformability measurements, and other tests were conducted in a gallery driven from the access tunnel in time to permit their integration into the final design process.

Rock mechanics for underground engineering begins with proper appreciation of the character of the rock. Rock that is able to bridge across an opening of 20 m or more without appreciable support could be considered competent. In such rocks, design can be aided by considerations of stress concentrations around the opening as deduced from the *theory of elasticity*.

When the rock is layered where bending and separation of strata are possible, the theory of *elastic beams and plates* can be invoked. For rock that presents time-dependent properties, such as rock salt, the *theory of linear viscoelasticity* provides useful concepts. In weak rocks, stresses around openings may reach the limit according to the criteria of failure, resulting in slow convergence (squeeze); in such rocks, a solution for stresses and displacements derived from the *theory of plasticity* provides a useful basis for engineering work. In jointed rock, only individual *limit equilibrium analysis* or studies with *numerical or physical models* may be appropriate. The discussions of competent rock, layered rock, and plastic rock that follow are intended only to provide simple models for guidance in engineering practice. These models can always be improved by using powerful numerical techniques, but the engineer cannot resort to such techniques for every question—he or she has to have some working tools to provide checks on computations, order of magnitude predictions, and sensitivity studies through parameter variation. This is the spirit in which the following theory is presented.

7.2 Openings in Competent Rock

In rock stressed below its elastic limit, that is, below about one-half of the compressive strength, and in which joints are widely spaced and tightly precompressed or healed, it is often acceptable to consider an opening as a long hole of constant cross section in an infinite volume. This is the plane strain[1] equivalent of a hole in a plate, and we can use the solution to the problem of a circular hole in a biaxially loaded plate of homogeneous, isotropic, continuous, linearly elastic material—the *Kirsch solution*. A point located at polar coordinate r, θ near an opening with radius a (Figure 7.3) has stresses σ_r, σ_θ, $\tau_{r\theta}$, given by

$$\sigma_r = \frac{p_1 + p_2}{2}\left(1 - \frac{a^2}{r^2}\right) + \frac{p_1 - p_2}{2}\left(1 - \frac{4a^2}{r^2} + \frac{3a^4}{r^4}\right)\cos 2\theta \qquad (7.1a)$$

$$\sigma_\theta = \frac{p_1 + p_2}{2}\left(1 + \frac{a^2}{r^2}\right) - \frac{p_1 - p_2}{2}\left(1 + \frac{3a^4}{r^4}\right)\cos 2\theta \qquad (7.1b)$$

$$\tau_{r\theta} = -\frac{p_1 - p_2}{2}\left(1 + \frac{2a^2}{r^2} - \frac{3a^4}{r^4}\right)\sin 2\theta \qquad (7.1c)$$

[1] The concept of plane strain is discussed in the derivation to Equation 7.1 in Appendix 4.

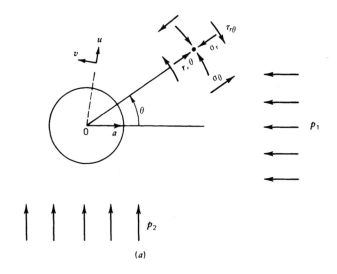

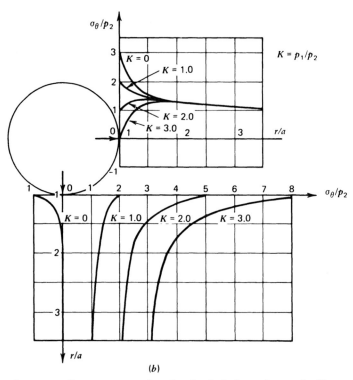

Figure 7.3 Stresses around a circular hole in an isotropic, linearly elastic, homogeneous continuum.

where σ_r is the stress in the direction of changing r, and σ_θ is the stress in the direction of changing θ.

Substituting the value $r = a$ in Equation 7.1 gives the variation of stresses on the walls of the opening. The radial and shear stresses are zero since this is a free surface. The tangential stress σ_θ varies from a maximum of $3p_1 - p_2$ at $\theta = 90°$ to a minimum of $3p_2 - p_1$ at $\theta = 0°$ (results used in Chapter 4). Away from the opening, the stress concentrations fall off quickly, as shown in Figure 7.3b and Table 7.1.

The Kirsch solution allows calculations of the potential influence of joints in the region of a tunnel. Assuming that a joint of given position and orientation introduces no change in the stress field, we compare the shear and normal stresses along its surface with the limiting values of shear stresses consistent with the criteria of peak shear strength presented in Chapter 5. This exercise defines a region of joint influence, which can be overlain on the actual or assumed geological section to isolate potential problem areas in the roof or walls.

Figure 7.4 shows regions of joint slip according to such an approach for three directions of jointing for an example in which $K = 2.33$ ($p_2/p_1 = 0.43$). The joints were assumed to obey Coulomb's law with $\phi_j = 31°$. The contour values give the number of times the lateral pressure of 1000 psi must be multiplied to produce slip on joints of the given orientation in the closed region inside the contour. The contour values must be multiplied by the shear strength intercept S_j of the joint and divided by 100. For example, a joint striking parallel to the tunnel and dipping at 30°, with a friction angle of 31° and a shear strength intercept of 50 psi, would slip throughout the region enclosed within the contour marked 0.50 if the horizontal stress were equal to 250 psi and the vertical stress were 108 psi.

An effective method for monitoring an underground opening is to measure relative displacements of points on the walls, for example, with a precision tape stretched between pairs of points, or with rods anchored at different depths in a borehole (a "multiposition extensometer"). To interpret such data, it is helpful to know the order of magnitude of displacements associated with elastic behavior. The displacements can be determined from the Kirsch solution, assuming conditions of plane strain:

$$u_r = \frac{p_1 + p_2}{4G} \frac{a^2}{r} + \frac{p_1 - p_2}{4G} \frac{a^2}{r} \left(4(1 - \nu) - \frac{a^2}{r^2} \right) \cos 2\theta \tag{7.2a}$$

and

$$v_\theta = - \frac{p_1 - p_2}{4G} \frac{a^2}{r} \left(2(1 - 2\nu) + \frac{a^2}{r^2} \right) \sin 2\theta \tag{7.2b}$$

Table 7.1 Stress Concentration σ_θ/σ_v in Roof ($\theta = 0$) and Wall ($\theta = 90°$) According to Kirsch Solution

σ_h/σ_v:	0		0.3		0.6		1.0	1.5		2.0		3.0	
θ:	0°	90°	0°	90°	0°	90°	All θ values	0°	90°	0°	90°	0°	90°
r/a													
1.0	−1.00	3.00	−0.10	2.70	0.80	2.40	2.00	3.50	1.50	5.00	1.00	8.00	0.00
1.0	−0.61	2.44	0.12	2.25	0.85	2.07	1.83	3.05	1.52	4.26	1.22	6.70	0.60
1.2	−0.38	2.07	0.25	1.96	0.87	1.84	1.69	2.73	1.51	3.77	1.32	5.84	0.94
1.3	−0.23	1.82	0.32	1.75	0.86	1.68	1.59	2.50	1.48	3.41	1.36	5.23	1.13
1.4	−0.14	1.65	0.36	1.60	0.85	1.56	1.51	2.33	1.44	3.16	1.37	4.80	1.24
1.5	−0.07	1.52	0.38	1.50	0.84	1.47	1.44	2.20	1.41	2.96	1.37	4.48	1.30
1.75	−0.00	1.32	0.40	1.32	0.80	1.33	1.33	1.99	1.33	2.81	1.36	3.97	1.33
2.0	+0.03	1.22	0.40	1.23	0.76	1.24	1.25	1.86	1.27	2.47	1.28	3.69	1.31
2.5	+0.04	1.12	0.38	1.13	0.71	1.14	1.16	1.72	1.18	2.28	1.20	3.40	1.24
3.0	+0.04	1.07	0.36	1.09	0.68	1.10	1.11	1.65	1.13	2.19	1.15	3.26	1.19
4.0	+0.03	1.04	0.34	1.04	0.65	1.05	1.06	1.58	1.08	2.10	1.09	3.14	1.11

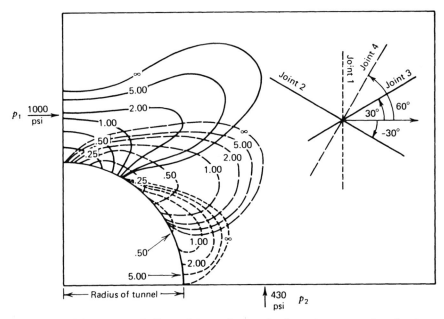

Figure 7.4 The extent of slip on joints of several orientations around a circular tunnel whose state of stress is given by the Kirsch solution (Equation 7.1). Contour values give the number of times the indicated boundary pressure must be multiplied to cause slip. Multiply the contour values by the joint cohesion $S_j/100$ psi.

in which u_r is the radial outward displacement and v_θ is the tangential displacement, as shown in Figure 7.3*a*, *G* is the shear modulus, and ν is Poisson's ratio.

It should be understood that the effect of gravity is not fully represented in the above equations. Gravity creates a vertical stress, represented by p_1 or p_2; but it also exerts a following force on loosened rock near the roof, a force that pursues the rock no matter where it might move. This effect cannot be modeled by any choice of loads on the boundary. One approach to introducing gravity in the computation of a tunnel would be to consider the weight of rock in the zone of joint slip, calculated according to the Kirsch solution stresses and assigned joint or material properties as in Figure 7.4. This added weight could be assigned to a system of supports such as rock bolts or shotcrete. The omission of gravity is one reason why the Kirsch solution does not offer a *size effect*, that is, the stresses on the walls predicted by Equation 7.1 are the same regardless of the diameter of the tunnel. This cannot be the whole truth for we know from experience that a small tunnel is more stable than a large one. Size effect can be introduced in real tunnels not only by including the additional load due to gravity acting on the rock near the tunnel, but also by introducing size effect in material strength. As a greater number of discontinuities are contained within a sample of rock, its strength must decrease. Accordingly, when the span of an opening is many times greater than the average spacing between discontinuities, the opening cannot be expected to stand without artificial support.

Openings of other shapes have been solved mathematically and solutions can be found in Muskhelishvili (1953). Elliptical and other idealized shapes are discussed by Jaeger and Cook (1976) and Obert and Duvall (1967). For three-dimensional problems, spheres and ellipsoid solutions are presented.

The inward radial displacement u of a point distance r from a spherical cavity of radius a excavated in a rock mass under hydrostatic initial stress p is

$$u = \frac{pa^3}{4r^2G} \tag{7.2c}$$

Comparing Equations 7.2c and 7.2a when $p_2 = p_1$ indicates that the radial displacement of a point on the wall of a spherical cavity under hydrostatic pressure is half that of a circular tunnel of equivalent radius, also under hydrostatic pressure, and assuming both rock masses behave elastically. This relationship is useful in approaching the relative response of instruments placed in long and short chambers, respectively.

In noncircular or nonspherical openings, stresses generally tend to concentrate at corners and concave bends of small radius and to decrease toward zero at convex bends. Unstressed rock suffers opening of joints and accelerated weathering and is often more burdensome than highly stressed rock underground at shallow or intermediate depth. Stress concentrations are usually high in the side walls and lowest where the line of action of the greatest initial stress intersects the opening. Stress concentrations will generally be the least troublesome if smooth shapes are used, without corners and reentrants, and if the major axis is aligned to the major principal stress, with the ratio of width to height proportional to K (Duvall, 1976). Table 7.2 gives some stress concentra-

Table 7.2 *Stress Concentrations around Openings under Vertical Stress Only $(K = 0)$*

Shape	Height/Width	Stress Concentration (σ_θ/σ_v)	
		Roof	Side
Ellipse	$\frac{1}{2}$	−1.0	5.0
Oval	$\frac{1}{2}$	−0.9	3.4
Rectangle (round corners)	$\frac{1}{2}$	−0.9	2.5
Circle	1	−1.0	3.0
Ellipse	2	−1.0	2.0
Oval	2	−0.9	1.6
Rectangle	2	−1.0	1.7

tions under vertical stress alone ($K = 0$) for ellipses, ovals, and rectangles, at extreme points (roof, and side). By superposition, the stress concentrations can be found corresponding to other values of K.

Multiple openings have been studied in models and results are presented by Obert and Duvall (1967). Under elastic conditions, two openings will interact with each other if separated by a thickness of rock less than two times the sum of their dimensions in the direction parallel to the separation. As the openings approach each other, the average stress in the pillar between them increases and approaches the maximum tangential stress. The maximum compressive stress concentration in the wall when $K = 0$ (i.e., only vertical stress) increases from 3, the value for one opening, to 4.2 for two openings separated by a distance equal to one-fifth of the opening width. In practice, multiple openings are usually designed on the basis of the average stress in the pillar $\bar{\sigma}_v$ given by the *tributary area* theory:

$$\bar{\sigma}_v = \frac{A_t}{A_p} \sigma_v \tag{7.3}$$

where A_t is the area supported by one pillar, A_p is the area of the pillar, and σ_v is the vertical stress at the level of the roof of the opening. For a square pillar (Figure 7.5a) A_t equals $(w_o + w_p)^2$ where w_o is the width of the rooms and w_p is the width of the pillars. Inclined joints in a square pillar will intersect the sides, reducing stability. For this reason, it is sometimes elected to use long rooms perpendicular to the strike of the most troublesome, steeply inclined joints. The most severe reduction in strength of a pillar occurs when the discontinuities strike parallel to the ribs and dip at an angle of $45 + \phi_j/2°$. Discontinuities at or close to this attitude should govern the orientations of ribs. In single chambers, it is usually desirable to choose the long axis oblique to the strike of all major discontinuity sets. (See section 7.8).

To determine the dimensions of a pillar or to evaluate the degree of safety of a given pillar configuration, the average pillar stress $\bar{\sigma}_v$ calculated with Equation 7.3 must be compared with the *pillar strength* σ_p. The latter is not simply the unconfined compressive strength of the material comprising the pillar q_u because shape and size effects introduce significant modifications from the breaking strength of unconfined compressive cylinders. Hustrulid (1976) reviewed size and shape corrections applicable to coal for which, due to the presence of numerous fissures, the effect of size is great; for example, the strength of a pillar 1 m high in coal is of the order of one-fourth the strength of a test cylinder 5 cm in diameter and 10 cm high (Bieniawski, 1968). On the other hand, the shapes of many underground pillars approach rectangular prisms having width-to-height ratios considerably more than one-half, the usual value for an unconfined compressive specimen; this produces a contrary correction, the strength of relatively short pillars being greater than the strength of relatively long pillars of the same volume.

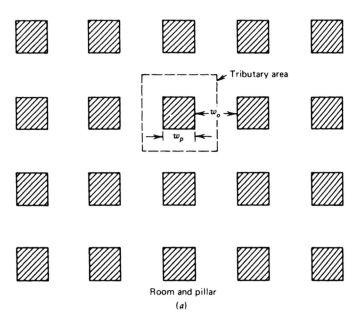

Room and pillar

(a)

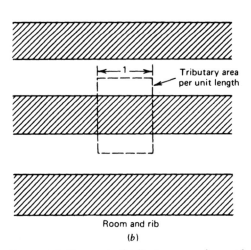

Room and rib

(b)

Figure 7.5 Concept of tributary area in regular arrangements of underground rooms.

An analysis of data reviewed by Hustrulid yields the following estimate of strength in compression for rectangular pillars of square cross section:

$$\sigma_p = \left(0.875 + 0.250 \frac{W}{H}\right) \left(\frac{h}{h_{crit}}\right)^{1/2} q_u \tag{7.4}$$

where σ_p is the strength of a pillar corrected for shape and size effects and assumed to have a height greater than h_{crit}; W and H are the width and height of the pillar, respectively; q_u is the unconfined compressive strength of the pillar material, performed on cylinders with height h equal to twice the diameter; and h_{crit} is the minimum height of a cubical specimen of pillar material such that an increase in the specimen dimension will produce no further reduction in strength. In experiments conducted by Bieniawski (1968), h_{crit} was 1 m (Figure 3.2.1). Equation 7.4 applies only for $h \leq h_{crit}$.

Using square pillars in a room and pillar panel presents a long roof span at the intersections of rooms. If roof stability is a problem, then long pillars will be warranted. Roof stability generally controls the width of rooms, whereas pillar strength controls the relative separation of rooms. Roof stability can be critical in horizontally layered rock.

7.3 Horizontally Layered Rock

When horizontally bedded rock lies above the roof, the thinner strata near the opening will tend to detach from the main rock mass and form separated beams. The stability of such beams is great if there is a horizontal stress and the span-to-thickness ratio is fairly small. Thin beds just above the opening will tend to fall down unless there is immediate support in the form of rock bolts or sets.

Figure 7.6 shows models of progressive failure of the roof of an opening with horizontally bedded rock. First, a relatively thinner beam in the immediate roof separates from the rock above, flexes downward, and cracks on its upper surface at the ends and on the lower surface in the middle. The end cracks occur first but are invisible underground. The inclined stress trajectories in the ends of the beam direct the crack propagation diagonally. Collapse of the first beam leaves cantilevers as abutments for the next beam so each layer above the roof has, in effect, a progressively smaller span. Continued failure and fall of beams eventually produces a stable, trapezoidal opening, a shape that could be selected for civil engineering applications in such rock. The beams are greatly strengthened by horizontal stress, up to about one-twentieth of the Euler buckling stress $(\pi^2 Et^2)/(3L^2)$ in which E is Young's modulus, t is the thickness of the

Figure 7.6 Models of roof behavior in horizontally layered rock. (*a*) and (*b*) show the deflection and cracking in the case of a thinner beam beneath a thicker beam.

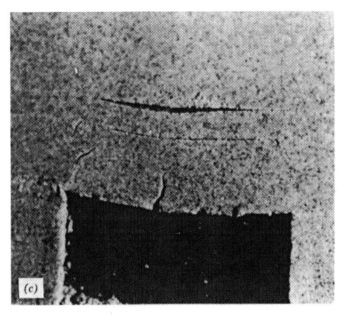

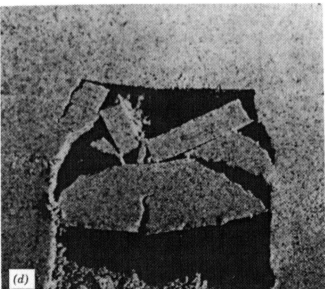

Figure 7.6 Models of roof behavior in horizontally layered rock. (*c*) and (*d*) show the deflection and failure in the opposite case—thick beneath thin. These models were made with a base friction machine. [See Goodman (1976).]

beam, and L is the span (Duvall, 1976). Assuming the roof acts like a clamped beam, the maximum tensile stress occurs at the top surface near the ends:

$$\sigma_{max} = \frac{\gamma L^2}{2t} - \sigma_h \qquad (7.5)$$

with the above constraint that $\sigma_h < \pi^2 Et^2/60L^2$. The maximum tensile stress in the center, at the bottom of the beam, is half the value given by Equation 7.5. To be conservative, σ_h can be assumed to be zero.

Visible deflection of the roof warns that a detached beam may have formed. Miners have been known to force a stick into a bow between the roof and floor, and draw a string taut between the ends so that any relaxation of the tension in the string will indicate continued downward deflection of the roof (or heave of the floor). A borehole periscope or television device should be used to inspect the roof for gaps between layers.

The maximum deflection of a clamped, elastic beam is given by

$$u_{max} = \frac{\gamma L^4}{32Et^2} \qquad (7.6)$$

For beams in a given rock type, where E and γ are constant from one layer to the next, load will be transferred from a thin beam into a thick beam if the thin beam lies above the thick beam. The stresses and deflections of the lower beam can be calculated by assigning it an increased unit weight γ_a given by

$$\gamma_a = \frac{E_{thick} t_{thick}^2 (\gamma_{thick} t_{thick} + \gamma_{thin} t_{thin})}{E_{thick} t_{thick}^3 + E_{thin} t_{thin}^3} \qquad (7.7)$$

This equation can be generalized for n beams, in which thickness decreases progressively upward. In the case where a thin beam underlies a thicker beam, a separation tends to form, as previously noted. If rock bolts are used, the bolts will have to stretch to permit the separation, and the load transference, which occurs naturally in the thin-over-thick case, will be achieved through the action of the rock bolts. In this case, the bolts must be designed to supply a force per unit area equal to Δq. The load per unit of surface area of each beam is $\gamma_1 t_1 + \Delta q$ and $\gamma_2 t_2 - \Delta q$ for the stiffer and less stiff beam, respectively. Substituting these loads in place of γt in Equation 7.6, and equating deflections of each beam (for $\sigma_h = 0$)

$$\frac{(\gamma_1 t_1 + \Delta q)L^4}{32E_1 t_1^3} = \frac{(\gamma_2 t_2 - \Delta q)L^4}{32E_2 t_2^3}$$

Solving for Δq,

$$\Delta q = \frac{\gamma_2 t_2 E_1 t_1^3 - \gamma_1 t_1 E_2 t_2^3}{E_1 t_1^3 + E_2 t_2^3} \qquad (7.8)$$

the stresses in the two layers are given by

$$\sigma_{max} = \frac{(\gamma t \pm \Delta q)L^2}{2t^2} \tag{7.9}$$

This type of load transference is termed a "suspension" effect by Obert and Duvall (1967) and Panek (1964). Added strength in the beam due to friction between the layers can be considered as follows. Let x be the coordinate parallel to the beam with the origin at one end. The shear force in the clamped beam (per unit of width) is

$$V = \gamma t \left(\frac{L}{2} - x \right)$$

The maximum shear stress, τ, at any section x is $3V/(2t)$, giving

$$\tau = \frac{3\gamma}{2} \left(\frac{L}{2} - x \right)$$

The maximum shear stress occurs at the ends, $x = 0, L$.

Consider a beam composed of layered rock with $\gamma_1 = \gamma_2$ and $E_1 = E_2$. If the friction angle between the layers is $\phi_j(S_j = 0)$, the beam can be forced to behave as if it were homogeneous if all interlayer slip is prevented. Rock bolts can be installed to achieve this is their spacing is such that they supply an average force per unit area at every x satisfying

$$p_b \tan \phi_j \geq \frac{3\gamma}{2} \left(\frac{L}{2} - x \right) \tag{7.9a}$$

Considering both friction and suspension, with Δq given by Equation 7.9, if a uniform spacing of bolts is desired, the average force per unit area exerted by the bolt system should be at least

$$p_b = \frac{3\gamma L}{4 \tan \phi_j} + \Delta q \tag{7.9b}$$

7.4 Rock with Inclined Layers

Rock with horizontal layering, as we have seen, tends to open up in the roof of an underground opening, yet remains tightly compressed in the walls. When the strata are dipping, the zone of interbed separation and potential buckling shifts off center and the walls may be undermined by sliding. How extensively these rock failure mechanisms may progress depends, among other things, on the friction between the layers since neither bending nor sliding is possible without interlayer slip (Figure 7.7a).

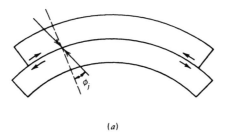

(a)

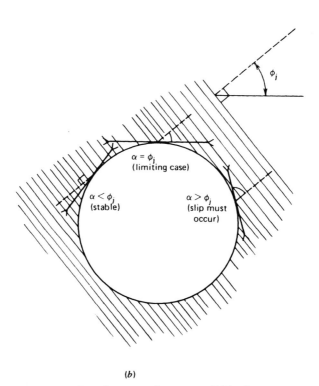

(b)

Figure 7.7 Requirements for compatibility between the flow of stress around an opening and limiting friction between layers delimit regions of slip and regions of stability.

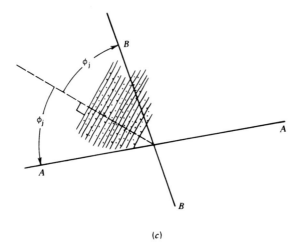

(c)

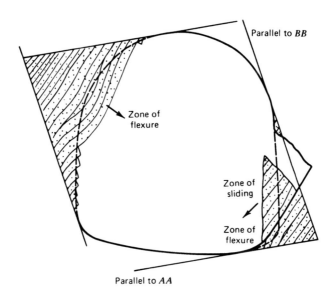

(d)

Figure 7.7 (continued)

When rock layers slide against one another, the interlayer forces become inclined ϕ_j degrees with respect to the normal to each layer. Consequently, if the beds are in static equilibrium, the resultant interlayer force cannot incline more than ϕ_j with the normals to beds (except that it may achieve the single value of 90° to the normal, that is, parallel to the beds, for then no shear resistance needs to be mobilized along the layers).

Consider an underground opening long with respect to its cross section and with initial principal stresses p_1 and p_2 in the plane of the cross section. At the surface of the opening in the absence of tunnel supports, the normal and shear stresses are zero. Therefore, the tangential stress is also the resultant force per unit area across the layers. In view of the above, at the periphery of an underground opening the tangential stress must lie less than ϕ_j degrees from the normal to the strata or lie exactly parallel to them. What happens when the tunnel walls meet the strata at an angle such that this is not true is that the layers must either mobilize cohesion or slip. The latter redistributes the stress, or changes the tunnel shape, or both (Figure 7.7b). The interlayer slip promotes sliding whenever the layers dip downward toward the opening and promotes flexure where they do not (Figure 7.7d). In other words, in regularly layered rock, the stress flows around the tunnel as if it had a shape different from that initially assumed.

The principles stated above suggest the simple construction shown in Figure 7.7c, d. To identify the zones of layer slip, with potential sliding and flexure, around a tunnel of any shape:

1. Draw the layers in their correct orientation in the cross section of the tunnel.
2. Construct lines AA and BB inclined ϕ_j with the normal to the layers.
3. Draw tangents to the tunnel periphery parallel to AA and BB.
4. Identify two opposed regions of interlayer slip delimited by these tangents; within these regions, a tangent to the tunnel surface is inclined more than ϕ_j with the normal to the strata.

Under the action of gravity, the interlayer slip regions may loosen progressively, destroying interlocks, until satisfactory support is provided. It may not be feasible to supply supports sufficiently fast or with sufficient pressure to prevent the initial formation of small slip regions. However, progressive enlargement of these regions by loosening of the rock under the action of gravity and atmospheric weathering, as depicted in Figure 7.7d, should be prevented by means of adequate support to minimize "overbreak"; in some cases the tunnel might collapse altogether.

Flexible supports, acting passively, would have to carry the weight of extensive slip regions corresponding to low values of friction representative of loosened rock masses. At the other extreme, prestressed supports could be designed to maintain the stress tangential to at least part of the opening by preventing interlayer slip. At a point like A in Figure 7.8, the radial support pressure p_b required to do this can be calculated from Bray's formula, given in Problem 10 of chapter 6:

$$p_b = (N_1 p_1 + N_2 p_2) \left\{ \frac{\tan |\psi|}{\tan (|\psi| + \phi_j)} \right\} \tag{7.10}$$

where ψ is the angle between the layers and the surface of the tunnel

p_1 and p_2 are the larger and smaller initial stresses in the plane of the cross section of the opening

N_1 and N_2 are the tangential stress concentrations at point A, meaning that before any supports are installed

$$\sigma_{\theta,A} = N_1 p_1 + N_2 p_2$$

and ϕ_j is the angle of friction between the layers of rock.

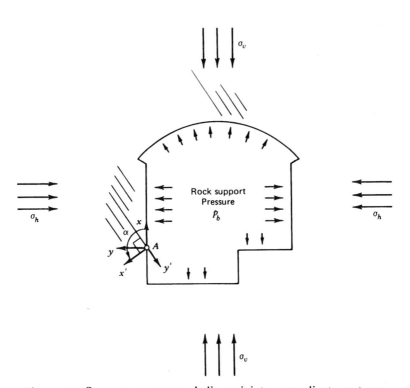

Figure 7.8 Support pressure and slip on joints—coordinate systems.

For a given value of ϕ_j at a set of points all around the surface of a tunnel, Equation 7.10 might be used to calculate the support pressure required theoretically to prevent any slip. N_1 and N_2 for each of these points can be determined from physical or numerical models or, for tunnels of regular shape, from closed form solutions. In the case of circular tunnels, N_1 and N_2 are determined by the second of Equations 7.1 with $r = a$, giving $N_1 = 1 - 2 \cos 2\theta$ and $N_2 = 1 + 2 \cos 2\theta$ (θ being measured from the line of action of p_1 as in Problem 7.12). Equation 7.10 can also be used to investigate the progressive loosening of rock by comparing results with various values of ϕ_j.

Deterioration of the rock around a tunnel, following slip on discontinuities, may result from the fallout or inward movement of rock wedges, driven by gravity or by initial stresses ("ground pressure"). Cording and Deere (1972) and Cording and Mahar (1974) relate experiences with wedge falls in tunnels and compare the results with Terzaghi's empirical formulas (Terzaghi, 1946).

When a tunnel is driven in weak rock, or at considerably depth, it may suffer failure of the wall rock itself along new fracture surfaces, causing progressive closure. This is considered in Section 7.5.

7.5 Plastic Behavior around Tunnels

When the tangential stress around an opening is greater than about one-half the unconfined compressive strength, cracks will begin to form. There is usually some rock breakage due to construction and a zone of relaxation around the skin of the opening but the new cracks are conspicuous in forming slabs parallel to the periphery. At great depth, such rock failure can cause violent "bursts."

Weak rocks like shale reach the condition for rock cracking at small depths. For example, a shale with compressive strength equal to 500 psi and with K equal to 2 will have sufficient stress for new cracking around a circular tunnel at a depth of only 50 ft. In such rocks, moreover, new cracking may initiate further loosening as water and air cause accelerated weathering. The gradual destruction of rock strength drives the zone of broken rock deeper into the walls, creating loads on the tunnel support system that may close the whole tunnel. Commonly, the supports experience a gradual buildup in pressure known as "squeeze." The severity of squeezing is related to the ratio of initial stress to unconfined compressive strength, and the durability of the rock.

As shown in Figure 7.9, two types of behavior might be experienced in squeezing ground. In rock that tends to arch and in which the supports are able to provide sufficient loads when needed to halt progressive deterioration of the tunnel, the inward displacement of the walls will decrease with time and approach an asymptote. If the supports are erected too late, or if the rock supplies a rock load too large for them to withstand, the displacements will accelerate

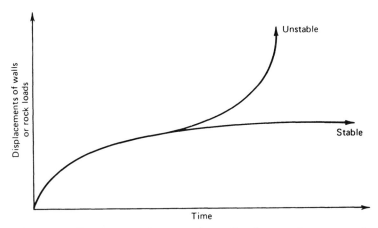

Figure 7.9 Convergence between the walls of a tunnel corresponding to stability and instability.

after some time; without proper engineering response, the tunnel will fail. In such rocks, it is essential to measure the displacements of the rock periphery at frequent stations and to plot the data promptly.

To gain a better understanding of the mechanics of a squeezing tunnel, and to provide an analytical framework for selecting appropriate support systems, we will consider a theoretical model proposed by John Bray (1967).[2] Assume that the construction of the tunnel creates intolerable stress conditions that result in failure of the rock according to the Mohr-Coulomb theory. To permit the analysis of the extent of failure—the "plastic zone"—the simplifying assumption is made that the state of stress is axisymmetric, that is, $K = 1$. Within

---

[2] The theory for *stresses* assuming plastic behavior according to the Coulomb equation has been treated for circular tunnels by a number of authors, including (1) H. Labasse (1948, 1949) Les pressions de terrains . . . , *Revue Universelle des Mines, Series 9,* Vols. V and VI; (2) H. Kastner (1949), Über den echten Gebirgsdruck beim Bau tiefliegender Tunnel *Österreich Bauzeitschrift,* Vol. 10, No. 11; (3) J. A. Talobre (1957), *La méchanique des Roches* (Dunod); (4) T. A. Lang (1962). Notes on rock mechanics and engineering for rock construction (unpublished); (5) N. Ikeda, T. Tanaka, and I. Higuchi (1966), The loosening of the rock around the tunnel and its effect on steel support, *Qtly. Report RTRI,* Vol. 7, No. 4; (6) John Bray (1967), A study of jointed and fractured rock—part II, *Felsmechanik und Ingenieurgeologie,* Vol. V, No. 4; (7) N. Newmark (1969), Design of rock silo and rock cavity linings, *Tech. Report,* Contract 155, Air Force Systems Command, Norton Air Force Base; (8) A. J. Hendron and A. K. Aiyer (1972), Stresses and strains around a cylindrical tunnel in an elasto-plastic material with dilatancy, *Corps of Engineers Omaha, Technical Report No. 10;* (9) Ladanyi, B. (1974) Use of the long term strength concept in the determination of ground pressure on tunnel linings, *Proc. 3rd Cong. ISRM* (Denver), Vol. 2B, pp. 1150–1156. Solutions for plastic *displacements* were pioneered by Bray, Newmark, Hendron and Aiyer, and Ladanyi.

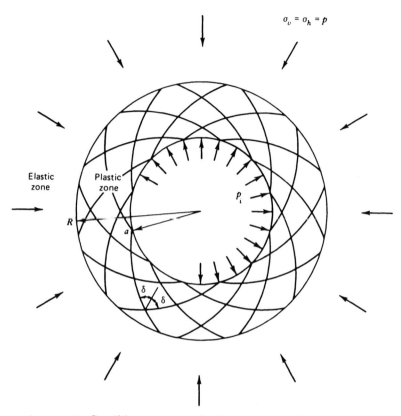

Figure 7.10 Conditions assumed for Bray's elastic-plastic solution.

the plastic zone, which extends to radius R, Bray assumed the fractures were log spirals inclined at δ degrees with the radial direction, as predicted by a strict application of the Mohr-Coulomb theory (see Figure 7.10). This is not appropriate for analysis of many rocks in which, as noted, the cracks will form slabs parallel to the walls and floor ("ring cracks"). In shales and clays, however, Bray's assumption of log spirals is considered to be acceptable. For minimum strength, the appropriate value of δ is $45 + \phi/2$ but the quantity δ will be left as an independent parameter of the solution. It proves useful to define a quantity Q given by

$$Q = \frac{\tan \delta}{\tan(\delta - \phi_j)} - 1 \tag{7.11}$$

Assuming that the broken rock inside the plastic zone contains log spiral surfaces with shear strength characteristics $\tau_p = S_j + \sigma \tan \phi_j$, the radius R of the plastic-elastic zone boundary is given by

$$R = a \left(\frac{2p - q_u + [1 + \tan^2(45 + \phi/2)]S_j \cot \phi_j}{[1 + \tan^2(45 + \phi/2)](p_i + S_j \cot \phi_j)} \right)^{1/Q} \tag{7.12}$$

where p is the initial rock stress ($\sigma_v = \sigma_h = p$), q_u is the unconfined compressive strength of the intact rock, p_i is the internal pressure in the tunnel provided by the supports, and ϕ is the angle of internal friction of the intact rock.

Within the elastic zone, Bray's solution determines the radial and tangential stresses as

$$\sigma_r = p - \frac{b}{r^2}$$

and

$$\sigma_\theta = p + \frac{b}{r^2}$$

where

$$b = \left(\frac{[\tan^2(45 + \phi/2) - 1]p + q_u}{\tan^2(45 + \phi/2) + 1} \right) R^2 \tag{7.13}$$

The radial and tangential stresses in the plastic zone are defined by

$$\sigma_r = (p_i + S_j \cot \phi_j)\left(\frac{r}{a}\right)^Q - S_j \cot \phi_j$$

and

$$\sigma_\theta = (p_i + S_j \cot \phi_j) \frac{\tan \delta}{\tan(\delta - \phi_j)} \left(\frac{r}{a}\right)^Q - S_j \cot \phi_j$$

$$(7.14)$$

The displacements are important also for they provide an observational framework for the engineer. The radially inward displacement u_r is given by

$$u_r = \frac{1 - \nu}{E} \left(p_i \frac{r^{Q+1}}{a^Q} - pr \right) + \frac{t}{r} \tag{7.15}$$

where

$$t = \frac{1 - \nu}{E} R^2 \left[(p + S_j \cot \phi_j) - (p_i + S_j \cot \phi_j)\left(\frac{R}{a}\right)^Q \right] + \frac{1 + \nu}{E} b \tag{7.16}$$

(b was given in Equation 7.13.)

For example, consider a case with the following properties: the fractures are described by $\phi_j = 30°$, $S_j = 0$, $\delta = 45°$; the rock has properties $q_u = 1300$ psi, and $\phi = 39.9°$; the initial stress $p = 4000$ psi, and the support pressure $p_i = 40$ psi. Then $Q = 2.73$, $R = 3.47a$, and $b = 33{,}732a^2$.

The stresses (psi) in the plastic zone are defined by

$$\sigma_r = 40 \left(\frac{r}{a}\right)^{2.73}$$

and

$$\sigma_\theta = 149 \left(\frac{r}{a}\right)^{2.73}$$

while the stresses (psi) in the elastic zone are

$$\sigma_r = 4000 - 33{,}732a^2/r^2$$
$$\sigma_\theta = 4000 + 33{,}732a^2/r^2$$

Figure 7.11 shows how the stresses vary around the tunnel in this example and for a case in which the material is elastic everywhere (Kirsch solution). In the plastic case, this stress difference is intolerable and the tangential stress has relaxed to the maximum value consistent with the strength of the material, 149 psi. For some distance behind the tunnel wall, the tangential stresses are lower than those predicted by elastic theory; thereafter, they are higher. The zone of relatively highly stressed rock behind the tunnel wall can sometimes be detected by seismic refraction measurements along the wall of the tunnel.

The plastic behavior of the region near the tunnel has the effect of extending the influence of the tunnel considerably farther into the surrounding rock. In the wholly elastic case, the tangential stresses would have fallen to only 10% above the initial stresses at a radius of 3.5 times the tunnel radius; in the elastic-plastic case considered in Figure 7.11, the elastic zone stresses are 70% greater than the initial stresses at this distance and 10 radii are required before the stress perturbation of the tunnel has fallen to 10%. Thus, two tunnels that do not interact with one another in elastic ground might interact in plastic ground.

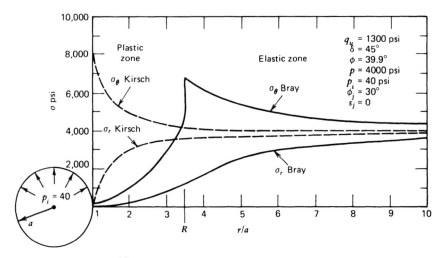

Figure 7.11 Stresses around the yielding tunnel in the example.

Displacements for the example can be discussed if values are given for E, ν, the elastic properties, and tunnel radius. Let $E = 10^7$ psi, $\nu = 0.2$, and $a = 96$ in. Then $t = 62.21$ and $u = 0.62$ in. This displacement is measured at the tunnel wall while the elastic displacement at the elastic-plastic boundary is 0.11 in. In the corresponding wholly elastic problem, the elastic displacement at the tunnel wall would be 0.046 in. These values of displacement are sensitive to the value of p_i. Had the internal pressure p_i been initially installed at 5 psi, instead of 40 psi, R would have been $7.44a$ and u_r would have been 1.55 in. Corresponding to p_i equal to 1 psi, R would have been $13.42a$ giving u_r equal to 4.26 in. Such a large value for R indicates that a substantial volume of rock exists in a loosened state around the tunnel. It would be prudent to assume that this material could continue to move under the influence of gravity and to allot an additional increment of reserve support capacity to hold it in place. As a simplification, we can add a support pressure equal to a portion of the weight of the overlying material within the loosened zone $c\gamma(R - a)$, where c is a constant ≤ 1. Then the total pressure that must be supplied by the supports is

$$p_{i,\text{total}} = p_i + \gamma(R - a)c \qquad (7.17)$$

Gravitational load would be felt at the roof and also at the sides due to Poisson effect. The load increment due to loosening of the rock (the "loosening pressure") will increase inversely with p_i because R/a increases inversely with p_i (Figure 7.12a). As a result, the total support pressure will exhibit a minimum when plotted against the initially installed support pressure (Figure 7.12b). Since the displacements also increase with R/a, the curve of p_i versus displacement will likewise display a minimum (Figure 7.12c) at a value of $u_r = u_{\text{crit}}$. The tunnel will be stable if the supports are installed with an initial pressure such that the equilibrium between rock and support is reached at a value of $u_r < u_{\text{crit}}$; any additional displacement of the tunnel wall lessens the required support pressure. Should the installation of supports be delayed, or should the supports used be too flexible, an initial equilibrium between rock pressure and support pressure might be reached at a value of the radial displacement greater than u_{crit}; this situation is less satisfactory since any additional increment of displacement would demand additional support pressure. As shown in Figure 7.12c, if the curve of $p_{i,\text{total}}$ versus u_r rises more steeply than that corresponding to the stiffness of the support, the tunnel will fail.

Bray's solution can be used to construct an estimate of a design curve like that of Figure 7.12c. For the case considered, assuming $c = 1$,[3] the three values of initially designed support pressure of 40, 5, and 1 psi produce final support pressures of 64.6, 58.7, and 104.5 psi, respectively, as shown in Table 7.3. If the initial support pressure were greater than 40 psi, and installed quickly with a reserve load-carrying capacity, the supports would begin to acquire additional

---

[3] This is probably an overconservative assumption.

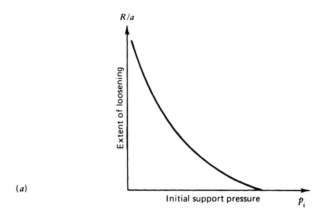

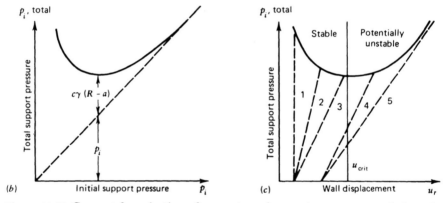

Figure 7.12 Concept for selection of supports and procedures: supports 1, 2, and 3 are safe; support 4 is potentially unsafe; support 5 is unsafe.

Table 7.3[a]

p_i (psi)	u_r (in.)	R/a	$R - a$ (ft)	$c\gamma(R - a)/144$ (psi)	$p_{i,total}$ (psi)
40	0.62	3.47	19.8	20.6	64.6
5	1.55	7.44	51.5	53.7	58.7
1	4.26	13.42	99.4	103.5	104.5

[a] $a = 8$ ft, $\gamma = 150$ lb/ft^3, $c = 1$.

Table 7.4 *Typical Support Pressures*[a]

Types of Support	Range of p_i	Delay Time until p_i Is Effective
Rock bolts	0–50 psi	Several hours
Shotcrete, 2–8″ thick	50–200 psi	Several hours
Steel sets	0–400 psi	One day to weeks
Concrete lining	100–800 psi	Several weeks to months
Steel lining	500–3000 psi	Months

[a] Load of set depends on manner of blocking and lagging.

load after installation and if they were sufficiently stiff the tunnel would come to equilibrium. In the case of a support pressure installed too slowly, or one that was too flexible, the ground would loosen into the potentially unstable region of Figure 7.12c and the tunnel might in time collapse. All this supposes that the properties ϕ_j and S_j in the plastic zone are unaffected by loosening. If there is a clay component in the rocks, weathering by slaking or swelling is possible, bringing ϕ_j to lower values. (This can be input in the analysis.) Available supports for tunnels provide a range in support capacities and stiffnesses as shown in Table 7.4 (see, in ch. 1 references, Hoek and Brown, 1980).

7.6. Use of the Geomechanics Classification

It does not require analysis to foresee that an insufficiently supported tunnel in imperfect rock will eventually cave. Knowing how long the workers can stay in the unsupported face region of an advancing tunnel permits the engineer and contractor to select an appropriate style of support and optimal lengths of drilling rounds. Although there is, as yet, no satisfactory wholly rational method for evaluating the time to failure—the "stand-up time"—of an unsupported rock span, a number of engineers have assisted engineering judgment in this matter through correlations with classifications of the rock. Lauffer[4] recognized that the stand-up time depends on both the rock condition and the "active span"; the latter is defined as the minimum of the tunnel width and the unsupported length at the face. Lauffer's chart relating log of active span to log of stand-up time was calibrated by Bieniawski (1976, 1984) in terms of the Geomechanics Classification introduced in Chapter 2, shown in Figure 7.13. The lower and upper curves delimit the range of prediction for the active span length that should fail after the given time. Contours of rock mass rating be-

[4] H. Lauffer (1958), Gebirgsklassiferung für den Stollenbau, *Geologie und Bauwesen* **24:** 46–51.

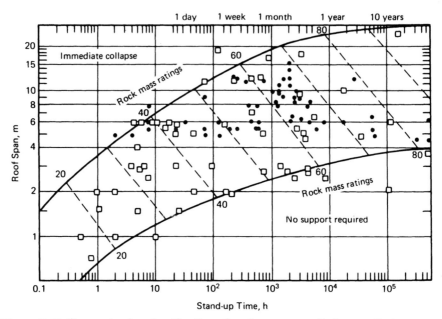

Figure 7.13 Geomechanics classification of rock masses applied to predicting tunneling conditions. The roof span is the length of the unsupported section at the face, or the width of the tunnel if greater. Reproduced from Bienawski (1984) *Rock Mechanics Design in Mining and Tunneling* (Balkema), with permission. The points denote case histories of roof falls; circles are for mines, squares are for tunnels.

tween these limiting curves separate fields. Thus, given the rock mass rating and the active span, Figure 7.13 offers a prediction of the time to failure. The chart is based on the set of points summarizing South African practice, which is somewhat less conservative than Austrian experience.

7.7 Time-Dependent Behavior of Tunnels

"Squeezing," we have noted, refers to the slow accumulation of load on supports. Nothing in the arguments concerning loosening of layered rocks or plastic behavior around tunnels introduces time explicitly; yet it is not hard to imagine that localized failure, crack propagation, and stress redistribution could occur progressively, manifesting their full effect only days or even weeks after excavation. Other phenomena may also cause tunnels to squeeze. Rock loads may change due to additional excavation. Effective stresses may increase due to drainage of water into the tunnel, or conversely they may decrease due to a rise in the groundwater table following the initial drawdown on tunnel driving. Rock may hydrate ("swell"), oxidize, or disintegrate or otherwise weather in response to the change in humidity and temperature consequent on excavation. And then, the rock may be genuinely viscous or viscoelastic in its stress-strain response; as noted in Section 6.5, bituminous, argillaceous, and saline deposits may creep at relatively low differential stress.

If the viscoelastic response is linear, we might be able to make satisfactory predictions of tunnel displacement rates using linear viscoelastic theory. However, when time dependency rises from changes in the *geometry* of the rock mass due to crack growth, as for Indiana limestone discussed in Chapter 6, the viscoelastic properties depend on stress and the response is nonlinear. Then, since there will be different stress states in each element of rock around the tunnel, the mass becomes nonhomogeneous, and simple solutions based on homogeneity may be misleading. For first approximations and for back figuring observed displacement, simple linear viscoelastic models may still be helpful, even for nonlinearly viscoelastic materials. In this spirit, we can consider several idealized systems.

AN UNLINED CIRCULAR TUNNEL
IN A BIAXIAL STRESS FIELD

Assume the tunnel is in a state-of-plane strain, as for the Kirsch solution to the elastic problem (Equation 7.1) and that it has been excavated in a Burgers material in distortion, which behaves elastically in hydrostatic compression (see Section 6.5). The secondary principal stresses in the plane perpendicular to the tunnel are p_1 and p_2. If the tunnel is unlined so that the boundary conditions on the wall impose zero or constant pressures, Equations 7.1 still hold for the stresses in the viscoelastic material. However, the material creeps and the strains and displacements will change with time. In place of Equation 7.2, the radial displacement u_r of a point at coordinates r, θ (Figure 7.2a) is described by

$$u_r(t) = \left(A - C + B\frac{d_2}{d_4}\right)\frac{m}{q} + \left(\frac{B(d_2/G_1 - d_1)}{G_1 d_3 - d_4} - \frac{A - C}{G_1}\right)e^{-(G_1 t/\eta_1)}$$

$$+ B\left(\frac{d_2(1 - m/\alpha) + d_1(m - \alpha)}{G_2(G_1 d_3 - d_4)}\right)e^{-(\alpha t/\eta_1)} + \frac{A - C + B/2}{\eta_2}t \quad (7.18)$$

where

$$A = \frac{p_1 + p_2}{4}\frac{a^2}{r}$$

$$B = (p_1 - p_2)\frac{a^2}{r}\cos 2\theta$$

$$C = \frac{p_1 - p_2}{4}\frac{a^4}{r^3}\cos 2\theta$$

$$m = G_1 + G_2 \qquad d_3 = 6K + 2G_2$$
$$q = G_1 G_2 \qquad d_4 = 6Km + 2q$$
$$d_1 = 3K + 4G_2 \qquad \alpha = \frac{3Km + q}{3K + G_2}$$
$$d_2 = 3Km + 4q$$

A useful special case of (7.18) corresponds to an incompressible rock mass ($v = \frac{1}{2}$). Then

$$u_r(t) = \left[A + B\left(\frac{1}{2} - \frac{a^2}{4r^2} \right) \right] \left(\frac{1}{G_2} + \frac{1}{G_1} - \frac{1}{G_1} e^{-(G_1 t/\eta_1)} + \frac{t}{\eta_2} \right) \qquad (7.19)$$

To illustrate the above equations, consider a circular tunnel 30 ft in diameter at a depth of 500 ft in rock salt having the following properties:

$$K = 0.8 \times 10^6 \text{ psi} \qquad \text{(bulk modulus)}$$
$$G_1 = 0.3 \times 10^6 \text{ psi}$$
$$G_2 = 1.0 \times 10^6 \text{ psi}$$
$$\eta_1 = 7.0 \times 10^8 \text{ psi min}$$
$$\eta_2 = 8.3 \times 10^{10} \text{ psi min}$$
$$\gamma_{\text{wet}} = 140 \text{ lb/ft}^3$$

We assume that the horizontal stress is twice the vertical stress (an assumption actually inconsistent with the choice of $\eta_2 < \infty$) so that $p_2 = 468$ psi (vertical) and $p_1 = 927$ psi. The displacements of points on the wall of the tunnel are plotted against time in Figure 7.14. There is a small elastic, instantaneous displacement, then a delayed elastic response tapering off after about 4 days, followed by slow, secondary creep. If the material around the tunnel can deform without rupture, secondary creep might continue for a long time. Otherwise, the strain will reach a sufficient magnitude to cause local rupture, changing the state of stress by the development of a plastic zone. If rock bolts are installed, the displacements of the walls will be only slightly reduced. This can be calculated approximately by superimposing on Equation 7.18 the displacements predicted by Equation 6.33, with a^2/r in place of a and the rock bolt pressure p_b in place of p assuming the bolts are long). Near the tunnel the secondary creep rate would then be reduced to

$$\dot{u}_r = \frac{(A - C + B/2) - (p_b/2)(a^2/r)}{\eta_2}$$

A structural lining would act quite differently.

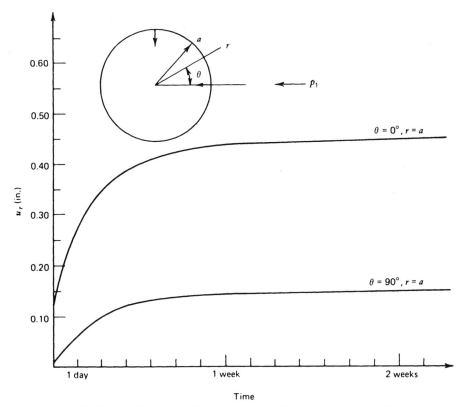

Figure 7.14 Radial displacement due to creep on the wall of the tunnel in the hypothetical example.

A LINED CIRCULAR TUNNEL IN A HYDROSTATIC STRESS FIELD

When a stiff lining is cast against the rock so that it remains in contact with the rock surface as the tunnel rock deforms, a displacement boundary condition is imposed. Now the Kirsch solution stress field no longer applies. If the rock behaves as a Burgers body, in time the pressure will build up on the lining while the stress difference in the rock declines. The final pressure on the lining can be approximated by assuming it to be loaded as a thick-walled cylinder by a uniform outside pressure equal to the initial stress in the rock. According to the values of η_1 and η_2, this may require years, or tens of years.

Let us consider the special case of a stiff elastic lining inside an incompressible viscoelastic material. Gnirk and Johnson (1964) discussed this problem for a Burgers material. For the present purposes, since the lining will not usually be constructed until the instantaneous elastic displacement has already occurred, it is sufficient to consider the rock to be an incompressible generalized Maxwell body (Figure 6.17c). Let η_1, η_2, and G_1 describe the time-dependent properties of the rock and ν' and G' represent the elastic properties of the lining. The rock tunnel has radius b while the lining has internal radius a. The pressure $p_b(t)$ that develops at the lining/rock interface is given by

$$p_b(t) = p_0(1 + Ce^{r_1 t} + De^{r_2 t})$$

(7.20)

where p_0 is the initial stress in the rock ($\sigma_1 = \sigma_2 = p_0$)

$$C = \frac{\eta_2}{G_1} r_2 \left(\frac{r_1(1 + \eta_1/\eta_2) + G_1/\eta_2}{(r_1 - r_2)} \right)$$

(7.20a)

$$D = \frac{\eta_2}{G_1} r_1 \left(\frac{r_2(1 + \eta_1/\eta_2) + G_1/\eta_2}{(r_2 - r_1)} \right)$$

(7.20b)

and r_1, r_2 are the real roots of

$$\eta_1 B s^2 + \left[G_1 B + \left(1 + \frac{\eta_2}{\eta_2'} \right) \right] s + \frac{G_1}{\eta_2} = 0$$

(7.20c)

in which

$$B = \frac{1}{G'} \left(\frac{(1 - 2\nu')b^2 + a^2}{b^2 - a^2} \right)$$

(7.20d)

The stresses and displacements in the lining ($a \leq r \leq b$) are

$$\sigma_r = p_b \frac{b^2}{b^2 - a^2} \left(1 - \frac{a^2}{r^2} \right)$$

(7.21a)

$$\sigma_\theta = p_b \frac{b^2}{b^2 - a^2} \left(1 + \frac{a^2}{r^2} \right)$$

(7.21b)

and

$$u_r = - \frac{b^2 r p_b (1 - 2\nu' + a^2/r^2)}{2G'(b^2 - a^2)}$$

(7.21c)

while the stresses and displacements in the rock ($r \geq b$) are

$$\sigma_r = p_0 \left(1 - \frac{b^2}{r^2} \right) + p_b \frac{b^2}{r^2}$$

(7.22a)

$$\sigma_\theta = p_0\left(1 + \frac{b^2}{r^2}\right) - p_b \frac{b^2}{r^2} \qquad (7.22b)$$

and

$$u_r = -\frac{b^2}{r} p_b\left(\frac{(1 - 2\nu')b^2 + a^2}{(2G')(b^2 - a^2)}\right) \qquad (7.22c)$$

In Equations 7.21 and 7.22, p_b varies with time according to (7.20).

As an example, suppose a lining 2 ft thick were placed inside a circular tunnel 30 ft in diameter in evaporite rocks with $p_1 = p_2 = p_0 = 1000$ psi. The rock is characterized by $G_1 = 0.5 \times 10^5$ psi, $G_2 = 0.5 \times 10^6$ psi, $\eta_1 = 5 \times 10^{10}$ psi/min, $\eta_2 = 1 \times 10^{13}$ psi/min, and $K = \infty$ ($\nu = \frac{1}{2}$). The elastic constants of the concrete are $\nu' = 0.2$ and $E' = 2.4 \times 10^6$ psi, giving $G' = 1 \times 10^6$ psi. Substituting $G_2 = 0.5 \times 10^6$ psi in Equation 7.2 with $p_1 = p_2 = 1000$ psi determines the instantaneous elastic displacement of the unlined tunnel to be $u_r = 0.18$ in. Introducing the assigned values of G_1, η_1, and η_2 in Equations 7.20 to 7.22 yields the displacements and stresses listed in Table 7.5. Figure 7.15a shows the time-dependent displacements of the rock surface with and without a lining.

The amount of displacement of the lined tunnel is relatively small: $u_r = 0.44$ *in.* after 10 years. However, since the concrete is stiff, the maximum compressive stress becomes large enough to crush the concrete in about a half a year and theoretically reaches 6365 psi in 10 years (Figure 7.15b). One solution for squeezing tunnels is to use a soft or yielding support system, for example, with crushable wood blocking or porous concrete. If the modulus of elasticity

Table 7.5 *Displacements and Stresses for the Example*

Time (t)	Rock Displacement		Maximum Stress in Concrete (psi)	Stresses on the Rock Surface	
	Unlined (in., total)	After Placement of Lining (in.)		σ_r (psi)	σ_θ (psi)
0	0.180	0	0	0	2000
1 day	0.183	0.003	43	5	1995
1 week	0.198	0.018	293	36	1964
28 days	0.251	0.066	1093	136	1864
56 days	0.320	0.121	1997	248	1751
$\frac{1}{2}$ year	0.598	0.273	4491	559	1441
1 year	0.921	0.353	5799	721	1278
2 years	1.360	0.383	6292	783	1217
10 years	2.018	0.387	6365	792	1208

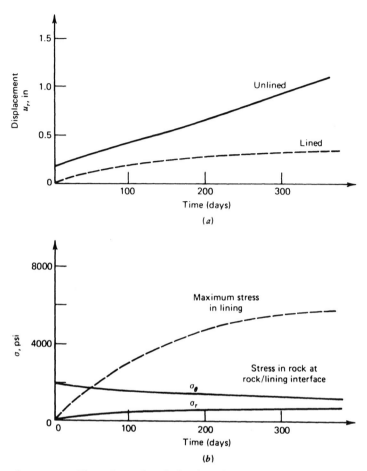

Figure 7.15 Time-dependent behavior of the tunnel in the example. (*a*) Inward movement of the rock wall in the lined and unlined cases. (*b*) Variation of stress with time in the lined tunnel.

of the lining is half the value input in the example, the maximum stress after 10 years is 5268 psi with a displacement of 0.64 in. at the rock-lining interface. An elastic modulus for the lining one-tenth that used in the example reduces the maximum lining stress to 2207 psi with a displacement of 1.34 in. at the rock surface. In comparison the displacement of the unlined tunnel after 10 years is 2.02 in.

7.8 Underground Openings in Blocky Rock—"Block Theory"

Excavations cut into rock masses with several sets of discontinuities may liberate rock blocks of various size. The potential movements of the most critically located of these may then undermine neighbor blocks, and the ensuing block falls and slides can menace the integrity of the engineering scheme. If the excavation is unsupported, block movements may unacceptably alter the excavation perimeter and the blocks may cause property damage and personal injury. If the excavation has been supported, the block movement tendency will transfer loads to the support system, which could fail if they have not been designed specifically to handle these loads. Since the rock itself is usually quite strong, it is mainly the potential block falls and slides that need to be addressed by the designer, and if these are adequately handled, or found to be unlikely, the excavation stability will be assured.

Suppose that a block of rock is isolated by the intersection of discontinuities and excavation surfaces. No matter how many faces it has, the block can move initially in only a few ways: by falling, by sliding on one face, or by sliding on two faces (or by combined sliding and rotation). All of these motions require that certain faces open. Thus, the first warning of block movement is the widening of particular joints. On the other hand, if the potentially dangerous blocks are found prior to movement and their stability is assured, then no block movements will occur anywhere. This is the principle of "block theory" (Goodman and Shi, 1984). The most dangerously located blocks are called "key blocks." The theory establishes procedures for describing and locating key blocks and for establishing their support requirements. By using the procedures it is possible to plan optimum reinforcement schemes, and to select excavation orientations and shapes that minimize or completely eliminate the need for artificial supports.

The shapes and locations of key blocks are fully three dimensional. In some types of engineering analyses, simpler two-dimensional configurations are adequate. For example, a layered mine roof can be analyzed as a beam rather than as a plate; or a potential shear failure can be analyzed as cylindrical rather than spherical. In the case of jointed rock masses, however, two-dimensional analysis would be an unwise simplification. In neglecting the third dimension, the savings realizable through geometry are abandoned, which is particularly foolish since three-dimensional analysis using block theory is quite easy. Three-dimensional block geometry allows one to find safe spatial directions for minimally supported excavations even when some joints are very weak. To simplify explanations, a series of two-dimensional illustrations are examined initially; then three-dimensional analysis is introduced, with extensive use of stereographic projection.

TYPES OF BLOCKS

Figure 7.16 identifies six types of blocks around an excavation. Type VI is a joint block, having no faces on the excavation perimeter, that is, no free faces. Type V has a free face but the block is infinite. Unless there are new cracks formed around the excavation, neither of these block types can be key blocks. The same is true for block IV, which has a tapered shape; there is no direction toward the excavated space in which the tapered block can move without pushing into its neighbors.

All of the other blocks are finite and *removable*. Whether they will move depends not only on geometry but on the direction of the resultant force and the magnitudes of the friction angles on the faces. Block III is safe under the action of gravity. Type II blocks are also safe by virtue of friction. The one in the roof has parallel sides so the block can move only in one direction, namely parallel to these sides; this restriction on the freedom to displace greatly increases the shear resistance on its faces, as discussed by Goodman and Boyle (1986). The type II block in the wall has a flat base so it is unlikely to move if the friction

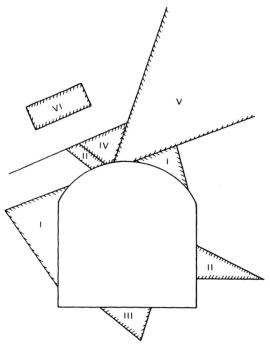

Figure 7.16 Types of blocks: I key blocks; II potential key blocks; III safe removable block; IV tapered block; V infinite block; VI joint block.

angle is any reasonable value (unless water or tractions acting on the faces, or accelerations rotate the direction of the resultant force away from the vertical toward the free space). Type I blocks will probably move unless supported as soon as the excavation succeeds in isolating them as individual blocks. The Type I block in the roof will fall out and the one in the wall will slide. These are the *key blocks*. The type II blocks are *potential key blocks*.

Block theory provides a system for dividing all the blocks into these groupings. The first decision separates the nonremovable blocks (IV, V, and VI) from the removable blocks (I, II, and III) by means of *Shi's theorem*. A "mode analysis," taking into account the direction of the sliding and falling tendencies, given the direction of the resultant force, then distinguishes type III blocks from type II and I blocks. Finally a limit equilibrium analysis, invoking friction on the block faces, establishes the key blocks and determines the support requirements. The basic analyses are dependent on the relative orientations of the joints and not on the specific block perimeters, so the infinity of block shapes that are created by intersecting discontinuities are all represented by a manageable number of analyses.

REMOVABILITY OF BLOCKS—SHI'S THEOREM

A geometric property of finite blocks enables finiteness and removability of blocks to be judged very simply. A finite block, in two dimensions, is shown in Figure 7.17. If the bounding faces are all moved without rotation toward the

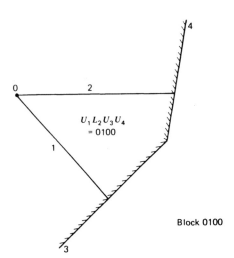

Figure 7.17 A removable block with two joints and two free surfaces (two-dimensional example).

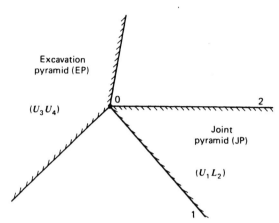

Figure 7.18 A joint pyramid and excavation pyramid for a two-dimensional removable block.

center of the block, the block becomes successively smaller until it shrinks to a single point. This cannot be done with an infinite block. The block in Figure 7.17 consists of the intersection of four half spaces: U_1, the half space above face 1; L_2, the half space below face 2; U_3, the half space above face 3; and U_4, the half space above face 4. The first two faces are formed by joint planes and the last two by free surfaces, that is, by excavation planes. In Figure 7.18, all of these faces have been moved without rotation to pass through a single point O. The intersection U_1L_2, is denoted the *joint pyramid*, and abbreviated JP; the JP is an angle with vertex at O. Similarly, the intersection U_3U_4 is called the *excavation pyramid* and abbreviated EP. This intersection of free half spaces is also an angle at O. Since the block in question is finite, JP and EP have no intersection. *Shi's theorem establishes that a block is finite if and only if JP and EP have no intersection.* In this two-dimensional illustration, the regions in question are angles in a plane. In three dimensions, these regions become pyramids with vertex at the origin.

USE OF STEREOGRAPHIC PROJECTION TO ESTABLISH REMOVABILITY OF BLOCKS

Stereographic projection (Appendix 5) simplifies the discussion of block theory in three dimensions. It reduces by one the dimensions of the geometric feature to be projected; a line passing through the center of the reference sphere projects as a point, while a plane passing through the center of the reference sphere projects as a great circle. Suppose a plane of interest dips α below horizontal in direction β measured clockwise from north. The center of the great circle that projects this plane is at C, whose distance from the center of the reference circle of radius R is given by

$$OC = R \tan \alpha \tag{7.23}$$

while the radius of the great circle is r given by

$$r = R/\cos \alpha \tag{7.24}$$

When the focus of the stereographic projection is at the bottom of the reference sphere, the distance OC is measured in the direction of the dip vector, along azimuth α. In this case, the region inside the reference circle represents all the lines through the center of the reference sphere that are directed into the upper hemisphere. Similarly, *the region inside the circle of radius r about C represents the complete set of lines through the center of the reference sphere that are directed into the upper half space of the plane represented by that circle, that is, plane α/β.*

Figure 7.19 presents an example of the stereographic projection of a joint and its two half spaces. The joint dips 30° to the east ($\alpha = 30°$, $\beta = 90°$). If we

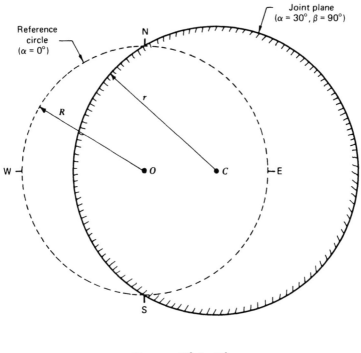

Given: $\alpha = 30°$, $\beta = 90°$
Choose $R = 5$, then:
   $r = 5/\cos 30 = 5.77$
   $Oc = 5 \tan 30 = 2.89$
   (Lower focal point)

Figure 7.19 Construction of a great circle given the dip and dip direction of a plane in the upper hemisphere (lower focal point) stereographic projection.

arbitrarily select $R = 5$, then $OC = 2.89$ and $r = 5.77$. (Changing R varies the dimensions of the drawing but not any of the angular relationships.) Drawing a circle at C with radius r determines the stereographic projection of the inclined plane $\alpha/\beta = 30/90$. The portion of this circle crossing the region inside the reference circle represents the lines in this plane that are directed into the upper hemisphere. The portion of the circle that lies outside the reference circle, similarly, represents the projections of the opposites to these lines, that is, the lines in the plane that are directed into the lower hemisphere. The space inside the circle about C contains all the lines passing through 0 directed into the upper half space of the plane 30/90. And the points outside the circle about C, that is, the rest of the projection plane, represent all the lines passing through 0 that are directed into the lower half space. If the circle about C is the projection of discontinuity set 1, then the region inside this circle is U_1 and the region outside of this circle is L_1.

Now consider a joint system consisting of three joint sets, as given in Figure 7.20. Set 1, bedding joints, are represented by the plane previously drawn, with $\alpha/\beta = 30/90$. Similarly, set 2 is a set of sheared surfaces ("shears") parallel to plane 60/45, and set 3, a true "joint" set, is parallel to plane 20/330. In Figure 7.20, these three planes have been projected, yielding three great circles. Their intersections generate eight spherical triangles. Consider point O, in the center of the reference sphere. This point is simultaneously inside each circle; therefore the line it represents is directed into the upper half space of each of the three joint sets. Let the digits 0 and 1 represent respectively the upper half space and lower half space of a joint and order the digits according to the order of the numbering of the joint sets. Accordingly the spherical triangle of point O has been labeled 000. The point $C2$, on the other hand, is inside only great circle 2 and lies outside great circles 1 and 3. Therefore, the spherical triangle of point $C2$ has been labeled 101.

The eight spherical triangles of Figure 7.20 are the three-dimensional analogs of the angle U_1L_2 in Figure 7.18. They are, in other words, the joint pyramids (JPs) of the system. Shi's theorem determines that a block is removable if and only if it has a JP on the stereographic projection and that JP has no intersection with the excavation pyramid (EP).

APPLICATION TO UNDERGROUND CHAMBERS

Each excavation face and the various excavation edges and corners have particular EPs. Consider, for example, the horizontal roof of an underground chamber. In Figure 7.18, the two-dimensional example, we saw that the excavation pyramid is the angle between the half spaces of the excavation planes that include the block. Any block in the roof of the chamber will lie above the half space of the roof plane. Therefore, in the case of the roof, the excavation pyramid is the half space above the roof. It is therefore the region inside the reference circle.

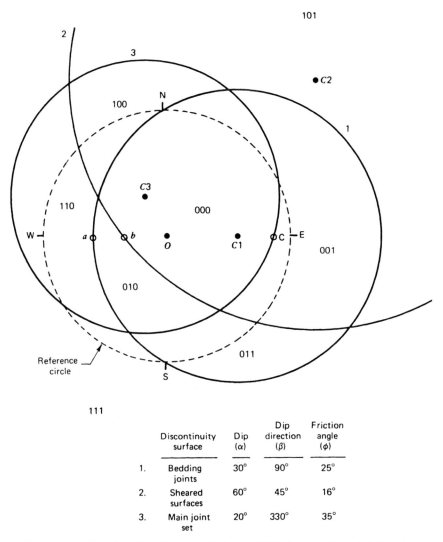

Figure 7.20 Construction of great circles and JP's for the given joint system.

	Discontinuity surface	Dip (α)	Dip direction (β)	Friction angle (ϕ)
1.	Bedding joints	30°	90°	25°
2.	Sheared surfaces	60°	45°	16°
3.	Main joint set	20°	330°	35°

Having identified the JPs and the EP for the roof of an underground cham-ber, we apply Shi's theorem to find the JPs that have no intersection with the EP. A quick search establishes the fact than only JP 101 satisfies this require-ment. (In Figure 7.21, all other JPs have been removed.) This construction proves rigorously that only blocks formed by the intersection of the roof with the lower half spaces of joints 1 and 3 and the upper half space of joint set 2 will be removable from the roof of the chamber.

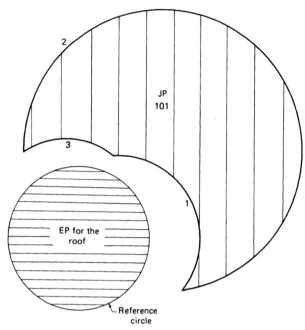

Figure 7.21 The only JP that has no intersection with the EP for the roof.

The next step is to draw the face of the removable block looking up at the roof. It is easier, perhaps, to first draw the free face in a conventional plan view, looking down. In Figure 7.22, the strikes of each of the joint sets have been drawn from the known values of β. The upper half space of an inclined plane is the side of the strike line that contains the dip vector, as marked in Figure 7.22a. Since the removable blocks belong to JP 101, simultaneously in the lower half spaces of joints 1 and 3 and in the upper half space of joint 2, the block must have the free face as drawn in Figure 7.22b. Then rotate about the EW horizontal line and turn north to south to obtain the drawing of the face as it appears looking up at the roof (Figure 7.22c). This drawing may be taken into the field to identify dangerous blocks as they become partly isolated by excavation; such blocks may then be supported before they become completely isolated.

Now consider removable blocks of a vertical wall of an underground chamber. For example, consider the east-west trending south wall of a chamber. Since a vertical plane dips at $\alpha = 90°$, by Equation 7.24 the radius of its great circle is infinite; thus the stereographic projection of a vertical plane is a straight line. In the case of the south wall of an underground opening, the rock is on the south side and the space is on the north side of the wall. Therefore the

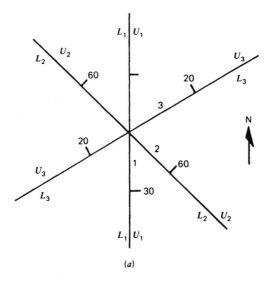

(a)

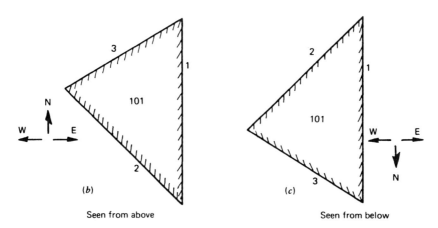

Seen from above   Seen from below

Figure 7.22 Construction for the trace of a convex removable block in the roof plane.

EP of the south wall is the region below the east-west line drawn through O, as shown in Figure 7.23. This EP together with the complete system of JPs of Figure 7.20 will show immediately that only JP 100 can yield removable blocks of this wall. (It can also be verified that only JP 011, the "cousin" of 100, will yield removable blocks in the north wall of the chamber.)

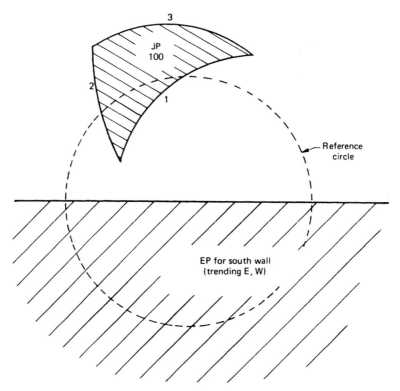

Figure 7.23 The only JP that has no intersection with the EP for the south wall.

To draw the free face of a removable block belonging to JP 100, we need to find the inclinations of the joint sets in the east-west trending vertical wall. Figure 7.24 shows a general procedure for measuring these inclinations from the construction. (The angles could also be read using a stereonet.) The inclinations are the lines represented by points a, b, and c where joint set great circles 1, 2, and 3, respectively, cross the east-west diameter of the reference circle. These are the "apparent dips" of the joint sets in the east-west wall. The traces of these joint planes in the wall are 30° above west, 53° above west, and 9° above east. "Above west" means that the angle of the trace is measured upward from the west end of a horizontal line; we know it is above west if the stereographic projection point is in the west side of the reference circle. Knowing that the critical JP is 100, we can draw the free face of a removable block as shown in Figure 7.25. Figure 7.25a is a drawing of the traces of the joint sets as seen looking north; the upper and lower half spaces of each joint set are marked. Figure 7.25b then determines the free face of a removable block of JP

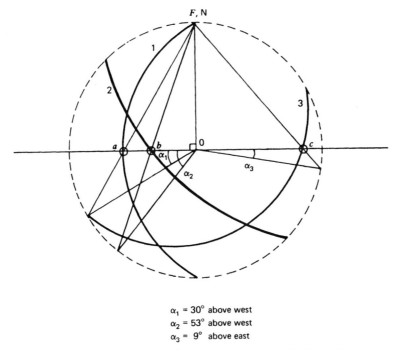

$\alpha_1 = 30°$ above west
$\alpha_2 = 53°$ above west
$\alpha_3 = 9°$ above east

Figure 7.24 Construction for the edges of a removable block in the south wall.

100 from the half space combinations, as seen looking north. Then, by rotating the figure about a vertical line, turning east into west, we can view the south wall from inside the chamber, looking south. This drawing, Figure 7.25c, can be used to identify potentially dangerous blocks as they are approaching isolation by excavation.

Now suppose we are able to view a joint trace map of the south wall, as in Figure 7.26a. The intersection of joint traces creates a large number of polygons. The previous analysis determines which of these are the faces of removable blocks. The free face of a removable block has the shape of the figure drawn in the bottom margin. It is possible to recognize this pattern in the trace map. Figure 7.26b shows the faces of all the removable blocks. If these are supported, nothing else can move and the entire wall must be safe. Formal procedures for finding removable blocks of more complex trace maps with more complex joint systems and with generally inclined excavations are presented by Goodman and Shi (1985). The engineer or geologist making use of these methods has the option of using manual constructions with the stereographic projection, as is done here, or switching on a number of interactive

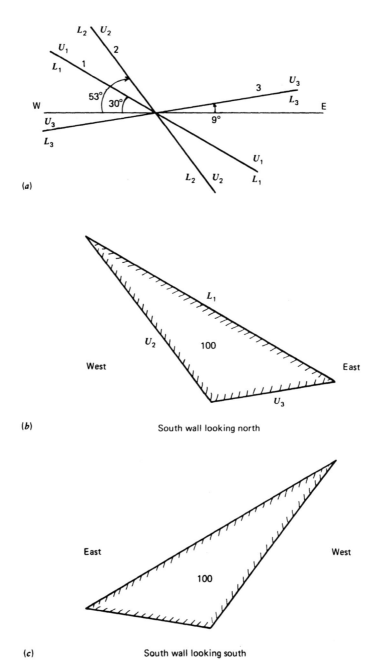

(a)

(b)   South wall looking north

(c)   South wall looking south

Figure 7.25 Construction for the edges of a convex removable block in the south wall.

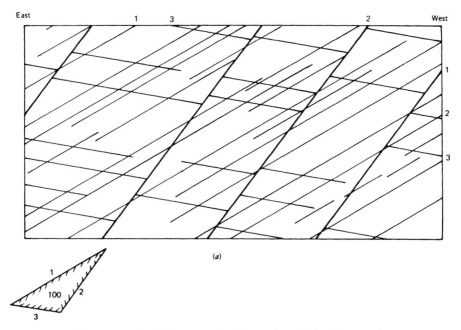

Figure 7.26 (*a*) Joint traces in the south wall, looking south.

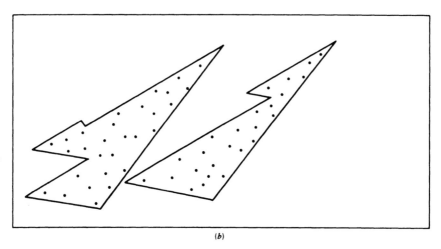

Figure 7.26 (*b*) Removable blocks of the south wall (looking south) determined for JP 100 from Fig. 7.26(*a*).

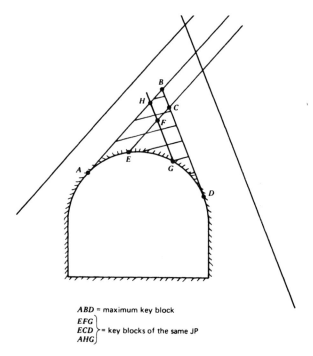

ABD = maximum key block

EFG ⎫
ECD ⎬ = key blocks of the same JP
AHG ⎭

Figure 7.27 The maximum key block of a tunnel corresponding to a given JP.

computer programs.[5] Computer programs have also been developed to draw joint trace maps for simulated rock masses using joint statistics, for both planar and generally curved excavation surfaces.

APPLICATION TO TUNNELS

In the case of a tunnel, the surface of the excavation is the union of a family of planes parallel to the tunnel axis. Therefore, almost every JP can produce a removable block somewhere around the interior of the tunnel. However, these removable blocks are restricted to particular portions of the tunnel surface. For example, consider the joint planes seen in the two-dimensional section in Figure 7.27. If we specify that the blocks must lie simultaneously in the lower half space of each joint plane of this drawing, then no block can be larger than the region *ABD*; real blocks may be smaller. In the absence of information about the spacing and extent of the joints, it would be prudent to design on the basis

[5] Programs for block theory by Gen hua Shi are available from Discontinuous Systems Research, 715 Arlington, Berkeley, CA 94707.

of *the maximum removable block, ABD.* Given the tunnel cross section, each JP (except those that contain the direction of the tunnel axis) has a maximum removable block. We now show how to determine it.

In Figure 7.28*a*, one particular JP, 010, is shaded. The tunnel is horizontal, bearing N 21° E; the projection of its axis is point *a*, or its opposite, -*a*. The corners of the JP are intersections of joint sets and each has an orthographic projection in the vertical plane perpendicular to *a*. These orthographic projections of the JP corners produce three points along the tunnel section, two of which are the projections of the extreme limits of the maximum removable block for JP 010 as seen in the tunnel cross section. To determine the inclinations of the extreme limits of the maximum removable block we construct two great circles through *a* and -*a* and a corner of the JP such that the JP is entirely enveloped. The traces of the limit planes thus constructed are represented by the points where they cross the plane of the tunnel section. The inclinations of these traces may be measured as shown in Figure 7.28*a*, making use of a property of stereographic projection, or they may be found using the stereonet. If a corner lies outside the reference circle, its opposite will lie within it. The

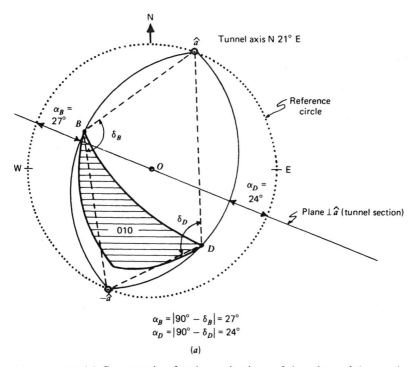

$$\alpha_B = |90° - \delta_B| = 27°$$
$$\alpha_D = |90° - \delta_D| = 24°$$

(*a*)

Figure 7.28 (*a*) Construction for the projections of the edges of the maximum removable block in the tunnel section corresponding to JP 010.

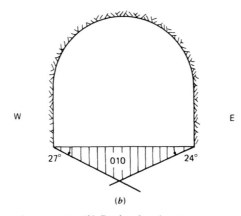

(*b*)

Figure 7.28 (*b*) Projection in the tunnel
section of the maximum removable block
corresponding to JP 010.

limit plane must pass through both the corner and its opposite. If the corner has
distance x from the center of the reference circle, its opposite has distance R^2/x
from the center in the opposite direction (where R is the radius of the reference
circle). In the case of Figure 7.28*a*, the JP lies *inside* each of the enveloping
great circles, so the JP is *above* both of them. The angles of inclination of the
orthographic projections of the corners in the tunnel section are transferred to
the tunnel section in Figure 7.28*b*, and the removable block is found as the
region simultaneously above each. In this case, the maximum removable block
is not a key block if gravity is the main contributor to the resultant force
direction in the block.

Figure 7.29*a* shows a similar construction but for JP 001. The corner I_{12} of
this JP is off the paper so its opposite $-I_{12}$ has been drawn. (The opposite is the
upper hemisphere line that is perpendicular to both of the normals n_1 and n_2 and
therefore perpendicular to the great circle connecting normals n_1 and n_2.) Great
circles through each corner in turn intersect the tunnel section at the points
shown and the limiting great circles are established as those through I_{23} and I_{13},
intersecting the tunnel section respectively 24° above east and 83° above west.
The JP is outside the first of these and inside the second so JP 001 is below the
first and above the second. Figure 7.29*b* transfers this information to the tunnel
section.

Having performed an analysis of maximum key blocks for two JPs, we now proceed to find all the rest. Since JPs 000 and 111 contain the tunnel axis, they have no maximum key block areas in the tunnel section. The maximum key block regions of all the other JPs are shown on Figure 7.30, where each tunnel section is drawn in the curved polygon corresponding to the JP in question. We

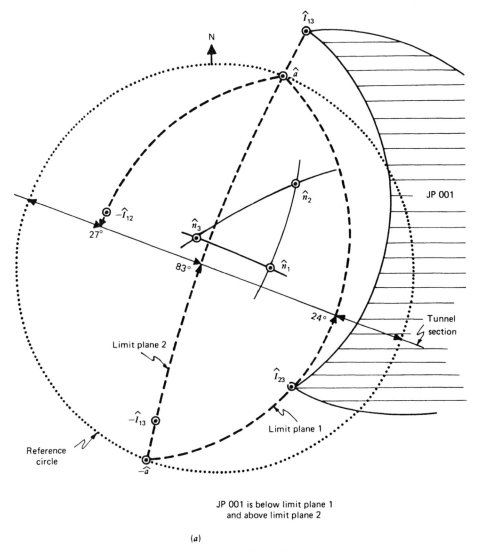

JP 001 is below limit plane 1
and above limit plane 2

(*a*)

Figure 7.29 (*a*) Construction for the projections of the edges of the maximum removable block in the tunnel section, corresponding to JP 001.

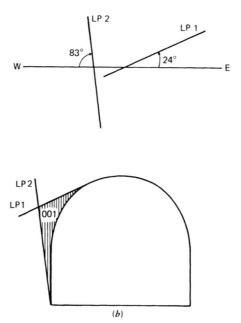

Figure 7.29 (*b*) Projection in the tunnel
section of the maximum removable block
corresponding to JP 001.

see at a glance that under gravity loading JPs 101, 100, and 001 may generate
blocks that require support. Although these drawings are two dimensional, the
three-dimensional maximum keyblocks can be drawn corresponding to each of
these joint pyramids and subjected to a limit equilibrium analysis. The direction
of sliding under gravity is determined by a mode analysis, based upon the
direction of the resultant force and the description of the joint pyramid, as
discussed by Goodman and Shi (1985). Then when friction angles are input on
each face, the support force vector can be computed for each maximum key
block using methods of analysis discussed in the next chapter. It may be that
none of the blocks require support; on the other hand, in certain tunnel directions, the same set of joint planes and friction angles may create the need for a
large support force. This is largely due to the changing size of the maximum key
blocks as the direction of the tunnel is varied.

CASE HISTORY—"TUNNEL SUPPORT SPECTRUM"

The effect of tunnel direction on rock mass support requirements will be illustrated for the rock mass of Figure 7.20 by studying the support force for

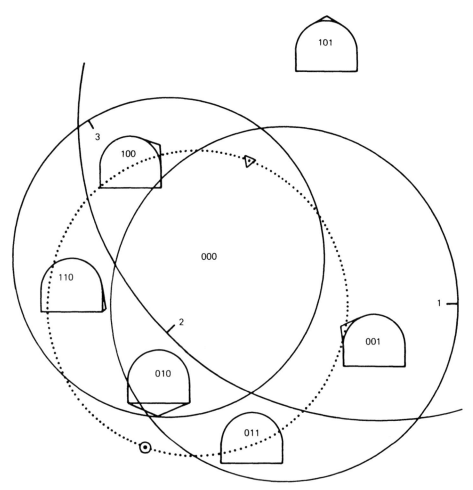

Figure 7.30 Tunnel sections showing maximum removable blocks for each JP superimposed on the stereographic projection of the JPs.

equilibrium of the most critical block as a function of a complete sweep of tunnel directions. All the examples considered thus far correspond to a horizontal tunnel axis in direction N 21° E. Considering only horizontal tunnels, for the time being, we can observe the relative size of the most critical key block, for five tunnel directions, in Figure 7.31. The maximum key blocks become dramatically larger as the direction 315° (N 45° W) is approached. Limit equilibrium analysis of this system, for a horseshoe tunnel 6 m wide and 5.4 m high, yields required support forces for the maximum key block varying from a low of less than 1 metric ton per meter length of tunnel, to a maximum of 33 metric

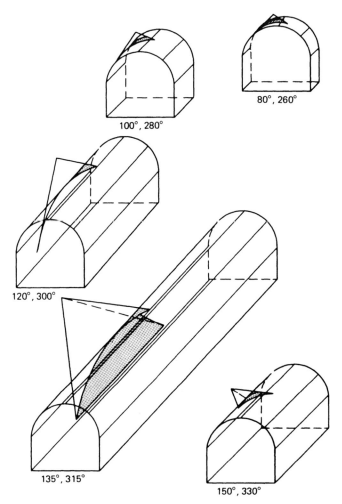

Figure 7.31 The maximum key blocks of the most critical joint pyramids, taking into account limit equilibrium analyses, for tunnels in different directions (indicated by the tunnel azimuth values given).

tons per meter of tunnel, as plotted in Figure 7.32. The sharpness of the peak in this diagram suggests the name "tunnel support spectrum" for this type of presentation.

A microcomputer program was used to perform a complete analysis of key block support requirements over the complete set of tunnel directions in space. The input to this program consists of the attitudes of the sets of joint planes and

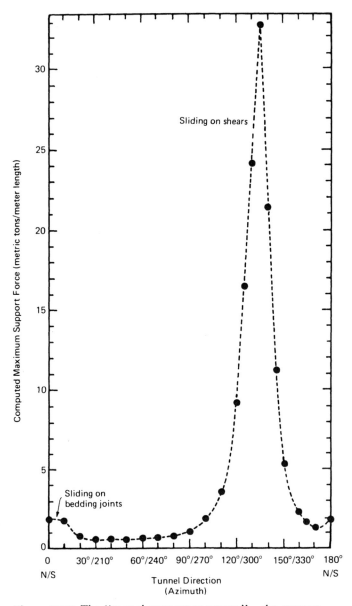

Figure 7.32 The "tunnel support spectrum"—the support force, directed opposite to the sliding direction, to reach limiting equilibrium for the most critical removable block for horizontal tunnels in all directions—for a horseshoe tunnel 6m wide by 5.4 m high.

their friction angles as well as the shape and dimensions of the tunnel section. The output is a contoured "equal area projection" of support force requirements for all tunnel/shaft directions in space, and a list of all sliding modes and sliding force directions. The equal area projection is a distortion of stereographic projection that yields equal areas for "squares" of the stereonet no matter where they project.

Figure 7.33 shows such a diagram for the problem being discussed. It demonstrates that the worst tunnel orientations are horizontal to azimuth 315° (N 45° W), and 23° above horizontal to azimuth 305° (N 55° W). Most tunnel/ shaft directions that are not near these orientations have very much smaller support needs.

Another example is presented in Figure 7.34, where a fourth joint set has been added to the three previously considered; the additional joint set has dip and dip direction equal to 75 and 190°, respectively, and has been assigned a

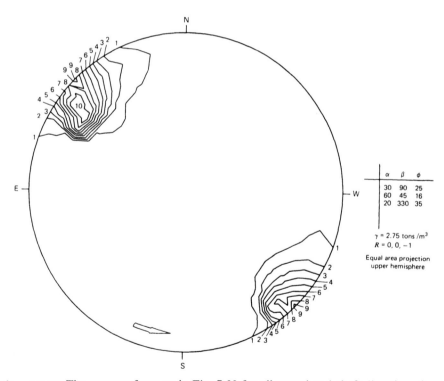

Figure 7.33 The support force as in Fig. 7.32 for all tunnel and shaft directions for the rock mass with three joints sets; multiply the contour values by 3.3 metric tons per meter of tunnel length.

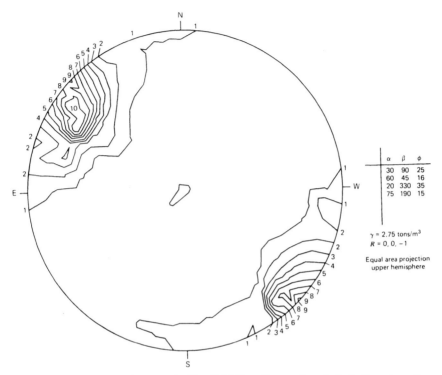

α	β	φ
30	90	25
60	45	16
20	330	35
75	190	15

γ = 2.75 tons/m³
R = 0, 0, −1

Equal area projection
upper hemisphere

Figure 7.34 The support force as in Fig. 7.33 for all tunnel and shaft directions for the rock mass with four joint sets; multiply the contour values by 3.3 metric tons per meter of tunnel length.

friction angle of 15°. The tunnel/shaft support force contours are only slightly less polarized, with vertical shafts now showing increased support needs. The orientations of the tunnel requiring greatest force and the magnitude of this support force is the same as that in Figure 7.33, with three joint sets. The magnitude of the support force for the worst tunnel is 32 tons per meter of tunnel length.

SUMMARY

For hard, jointed rock, the design of tunnel supports should be based on the force required to stabilize potential key blocks formed by the intersection of preexisting joint planes and the tunnel surface. Describing and analyzing these blocks necessitates a three-dimensional approach. Block theory is a convenient way to determine the most critical blocks, given any tunnel direction, shape, and dimensions. The information required to describe the rock mass consists simply of the orientations of the joint planes and their friction angles.

This section introduced the theoretical basis and graphical procedures for application of block theory to the computation of tunnel support forces. These procedures offer the possibility to optimize tunnel and shaft design with respect to support requirements. The information about the rock required to perform this analysis is minimal and the computations are not tedious, particularly since microcomputer programs are available. Block theory offers potential cost savings in tailoring the layout and design of tunnels and shafts to the geometric properties of jointed rock masses.

References

Benson, R. P., Conlon, R. J., Merritt, A. H., Joli-Coeur, P., and Deere, D. V. (1971) Rock mechanics at Churchill Falls, *Proceedings, Symposium on Underground Rock Chambers* (ASCE), pp. 407–486.

Bieniawski, Z. T. (1968) op. cit., Chapter 3.

Bieniawski, Z. T. (1975a) op. cit., Chapter 3.

Bieniawski, Z. T. (1975b) Case studies: Prediction of rock mass behavior by the geomechanics classification, *Proceedings, 2nd Australia-New Zealand Conference on Geomechanics* (Brisbane), pp. 36–41.

Bieniawski, Z. T. (1976) op. cit., Chapter 2.

Bieniawski, Z. T. (1984) op. cit., Chapter 1.

Bray, J. W. (1967) A study of jointed and fractured rock, II. Theory of limiting equilibrium, *Felsmechanik und Ingenieurgeologie (Rock Mechanics and Engineering Geology)* 5: 197–216.

Coates, D. F. (1970) See references, Chapter 1.

Cording, E. J. and Deere, D. V. (1972) Rock tunnel supports and field measurements, *Proceedings, 1st Rapid Excavation and Tunneling Conference* (AIME), Vol. 1, pp. 601–622.

Cording, E. J. and Mahar, J. W. (1974) The effect of natural geologic discontinuities on behavior of rock in tunnels, *Proceedings, 1974 Excavation and Tunneling Conf.* (AIME), Vol. 1, pp. 107–138.

Dube, A. K. (1979) Geomechanical evaluation of tunnel stability under falling rock conditions in a Himalayan tunnel, Ph.D. Thesis, University of Roorkee, India.

Duvall, W. (1976) General principles of underground opening design in competent rock, *Proceedings, 17th Symposium on Rock Mechanics* (University of Utah), Paper 3A1.

Gnirk, P. F., and Johnson, R. E. (1964) The deformational behavior of a circular mine shaft situated in a viscoelastic medium under hydrostatic stress, *Proceedings, 6th Symposium on Rock Mechanics*, University of Missouri (Rolla), pp. 231–259.

Goodman, R. E. and Boyle, W. (1986) Non-linear analysis for calculating the support of a rock block with dilatant joint faces, *Felsbau* 4: 203–208.

Goodman, R. E. and Shi, G. H. (1985) *Block Theory and Its Application to Rock Engineering*, Prentice-Hall, Englewood Cliffs, NJ.

Hoek & Brown (1980) op. cit. chap 1.

Holland, C. T. (1973) Pillar design for permanent and semi-permanent support of the overburden in coal mines, *Proceedings, 9th Canadian Rock Mechanics Symposium.*

Hustrulid, W. A. (1976) A review of coal pillar strength formulas, *Rock Mech.* **8:** 115–145.

Indraratna, B. and Kaiser, P. K. (1987) Control of tunnel convergence by grouted bolts, *Proc. Rapid Excav.* and *Tunneling Conf.* (RETC), New Orleans

Jaeger, J. and Cook, N. G. W. (1976) See references, Chapter 1.

Jethwa, J. L. (1981) Evaluation of rock pressure in tunnels through squeezing ground in the lower Himalayas, Ph.D. thesis University of Roorkee, India.

Jethwa, J. L. and Singh, B. (1984) Estimation of ultimate rock pressure for tunnel linings under squeezing rock conditions, *Proceedings, ISRM Symposium on Design and Performance of Underground Excavations* (Cambridge), pp. 231–238 (Brit. Geotech. Soc., London).

Kaiser, P, K. and Morgenstern, N. R. (1981, 1982) Time-dependent deformation of small tunnels. I, Experimental facilities; II, Typical test data; III, Pre-failure behaviour, *Int. J. Rock Mech. Min. Sci.* I, **18:** 129–140; II, **18:** 141–152; III, **19:** 307–324.

Kastner, H. (1962) *Statik des tunnel—und stollenbaues.* Springer-Verlag, Berlin.

Korbin, G. (1976) Simple procedures for the analysis of deep tunnels in problematic ground, *Proceedings, 17th Symposium on Rock Mechanics* (University of Utah), Paper 1A3.

Ladanyi, B. (1974) Use of the long term strength concept in the determination of ground pressure on tunnel linings. *Proc. 3rd Cong. ISRM* (Denver), Vol. 2B, pp. 1150–1156.

Lang, T. A. (1961) Theory and practise of rock bolting, *Trans. Soc. Min. Eng.,* AIME **220:** 333–348.

Lang, T. A. and Bischoff, J. A. (1981) Research study of coal mine rock reinforcement, A report to the U.S. Bureau of Mines, Spokane (available from NTIS, #PB82-21804).

Lang, T. A., Bischoff, J. A., and Wagner, P. L. (1979) Program plan for determining optimum roof bolt tension—Theory and application of rock reinforcement systems in coal mines; A report to the U.S. Bureau of Mines, Spokane (available from NTIS, #PB80-179195).

McCreath, D. R. (1976) Energy related underground storage, *Proceedings, 1976 Rapid Excavation and Tunneling Conf.* (AIME), pp. 240–258.

McCreath, D. R. and Willett, D. C. (1973) Underground reservoirs for pumped storage, *Bull. Assoc. Eng. Geol.* **10:** 49–64.

Muskhelishvili, N. I. (1953) *Some Basic Problems of the Mathematical Theory of Elasticity,* 4th ed., translated by J. R. M. Radok, Noordhof, Groningen.

Obert, L. and Duvall, W. (1967) See references, Chapter 1.

Panek, L. A. (1964) Design for bolting stratified roof, *Trans. Soc. Min. Eng.,* AIME, Vol. 229, pp. 113–119.

Peck, R. B., Hendron, Jr., A. J., and Mohraz, B. (1972) State of the art of soft ground tunneling, *Proceedings 1st Rapid Excavation and Tunneling Conference* (AIME) **1:** 259–286.

Stephenson, O. (1971) Stability of single openings in horizontally bedded rock, *Eng. Geol.* **5:** 5–72.

Szechy, K. (1973) *The Art of Tunnelling,* 2d ed., Akademiado, Budapest.

Terzaghi, K. (1946) Rock defects and loads on tunnel supports, in R. V. Proctor and T. L. White, *Rock Tunneling with Steel Supports,* Commercial Shearing and Stamping Co., Youngstown, OH.

Problems

1. Draw vectors to scale showing the normal and shear stresses in psi along the locus of a fault, at points *A*, *B*, and *C* in the following diagram. The fault strikes parallel to a circular tunnel, 15 ft in radius, and at its closest point it is 10 ft from the tunnel. It dips 60° as shown in the diagram. The tunnel is driven at a depth of 500 ft in granite. Assume $K = 1.0$.

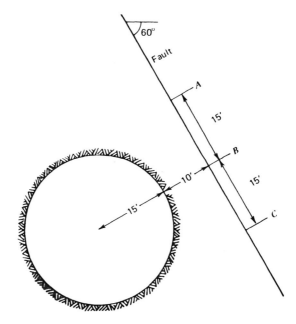

2. Calculate the maximum compressive stress around an elliptical underground opening, twice as high as it is wide, for values of K equal to $0, \frac{1}{3}, \frac{2}{3}, 1,$ 2, and 3. The tunnel is 1000 ft deep. Indicate in each case where the maximum stress occurs.

3. Calculate the deflection and stresses of each layer above the roof of a mine chamber 50 ft wide assuming elastic behavior. The layer closest to the opening is sandstone, 5 ft thick, with a unit weight of 160 lb/ft³. The layer above is shale, 2 ft thick, with unit weight of 140 lb/ft³. $E_{s.s} = 10^6$ psi; $E_{shale} = 5 \times 10^5$ psi.

4. Plot a curve of σ_r and σ_θ as a function of radius and calculate u_r at the tunnel wall using Bray's solution.

   (a) The properties are $\phi_j = 20°$; $S_j = 0$; $\delta = 55°$; $q_u = 500$ psi; $\phi = 35°$; $p = 4000$ psi; $p_i = 40$ psi; $a = 96''$; $E = 10^6$ psi; $\nu = 0.2$.

   (b) Find u_r and R for $p_i = 400$ psi, all other quantities being the same.

5. In an underground room 30 ft wide, limestone 3 ft thick is overlain by sandstone 6 ft thick.

   (a) Assuming the horizontal stress to be zero, what pretension force should be installed in rock bolts spaced on a square pattern in the roof with 3 ft spacing to achieve a "suspension effect" in the roof? (For the sandstone $E = 1 \times 10^6$ psi; for the limestone, $E = 0.3 \times 10^6$ psi; $\gamma = 150$ lb/ft³).

   (b) What is the corresponding maximum tensile stress in each layer?

6. Calculate and plot the variation of displacement with time along the horizontal diameter of a circular underground opening 50 ft in diameter between bench marks 10 ft deep in each wall. Assume that under hydrostatic pressure the rock exhibits only elastic, non-time-dependent volume change and in distortion behaves like a Burgers material with

$$G_1 = 0.5 \times 10^5 \text{ psi}$$
$$G_2 = 0.5 \times 10^6 \text{ psi}$$
$$\eta_1 = 8.3 \times 10^9 \text{ psi/min}$$
$$\eta_2 = 8.3 \times 10^{11} \text{ psi/min}$$
$$K = 1.0 \times 10^6 \text{ psi}$$

The initial vertical and horizontal stresses are 2000 and 4000 psi, respectively.

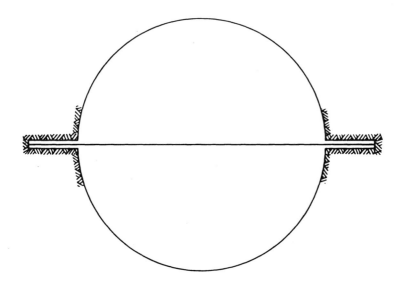

7. In Problem 6, suppose a rock bolt 15 ft long is installed 12 h after excavation along a radius 30° above the horizontal. Assuming that the rock bolt is insufficiently stiff to reduce the displacements of the tunnel, (a) calculate the strain in the bolt as a function of time; (b) if the bolt is made of steel and has a diameter of 1.25 in., calculate the force developed in the bolt as a function of time.

8. Calculate the radial displacement with time of a 40-ft-diameter tunnel at a depth of 1000 ft in rock salt with hydrostatic initial stresses and an internal pressure of 100 psi suddenly applied 24 h after excavation. The rock is elastic in hydrostatic compression and a Burgers body in distortion, with $K = 0.8 \times 10^6$ psi, $G_1 = 0.1 \times 10^6$ psi, $G_2 = 0.6 \times 10^6$ psi, $\eta_1 \times 10^8$ psi/min, and $\eta_2 = 10^{12}$ psi/min. $\gamma_{wet} = 150$ lb/ft³. (*Hint:* Superpose the solutions for dilatometer test in Chapter 6 and a tunnel in Chapter 7.)

9. A rock mass is rated 20 according to the Geomechanics rating. Plot a curve of stand-up time versus unsupported span in meters (see Figure 7.13).

10. A rock tunnel 5 m in diameter is driven with maximum unsupported length, at the face, equal to 4 m. Plot a curve of stand-up time versus rock mass rating for the tunnel.

11. Suppose you were given a set of identical rectangular prismatic blocks (thickness t, length s, and width b) and you wanted to construct an arched tunnel with them.
    (a) Calculate the shape and dimensions of the widest safe tunnel. (*Hint:* Any block i can extend a distance x_i over the block below it (see diagram) and the complete arch may be calculated as a symmetrical arrangement of such cantilevered blocks.)

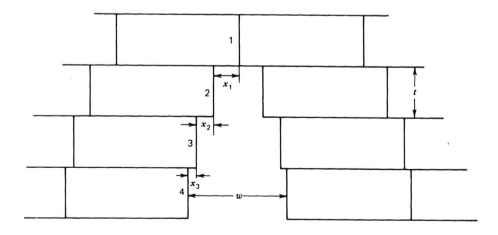

(b) Discuss the influence of horizontal stress in the mass on the stability of such a tunnel.

(c) Discuss the influence of limited tensile strength on the stability of the system.

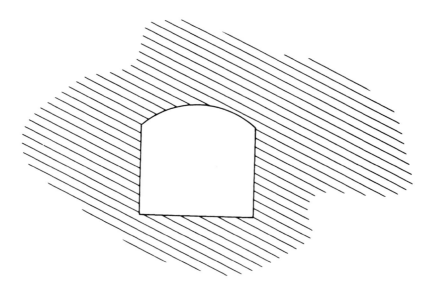

12. A long underground chamber, shown in cross section, is to be constructed in rock inclined 25° toward the right. (a) Assuming the friction angle ϕ_j is 50°, use the construction of Figure 7.7 to locate zones of potential distress around the opening. (b) Repeat this procedure with $\phi_j = 20°$ to investigate the effect of rock deterioration on the flow of stress around the tunnel.

13. (a) Derive a formula permitting calculation of the maximum inclination of σ_1 with the normal to layers, given values of σ_3, σ_1, and ϕ_j.

(b) In a bedded rock having $\phi_j = 20°$ and $\sigma_1 = 1$ MPa, how large must σ_3 be in order that σ_1 have an inclination of 30° with the normal to layers?

(c) How can this exercise be used to evaluate support requirements for the tunnel in Problem 12?

14. A circular tunnel will be constructed in regularly layered rock dipping 45° to the left with initial stresses $p_1 = 1.5$ MPa acting horizontally and $p_2 = 1.0$ MPa acting vertically (see figure). Suppose it were possible to apply a radial support pressure p_b inside the tunnel so soon after excavation that slip could be prevented along the rock layers. Calculate the value of p_b required to achieve this result at points around the tunnel surface defined by $\theta = 0$, 15, 60, 90, 120, and 180°. Assume $\phi_j = 30°$.

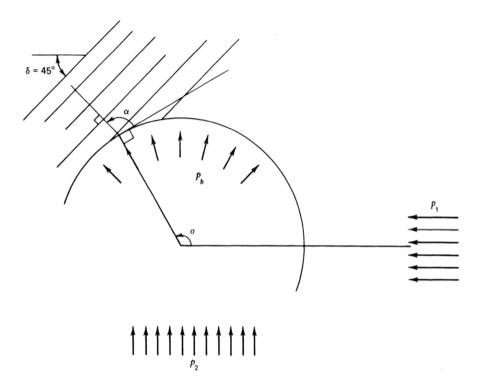

15. (a) Derive a formula to express the inward radial displacement (U_a) of a circular tunnel at radius a due to bulking of the annulus of rock between radii a and b ($b > a$). The bulking factor k_B is the increase in volume caused by rock breakage and decompression divided by the original volume before rock expansion. Assume that the outer circle of the annulus is stationary ($U_b = 0$).
    (b) Solve the relationship to express the rock bulking factor k_B as a function of U_a, a, and b where $U_b = 0$.
    (c) Derive a formula for k_B as above for the more general case with $U_b \neq 0$.

16. A circular tunnel of 2.12 m radius in phyllite is squeezing. Extensometers along a radius determine the radial displacement U_r with time as follows (data are from Jethwa (1981)):

t (days)	r (m)	U_r (mm)
20	2.12	75
	4.5	49
	7.0	30
	9.4	18
100	2.12	135
	4.5	93
	7.0	65
	9.4	49
800	2.12	253
	4.5	180
	7.0	142
	9.4	117

Using the results of Problem 15c, calculate the volume change factor k_B in each of the three rings between extensometer stations (2 to 4.5 m, 4.5 to 7.0 m, and 7.0 to 9.4 m) as a function of time.

17. Estimate the radius R of the destressed zone for the extensometer data of Problem 16 using the following two methods:
    (a) Dube (1979) showed that $R \simeq 2.7$ times the radius r_c separating the expanding portion from the contracting portion of the destressed zone. This radius can be interpolated from your answer to Problem 16.
    (b) Jethwa (1981) showed that R is the value of r at which the lines u versus $\log r$ intersect the curve for elastic displacement of a circular opening of radius r.

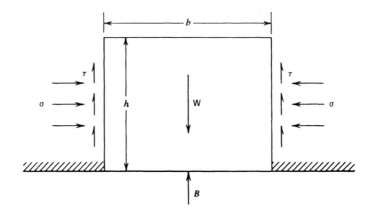

Calculate $u_{elas} = [(1 + \nu)/E] \, pr$ with $p = 0.4$ MPa, $\nu = 0.2$, and $E = 500$ MPa, and plot the curve u_{elas} versus log r. Plot the data sets u versus log r for each time and determine the intersections with $u_{elas}(r)$.

18. (a) Derive a formula for the ratio B/w of support forces to weight to stabilize the roof block under conditions of limiting equilibrium. The friction angles ϕ_j are the same on all joints (see figure). The unit weight is γ. The joints are smooth.

    (b) Solve (a) to find the maximum width b of a block such that no support is required.

19. In Problem 18, for limiting equilibrium one assumes that full friction is mobilized. This requires some block displacement, however. Since the sides are parallel, symmetry prevents any dilatant displacement. Assuming a dilatancy angle i on each joint and no initial normal stress, find the ratio of support force to block weight required for equilibrium as a function of block displacement u. (*Hint:* Assume rigid wall rock; calculate the strain of the rock block that would occur if dilatancy were allowed. Find the normal stress increment to accommodate this much strain.)

20. Restriction of dilatancy on the vertical joints of Problem 18 corresponds to requiring a horizontal path across Figure 5.17b. Thus if the initial normal stress corresponded to a of this figure, the displacement path would follow

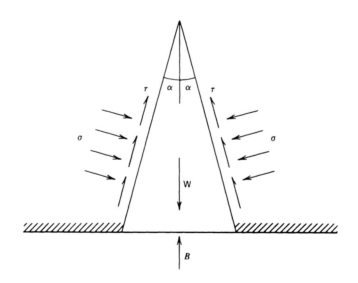

line *3, 4, 5, 6* of Figure 5.17*b*. The consequent values of σ and τ would then be determined as functions of displacement u and could be entered in the equilibrium expression to determine B/W as a function of u (see Goodman and Boyle, 1986). Suppose the block in question were a symmetrical wedge in the roof, as shown in the figure. What would be the corresponding displacement path across Figure 5.17*b*?

21. A circular tunnel has a symmetrical wedge of rock in the roof. Instead of falling out as a whole block, it breaks in two and the upper piece remains in place as the lower piece falls. What explanation can you give for this behavior?

22. (a) Calculate the abutment reactions H and V and their location for limiting equilibrium of the three hinged beam (see figure) in the roof of an underground opening.
    (b) Find the limiting value of settlement Δy at which an instability develops. What is the corresponding value of the horizontal displacement Δx of each abutment?

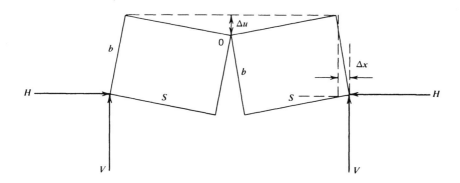

23. Assume that both ends of a tensioned rock bolt produce compressive stress inside a cone of 45° centered about the bolt axis (see figure). Determine the length-to-spacing ratio ℓ/s to provide a continuous zone of compressive stress:
    (a) Of minimum thickness $\ell/2$ in the middle of a straight beam of rock.
    (b) Of minimum thickness $a/2$ in a curved beam with inside radius a (let $s = r\theta$).

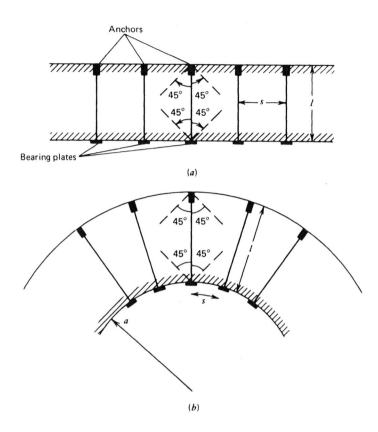

(a)

(b)

24. A system of rock bolt reinforcement for a flat-bedded roof is shown in the figure. What is the purpose of the angled bolts?

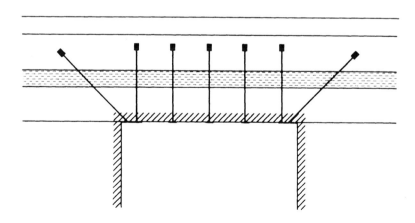

25. (a) Determine the joint pyramid codes for all blocks that are removable from the north wall of an underground chamber in a rock mass having the following joint sets:

Set	Dip	Dip Direction
1	30°	70°
2	50°	140°
3	60°	270°
North wall	90°	0°

   (b) Do the same for the south wall.
   (c) Draw the visible face of the critical block in the north wall as seen from inside the excavation (looking north).

26. A circular tunnel will be driven horizontally to the N 20° E (azimuth 20°) in the rock mass of Problem 25. On a section of the tunnel, looking in the direction of tunnel driving, draw the maximum key block region for JP 101.

Chapter 8

Applications of Rock Mechanics to Rock Slope Engineering

8.1 Introduction

Surface excavations in rock range from small rock walls ripped by bulldozers to make room for a shopping center, to enormous open pits up to 1 km deep and 1 km^3 in volume. This chapter deals with methods for planning the orientation, dimensions, and appurtenant features of such excavations so that they will safely serve their intended purpose.

For transportation routes, numerous high rock cuts are required in mountainous terrain. Restricted curvature specifications tend to make cuts for canals and railways relatively higher and more numerous than cuts for highways. A certain amount of rock fall and slide movement is often unavoidable along rights of way for transportation routes because the cost of assuring complete safety for so many square meters of excavated rock surface would be unbearable. However, railroad cuts and urban highway cuts should be designed to be absolutely safe at all times, due to the great cost of an accident. In urban areas, cuts may be made nearly vertical to preserve adjacent land values, necessitating installation of permanent rock support.

When rock cuts are to house installations, such as buildings, powerhouses, or portals to underground workings, the cost of a failure would be far greater than the cost of the excavation itself. Accordingly, great care and attention will be warranted in engineering such excavations, as if they were structures themselves. Drainage, supports, instrumentation, and construction sequence may be specified.

This is also true of cuts for spillways of earth dams and for surface penstocks; in either case, a slope failure could trigger a disaster.

Open pit mines are large rock excavations that are usually intended mainly to strip away overburden material from ore. Their design—that is, the choice of angles for slopes, widths of benches, and overall shape—is now integrated with other mining cost factors to achieve maximum profit. Too flat a slope will mean extra excavation and extra waste rock; but too steep a slope will increase the number of lost-time haulage road blockages and accidents. Most of the slopes of an open pit mine are temporary since the pit is ever enlarging. Simple instrumentation and quick response to signs of instability have allowed mining companies to work safely with slopes that would be judged too steep for civil engineering excavations of comparable size.

In soft rocks like shales, hydrothermally altered zones, and deeply weathered granites, design of safe slopes is an extension of soil mechanics theory since such materials tend to fail by slumping or sliding through the body of rock itself. In most hard rocks and in some of the softer rocks as well, preexisting discontinuities control the avenues of rock movement so modes of slope failure occur that are not usual in soils. Special methods for analyzing these structurally controlled failure modes have been devised by workers in rock mechanics.

It turns out to be cheaper to choose a safe angle for a rock slope than to try to hold it up on an unnatural slope with artificial supports, for when a high rock face starts to move, the forces required to hold it back are enormous. Choosing safe angles for rock slopes requires that shear strength characteristics of the controlling discontinuity surfaces be evaluated, often necessitating laboratory and field shear testing programs. However, if the strike of the cutting can be altered to suit the structural properties of the rock mass, it is often possible to choose an orientation for the excavation such that rock failure cannot occur, regardless of friction angle of the discontinuities. This is true because of the highly directional characteristics of failure modes along structural weakness planes.

8.2 Modes of Failure of Slopes in Hard Rock

Hard rock is usually so strong that failure under gravity alone is possible only if discontinuities permit easy movement of discrete blocks. In regularly bedded or foliated rock, cut by joints, there are many possibilities for block movement along weakness planes and a large variety of behavioral modes are exhibited. With an appreciation of the mode of failure, it is possible to evaluate the probability of failure, or the factor of safety, and to engineer a remedy if the degree of risk is unacceptable. When there are multiple sets of discontinuity planes intersecting in oblique angles, kinematic model studies may be helpful in anticipating the most likely pattern of slope failure. Failures involving movement of rock blocks on discontinuities combine one or more of the three basic modes—*plane sliding, wedge sliding,* and *toppling.*

A *plane slide* forms under gravity alone when a rock block rests on an inclined weakness plane that "daylights" into free space (Figure 8.1*a*). The inclination of the plane of slip must be greater than the friction angle of that

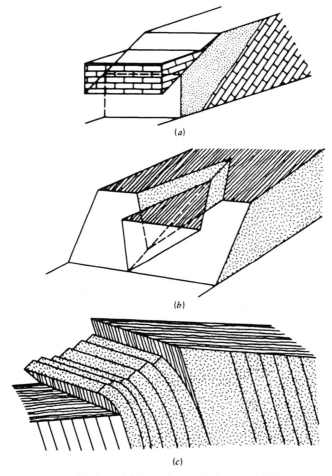

(a)

(b)

(c)

Figure 8.1 Modes of failure for rock slopes. (*a*) Plane slide. (*b*) Wedge slide. (*c*) Toppling.

plane. The conditions for failure reside dormantly in the slope until excavation or rock movement removes the barrier to block translation. Movement of a block like that shown in Figure 8.1*a* supposes that the restraint to sliding has been overcome not only along the surface of sliding but along the lateral margins of the slide as well. In soft rocks, like shale, the side restraint can be released by rupture of the rock itself if the base of sliding is inclined considerably steeper than the friction angle. In hard rocks, plane sliding can occur only if there are other discontinuities or valleys transverse to the crest of the slope releasing the sides of the block. In Figures 8.2*a* and *b*, the release is provided by joints.

Wedge slides (Figures 8.1*b* and 8.2*c*) can occur when two planes of weakness intersect to define a tetrahedral block. Slip can occur without any topographic or structural release features if the line of intersection of two discontinuities daylights into the excavation. It was the movement of a large wedge that undermined the Malpasset Dam in France (1959) causing its complete destruction and much loss of life (see Figure 1.5).

Toppling failure, Figures 8.1*c* and 8.3, involves overturning of rock layers like a series of cantilever beams in slates, schists, and thin-bedded sediments inclined steeply into the hillside. Each layer tending to bend downhill under its own weight transfers force downslope. If the toe of the slope is allowed to slide or overturn, flexural cracks will form in the layers above, liberating a large mass of rock. If there are frequent cross joints, the layers can overturn as rigid columns rather than having to fail in flexure. In either event, destructive slope movements must be prefaced by interlayer slip of a normal fault type (Figure 8.3*e*).

Many "higher modes" of failure are exhibited by complexly jointed and bedded rocks, in which plane sliding, wedge sliding, and toppling occur simultaneously or successively sometimes with failure through rock bridges by flexure, shear, or splitting (Figure 8.3*f*). Goodman and Bray (1977) show several examples of such modes involving toppling in conjunction with sliding. Failure through the intact material can also occur together with any of these modes, as previously noted. For example, "footwall slides" occur in high dip slopes following buckling of the strata near the toe (see Problem 8). Failure totally within intact rock, like slumping in clay soils, occurs in rock slopes only in very weak continuous rocks, in hydrothermally altered zones, and in highly weathered rocks. Pervasively fractured rocks may behave in a "pseudocontinuous" fashion because they exhibit so many combinations of failure modes that there is always a locus along preexisting cracks corresponding to the most critical failure locus of a continuous material. Slopes in such rocks have been analyzed using soil mechanics techniques (Hoek and Bray, 1977). When the rock is weakened by well-defined and regular sets of discontinuities, on the other hand, kinematic, rigid block analysis is preferable.

Figure 8.2 Slides controlled by discontinuities in hard rock. (*a*) Shallow sliding on bedding, restrained by a system of rock bolts (Colombia). (*b*) Loss of a bench in a slate quarry due to sliding on steeply dipping bedding and lateral release by vertical joints (North Wales). (*c*) Surfaces of a wedge failure (near Trondheim, Norway.)

Figure 8.3 Toppling failures. (*a*) A small topple in Clear Creek Canyon, in the Grand Canyon, developed in schist. (*b*) The toe of a large topple near the previous one.

Figure 8.3 Toppling failures. (c) A toppling failure in process in Alberta, causing raveling of rocks from the rock face (photo by Duncan Wyllie). (d) Detail of lower part of a topple in sandstone and shale (North Devon; England); there is no failure but a zone of concentrated flexural cracking and sharp bending.

Figure 8.3 Toppling failures. (*e*) The top of a large topple in North Wales, showing open tension cracks and obsequent slopes creted by normal fault type displacement along the slaty cleavage. (*f*) The toe of a topple in sandstone and shale strata (North Devon) showing sliding on flexural cracks and cracking of the columns.

8.3 Kinematic Analysis of Slopes

"Kinematics" refers to the motion of bodies without reference to the forces that cause them to move. Many rock cuts are stable on steep slopes even though they contain steeply inclined planes of weakness with exceedingly low strength; this happens when there is no freedom for a block to move along the weak surface because other ledges of intact rock are in the way. Should the blockage be removed by erosion, excavation, or growth of cracks, the slope would fail immediately. This section deals with an approach to slope design making use mainly of the directionality of the discontinuous rock mass to insure that there is always rock "in the way" of potential failure blocks. Only minimal reference is made to the strength parameters of the rock for the principal considerations are the orientations of the planar weaknesses in relation to the orientation of the excavation. (The subject of kinematics is discussed again, using block theory, in Section 8.7.)

Figure 8.4 shows the three basic line elements of a rock mass[1]: the *dip vector, ($\hat{D}_i$)* pointed down the dip of a weakness plane; the *normal vector, ($\hat{N}_i$)* (or "pole") pointed in the direction perpendicular to the plane of weakness; and the *line of intersection ($\hat{I}_{ij}$)* of weakness planes i and j. Recall that the dip vector is a line bearing at right angles to the strike and plunging with vertical angle δ below horizontal. The lower hemisphere stereographic projection will be used exclusively in this section, so the dip vector always plots inside the circle representing the horizontal plane. (The principles of stereographic projection are introduced in Appendix 5.) The lower hemisphere normal $\hat{N}$ plots 90° from the dip vector in the vertical plane containing the dip vector (Figure 8.4a).

The line of intersection $\hat{I}_{ij}$ of two planes i and j can be found as the point of intersection of the great circles of each plane (Figure 8.4b). Alternatively, $\hat{I}_{ij}$ is determined as the line perpendicular to the great circle containing normals $\hat{N}_i$ and $\hat{N}_j$. Since normals to planes are plotted in the joint survey, as discussed in Chapter 5, the latter technique of locating the intersections is useful in practice. Once all the line elements $\hat{D}$, $\hat{N}$, and $\hat{I}$ are plotted for a rock mass, the kinematic requirements for possible slope failure can be examined for a rock slope of any strike and dip.

Consider the case of plane sliding under gravity (Figure 8.5). Any block tending to slide on a single plane surface will translate down the slope parallel to the dip of the weakness plane, that is, parallel to $\hat{D}$. If the slope is cut at angle α with respect to horizontal, the conditions for a slide are simply that $\hat{D}$ be pointed into the free space of the excavation and plunge at an angle less than α (Figure 8.5a). Figure 8.5b shows a cut slope plotted as a great circle in the lower hemisphere. The kinematic requirements for plane sliding are satisfied if

---

[1] Letters with carets represent unit vectors.

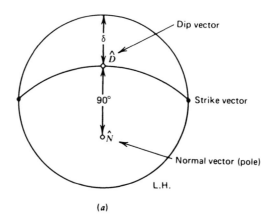

(a)

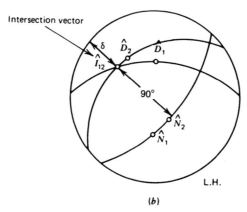

(b)

Figure 8.4 Stereographic projections of line elements relevant to analysis of rock slopes.

the dip vector of a possible surface of sliding plots in the ruled area above the cut slope circle. Plane 1 in this example would allow a slide whereas plane 2 would not. Given the dip vector of a potentially troublesome weakness surface, it is possible to determine the limiting safe angle (i.e., the steepest safe slope) corresponding to a cut of assigned strike by applying the simple construction of Figure 8.5c. For a rock cut having "strike 1" (in Figure 8.5c) the maximum safe angle α_1 is the dip of the great circle passing through "strike 1" and $\hat{D}_1$. Similarly, the steepest safe slope for a cut with strike 2 is α_2, the dip of the great circle through strike 2 and $\hat{D}_1$. For plane sliding on a single weakness plane, we see there is kinematic freedom for slip only in one-half of the set of possible cut orientations; cut orientations nearly parallel to the dip direction of the plane of weakness will be stable even when nearly vertical.

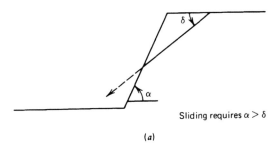

Sliding requires $\alpha > \delta$

(a)

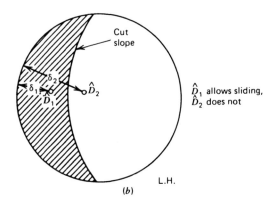

$\hat{D}_1$ allows sliding, $\hat{D}_2$ does not

(b)

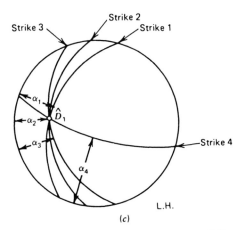

(c)

Figure 8.5 Kinematic test for plane sliding.

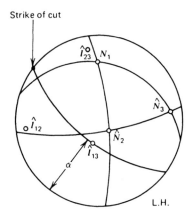

Figure 8.6 Example of a kinematic test for a wedge failure.

In the case of a wedge sliding along the line of intersection of two planes, exactly the same analytical procedure can be followed to find a maximum safe angle for slopes of assigned strike if in place of the line element $\hat{D}$, we substitute line element $\hat{I}$. Figure 8.6 gives an example of kinematic analysis of wedge failures for a rock mass comprised of three sets of joints. If a cut is made with strike shown, only wedges formed of planes 1 and 3, or planes 1 and 2 could potentially slide. If the cut is inclined at angle α, determined by the dip of the great circle passing through $\hat{I}_{13}$ and having the assigned strike for the cut, then only the wedge determined by planes 1 and 2 is capable of sliding. Moreover, since $\hat{I}_{12}$ plunges at a low angle, it is unlikely to cause trouble.

In the case of toppling failure, we have noted that interlayer slip must occur before large flexural deformations can develop. In a rock slope, the surface of the cut is the direction of major principal stress over the entire slope length. If the layers have angle of friction ϕ_j, slip will occur only if the direction of applied compression makes an angle greater than ϕ_j with the normal to the layers. Thus, as shown in Figure 8.7, a precondition for interlayer slip is that the normals be inclined less steeply than a line inclined Φ_j degrees above the plane of the slope. If the dip of the layers is δ, then toppling failure with a slope inclined α with horizontal can occur if $(90 - \delta) + \phi_j < \alpha$. On the stereographic projection, this means that toppling can occur only if the normal vector $\hat{N}$ lies more than ϕ_j degrees below the cut slope. Moreover, toppling can occur only if the layers strike nearly parallel to the strike of the slope, say within 30°.[2] Thus, toppling is a possibility on a regular, closely spaced discontinuity set if its

---

[2] The 15°, previously recommended by Goodman and Bray (1977), has been found to be too small.

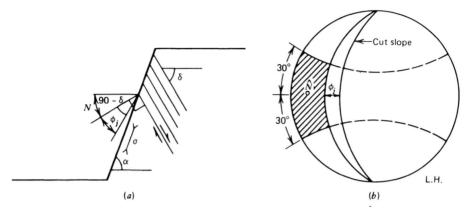

Figure 8.7 Kinematic test for toppling. (*a*) $(90 - \delta) + \phi_j < \alpha$. (*b*) $\hat{N}$ must plot in the shaded zone.

normal plots inside the ruled region of Figure 8.7*b*. This region is bounded by a great circle ϕ_j degrees below the cut slope and striking parallel to it, the horizontal great circle, and two small circles perpendicular to the strike of the cut and 30° from the center of the net.

In any case, the joint survey is likely to generate a multiplicity of discontinuities and many individual dip vectors, normal vectors, and lines of intersection. However, it is possible to reduce the number of lines to a manageable number for analysis by preparing two simple overlays, as shown in Figure 8.8. In the case of plane sliding under self-weight alone, failure can occur only if the surface of sliding dips steeper than ϕ_j. (For a very acute wedge, any roughness

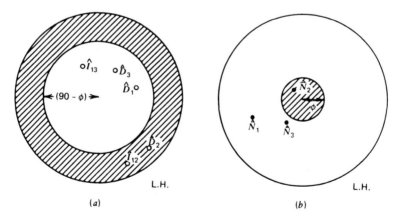

Figure 8.8 Use of kinematic tests to select important line elements for further analysis. (*a*) Retain $\hat{I}_{13}$, $\hat{D}_3$, and $\hat{D}_1$ for sliding. (*b*) Retain $\hat{N}_1$ and $\hat{N}_3$ for toppling.

on the planes of weakness adds considerable strength so, in fact, $\hat{I}$ can often incline considerably steeper than ϕ_j without wedge failure.) Draw a small circle of radius $90 - \phi_j$ at the center of the net as shown in Figure 8.8a. The ruled area outside this circle contains all lines plunging less than ϕ_j. Therefore, all $\hat{I}$ and $\hat{D}$ vectors within the ruled area can be eliminated from further consideration. Similarly, toppling can occur only on discontinuities whose normals plunge at an angle less than $90 - \phi_j$. So all $\hat{N}$ vectors inside the ruled area of Figure 8.8b can be eliminated from further analysis of toppling failure. This area is inside a circle of radius ϕ_j about the center of the projection.

As an example of a kinematic analysis, consider the design of a circular open pit mine in a rock mass presenting two sets of discontinuities in orientations shown in Figure 8.9a. Surfaces 1 strike N 32° E and dip 65° to the N 58° W; discontinuities 2 strike north and dip 60° E. The line of intersection of these two planes plunges 28° to the N 18° E. Assume ϕ_j equals 25°.

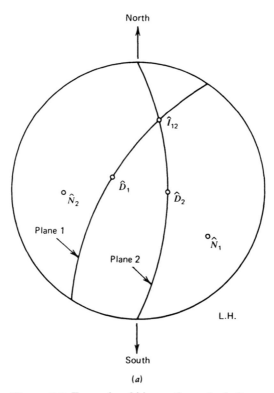

(a)

Figure 8.9 Example of kinematic analysis for a circular open pit. (a) Planar features of the rock mass and their representation by lines.

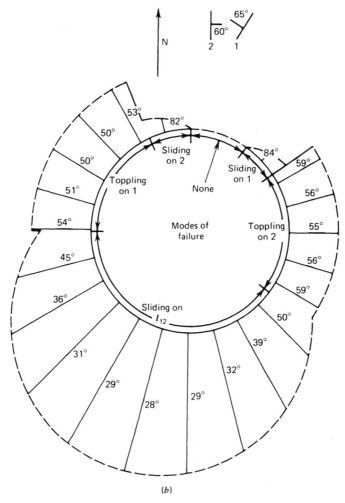

Figure 8.9 Example of kinematic analysis for a circular open
pit. (*b*) Steepest safe slopes every 15° around the pit. Lengths
are proportional to cot α.

In an open pit mine, there is a cut slope with strike corresponding to every
compass point. It might not be feasible, considering the geometry of access
roads into the bottom of the mine, to use different slopes in different parts of
the mine. But kinematic analysis will demonstrate that slopes of different
strikes have vastly different requirements for safety. By considering individu-
ally each of the line elements $\hat{D}_1$, $\hat{D}_2$, $\hat{I}_{12}$, $\hat{N}_1$, and $\hat{N}_2$, and applying the construction of Figure 8.5, maximum safe cut slopes were determined every 15°
around the pit (Table 8.1). The lowest value of α for each of the five modes of
failure is retained as the maximum safe value for design. Table 8.1 and Figure

Table 8.1 *Kinematic Tests—Circular Excavation*

Joint 1 strikes N 32° E
Joint 2 strikes NS

$\hat{D}_1$ 65° N 58° W
$\hat{D}_2$ 60° E
$\hat{I}_{12}$ 28° N 18° E

Strike of Cut	Direction of Dip of Cut	D_1 Analysis for Sliding Dip of Joint Set 1 Maximum Safe Cut Angle	D_2 Analysis for Sliding Dip of Joint Set 2 Maximum Safe Cut Angle
N	E	90°	60°
N 15° E	SE	90°	61°
N 30° E	SE	90°	64°
N 45° E	SE	90°	68°
N 60° E	SE	90°	74°
N 75° E	SE	90°	82°
E	S	90°	90°
S 75° E	SW	90°	90°
S 60° E	SW	90°	90°
S 45° E	SW	84°	90°
S 30° E	SW	78°	90°
S 15° E	SW	73°	90°
S	W	69°	90°
S 15° W	NW	66°	90°
S 30° W	NW	65°	90°
S 45° W	NW	66°	90°
S 60° W	NW	68°	90°
S 75° W	NW	71°	90°
W	N	76°	90°
N 75° W	NE	82°	81°
N 60° W	NE	89°	74°
N 45° W	NE	90°	68°
N 30° W	NE	90°	64°
N 15° W	NE	90°	61°

8.9*b* show these slope values and the critical modes of failure all the way around the pit. This analysis shows that vertical slopes are safe if oriented between east and S 60° E, whereas any cut steeper than 28° can potentially fail for a cut striking N 75° W. If the friction angle of the discontinuities is greater than 28°, $\hat{I}_{12}$ is no longer a concern and the slope can be steepened to at least 50°, at which inclination toppling becomes critical. If the discontinuities are widely spaced or irregular, then toppling is improbable and the slope can be steepened to at least 60°.

I_{12} Analysis for Sliding on Planes 1 & 2 Along I_{12} Maximum Safe Cut Angle	T_1 Analysis for Toppling on Joint 1 ($\phi = 25°$)	T_2 Analysis for Toppling on Joint 2 ($\phi = 25°$)	Applicable Mode	Maximum Safe Slope Angle
61°	54°	90°	T_1	54°
85°	51°	90°	T_1	51°
90°	50°	90°	T_1	50°
90°	50°	90°	T_1	50°
90°	53°	90°	T_1	53°
90°	90°	90°	D_2	82°
90°	90°	90°	None	90°
90°	90°	90°	None	90°
90°	90°	90°	None	90°
90°	90°	90°	D_1	84°
90°	90°	59°	T_2	59°
90°	90°	56°	T_2	56°
90°	90°	55°	T_2	55°
85°	90°	56°	T_2	56°
69°	90°	59°	T_2	59°
50°	90°	90°	I_{12}	50°
39°	90°	90°	I_{12}	39°
32°	90°	90°	I_{12}	32°
29°	90°	90°	I_{12}	29°
28°	90°	90°	I_{12}	28°
29°	90°	90°	I_{12}	29°
31°	90°	90°	I_{12}	31°
36°	90°	90°	I_{12}	36°
45°	90°	90°	I_{12}	45°

The kinematic analysis of sliding consists of finding the dip of a cut with given strike and containing $\hat{I}$ or $\hat{D}$. In structural geology terms, we determine the true dip α of a cut slope having apparent dip δ in a known direction. Let Σ be the angle between the strike of the cut and the bearing of $\hat{D}$ or $\hat{I}$, for plane sliding and wedge sliding respectively; then the maximum safe slope α is

$$\alpha = \tan^{-1} \frac{\tan \delta}{\sin \Sigma} \tag{8.1}$$

For example, in the open pit problem discussed previously, consider the portion of the pit where the slope strikes N 60° E and dips SE. Joint set 2 strikes N and dips 60° E; for this joint set, then, $\delta = 60°$ and $\Sigma = 30°$. Equation 8.1 gives $\alpha = 74°$.

8.4 Analysis of Plane Slides

A simple formulation of conditions for limiting equilibrium of a plane slide proves useful in back calculating actual failure cases. This is an important step in attempting to design a new excavation in a rock mass, for natural failures represent giant "test specimens." Due to the unknown importance of scale effects, it is far more suitable to rework field data using an appropriate model than to attempt a program of field tests, although the latter will also be useful to check assumptions and specific geological structures not represented in case histories. A suitable basis for assessing rock mass properties based on field cases is presented in *Rock Slope Engineering* by Hoek and Bray (1977).

Figure 8.10 shows the two cases of plane failure that need to be considered. Usually, a tension crack delimits the top of the slide at a point beyond the crest of the slope; occasionally the tension crack intercepts the slope face itself. In both cases, the depth of the tension crack is represented by the vertical distance z from the crest of the slope to the bottom of the crack. If the tension crack is filled with water to depth z_w, it can be assumed that water seeps along the sliding surface, losing head linearly between the tension crack and the toe of the slope. If the slide mass behaves like a rigid body, the condition for limiting equilibrium is reached when the shear force directed down the sliding surface equals the shear strength along the sliding surface; that is, failure occurs when

$$W \sin \delta + V \cos \delta = S_j A + (W \cos \delta - U - V \sin \delta)\tan \phi_j \qquad (8.2)$$

where δ is the dip of the surface of sliding
   S_j and ϕ_j are the shear strength intercept ("cohesion") and friction angle of the sliding surface
   W is the weight of the potentially sliding wedge
   A is the length (area per unit width) of the sliding surface
   U is the resultant of the water pressure along the sliding surface
   V is the resultant of the water pressure along the tension crack

(The toe of the slope is free draining.)
   Considering Figure 8.10, with tension crack to depth Z filled with water to depth Z_w,

$$A = \frac{H - Z}{\sin \delta} \qquad (8.3)$$

$$U = \tfrac{1}{2}\gamma_w Z_w A \qquad (8.4)$$

$$V = \tfrac{1}{2}\gamma_w Z_w^2 \tag{8.5}$$

and if the tension crack intercepts the crest of the slope (the usual case),

$$W = \tfrac{1}{2}\gamma H^2 \left\{ \left[1 - \left(\frac{Z}{H} \right)^2 \right] \cot \delta - \cot \alpha \right\} \tag{8.6a}$$

while if the tension crack intercepts the face,

$$W = \tfrac{1}{2}\gamma H^2 \left[\left(1 - \frac{Z}{H} \right)^2 \cot \delta (\cot \delta \tan \alpha - 1) \right] \tag{8.6b}$$

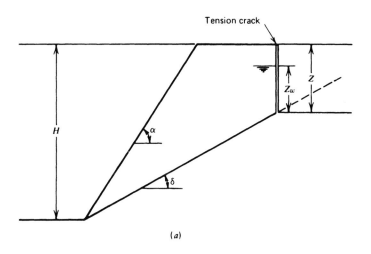

(a)

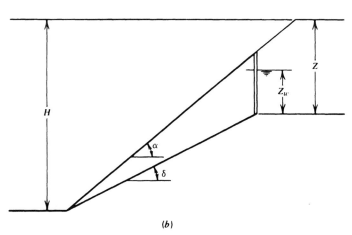

(b)

Figure 8.10 Geometry for analysis of plane failure. [After Hoek and Bray, (1977).]

A convenient way to use a failure case is to solve Equation 8.2 with the known geometry and presumed water conditions at the time of failure to yield a value for S_j, since this quantity is hard to measure in the laboratory. When a distribution of values for S_j has been determined in this way from case histories, Equation 8.2 can be used to generate a *slope chart* for design, in which H is plotted against cot α. A factor of safety F can be introduced for this purpose by multiplying the left side of Equation 8.2 by F. For generating a slope chart with factor of safety F, Equation 8.2 can be solved for cot α:

$$\cot \alpha = \frac{[a(F \sin \delta - \cos \delta \tan \phi) + U \tan \phi + V(\sin \delta \tan \phi + F \cos \delta)] - S_j A}{b(F \sin \delta - \cos \delta \tan \phi)}$$

(8.7)

in which the tension crack is assumed to intersect the slope crest,

$$a = \tfrac{1}{2}\gamma H^2 \left[1 - \left(\frac{Z}{H}\right)^2\right] \cot \delta$$

and

$$b = \tfrac{1}{2}\gamma H^2$$

Through variation of parameters in Equation 8.2, Hoek and Bray showed the following. A reduction in S_j affects steep slopes more than flat slopes. A reduction in ϕ_j reduces the factor of safety of high slopes more than low slopes. In addition, filling a tension crack with water reduces the stability of all heights and angles of slopes. Drainage is frequently found to be effective in stabilizing rock slopes that exhibit tension cracks and other signs of incipient movement.

8.5 Analysis of Plane Sliding on the Stereographic Projection

The facility with which three-dimensional relationships can be graphed and manipulated in stereographic projection makes this method attractive for problems of rock slope stability, especially for wedge failures that are fully three dimensional. The basic step in applying the stereonet to such problems is the recognition that friction between surfaces can be represented by small circles in the projection. According to the definition of the friction angle ϕ_j, a block will remain at rest on a planar surface if the resultant of all forces acting on the block is inclined with the normal to the surface at an angle less than ϕ_j (Figure 8.11a). If the block is free to move in any direction, the envelope of all allowable resultant forces on the block is therefore a cone of vertex angle $2\phi_j$ centered around the normal to the plane (Figure 8.11b). This cone of static friction

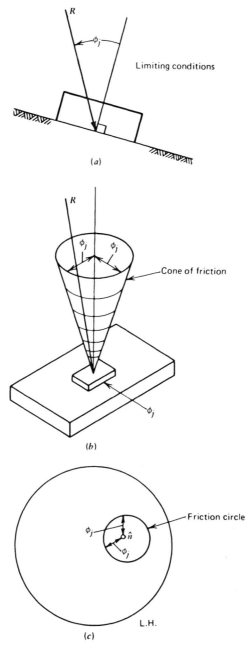

Figure 8.11 The friction circle concept. (a) Limiting conditions. (b) The cone of static friction. (c) Friction circle. **R** is "safe" if it plots inside the circle.

will be projected on the stereographic projection as a small circle of radius ϕ_j about the normal $\hat{n}$ (Figure 8.11c).

To project a small circle on the stereographic projection, plot two points on the diameter of the circle (p and q in Figure 8.12), bisect them to locate the center, and draw the circle with a compass. Do not make the mistake of placing the compass on the point $\hat{n}$, for it is a property of the stereographic projection that the geometric center of a small circle, representing a cone about an inclined axis in the reference sphere, migrates outward away from the projection of its axis.

The friction circle permits a quick and graphic examination of forces affecting the stability of a potentially sliding block. Forces can be entered in the projection as follows. Let $\mathbf{F}_1$ be a specific force acting on the block with magnitude $|F_1|$ and direction $\hat{f}_1$; that is, $\mathbf{F}_1 = |F_1|\hat{f}_1$. We can perceive the reference sphere as the locus of all unit vectors radiating from a point; $\hat{f}_1$ is one such unit vector. We can therefore represent $\hat{f}_1$ as a point on the projection. The magnitude $|F_1|$ will have to be noted down separately.

The direction of two forces, $\mathbf{F}_1$ and $\mathbf{F}_2$, are plotted in Figure 8.13. $\mathbf{F}_1$ is 20 MN plunging 30° to N 40° W. $\mathbf{F}_2$ is 30 MN plunging 40° to N 35° E. If the analysis of rotation is not considered, either force can be moved parallel to itself until it becomes coplanar with the other. The plane common to the two

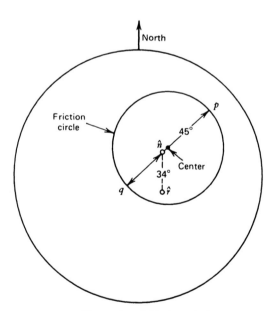

Figure 8.12 Example of friction circle analysis. $\hat{n}$ plunges 60° to N 50°E; $\phi_j = 45°$; $\hat{r}$ plunges 63° to S 55°E.

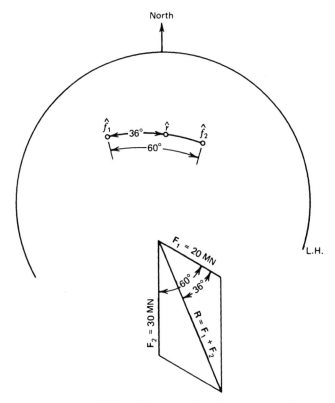

Figure 8.13 Addition of forces using the stereographic projection.

forces will then permit a vector resultant to be found, using the parallelogram rule. The stereonet allows us to find the common plane and the angle between two forces. Rotate the tracing containing $\hat{f}_1$ and $\hat{f}_2$ until they fall along the same great circle (denoted $\hat{f}_1\hat{f}_2$). Then measure the angle between $\hat{f}_1$ and $\hat{f}_2$ by counting the small circles between them. In this example, it is 60°. Then, in a separate diagram, add $\mathbf{F}_1$ and $\mathbf{F}_2$; as shown, their resultant $\hat{R}$ has direction $\hat{r}$, 36° from $\hat{f}_1$ in plane $\hat{f}_1\hat{f}_2$. The result of any number of vectors can be projected by repetition of this procedure and in this way the directions of the resultant of all forces acting on the block can be mapped.

If $\hat{r}$ lies within the friction circle, the block will remain at rest. For example, in Figure 8.12, $\hat{r}$ is safe as long as $\phi_j > 34°$. In the analysis we will use the conventions that $\hat{n}$ is the normal pointed out of the block, into the support, and the forces plotted are those acting on the block.

The forces that could enter into stability calculations include a block's self-weight, load transmitted from adjacent blocks, water forces, earthquake forces

Figure 8.14 The type of rock slope support often used for cuts in Venezuela, where decomposition of schistose rocks proceeds rapidly after excavation. The initial cut slope is coated with shotcrete and wire mesh and tied back with rock bolts or anchored cables.

or other dynamic loads, and support forces (Figure 8.14). The weight force plots in the center of the lower hemisphere:

$$\mathbf{W} = |W|\hat{w} \qquad (8.8)$$

The load of an adjacent block contains a normal force $\mathbf{F}_N$ parallel to the normal to the common face, and a shear force $\mathbf{F}_T$ parallel to the sense of shear motion on that face.

Water force $\mathbf{U}_1$ on the plane with normal $\hat{n}_1$ acts in direction $-\hat{n}_1$. If A is the area of plane 1 at the base of the block, the water force $\mathbf{U}_1$ is related to the average water pressure on the face $\mathbf{u}_1$ by

$$\mathbf{U}_1 = |\mathbf{U}_1|(-\hat{n}_1) = \mathbf{u}_1 \cdot A(-\hat{n}_1) \qquad (8.9)$$

An earthquake force can sometimes be treated as a "pseudostatic" force with constant acceleration $\mathbf{a} = \mathbf{K}g$. The inertial force is then

$$\mathbf{F}_I = \mathbf{K}g \, \frac{|\mathbf{W}|}{g} = \mathbf{K}|\mathbf{W}| \qquad (8.10)$$

$\mathbf{K}$ has dimensionless magnitude, and direction opposed to the earthquake acceleration. Since the latter will seldom be known, the most critical direction is usually selected.

The action of active supports (like pretensioned rock bolts) and passive supports (like retaining walls, grouted reinforcing bars, and dead weights) can also be plotted. Let the support force be

$$\mathbf{B} = |\mathbf{B}|\hat{b} \tag{8.11}$$

The best direction $\hat{b}$ can be found as the cheapest of a set of trial solutions, as shown in the following example.

In Figure 8.15, we reconsider the example plotted in Figure 8.12, in which $\hat{n}$

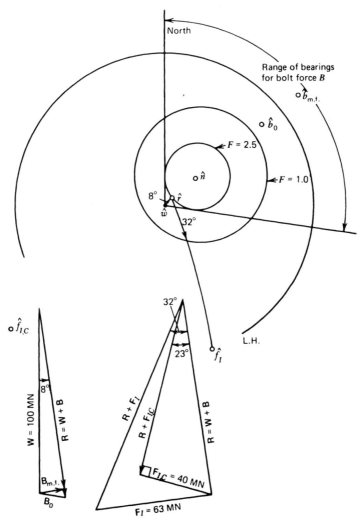

Figure 8.15 Application of the friction circle analysis to design of rock bolt support for a plane slide.

plunges 60° N 50° E and $\phi_j = 45°$. Suppose a potentially sliding block on this plane weighs 100 MN. We examine the following basic questions.

1. *What is the factor of safety of the block?* Factor of safety will be defined as

$$F = \frac{\tan \phi_{\text{available}}}{\tan \phi_{\text{required}}} \tag{8.12}$$

where $\phi_{\text{available}}$ is the friction angle adopted for design, and ϕ_{required} is the friction angle corresponding to limiting equilibrium conditions under a given set of forces; ϕ_{required} is the friction angle that causes the friction circle to pass through the resultant. In this example, ϕ_{required} is 30° giving $F = 1.73$.

2. *What rock bolt vector* **B** *should be used to raise the factor of safety to 2.5?* $F = 2.5$ gives $\phi_{\text{required}} = 22°$. The friction circle for $\phi_j = 22°$ is shown in Figure 8.15. Since **W** is vertical, the plane common to any bolting vector **B** and **W** will be a vertical plane, that is, a straight line through the center of the net. Any bearing for $\hat{b}$ between north and south 83° E can provide satisfactory support vectors. The actual direction chosen must integrate various factors including access, clearance for the drill, cost of drilling, and cost of steel. The *minimum support force* is provided by a specific vector oriented in order to rotate the resultant of **W** and **B** onto the 22° friction circle with minimum force. This direction is $\hat{b}_{\text{m.f.}}$, 8° above horizontal to the N 50° E. The force polygon shows the bolt force $\mathbf{B}_{\text{m.f.}}$ for direction $\hat{b}_{\text{m.f.}}$ is 13.9 MN. The *shortest length of bolts* is in direction $\hat{n}$, but the required support force will be considerably greater. Given cost data for steel and drilling, the optimum solution will be found. If the bolts are installed 10° below the horizontal, in direction $\hat{b}_0$ (Figure 8.15), the required bolt force $\mathbf{B}_0$ is 14.6 MN.

3. Assuming that bolts are installed in direction $\hat{b}_0$ and tensioned to a total force of 14.6 MN, *what acceleration coefficient will initiate block* sliding if the acceleration is horizontal in direction N 20° W, S 20° E? The inertial force that is most critical is directed to the S 50° W at an angle 23° + 90° from $\hat{r}$, that is, 15° upward ($\hat{f}_{IC}$ in Figure 8.15). However, we are given direction $\hat{f}_I$ horizontal to the S 20° E. Before the earthquake, the resultant force is in direction $\hat{r}$. During the earthquake, the resultant moves along the great circle common to $\hat{r}$ and $\hat{f}_I$ and the block starts to slip when the resultant has rotated through 32°. The force polygon shows the required inertia force KW is 63.4 MN, giving $Kg = 0.63$ g. The minimum yield acceleration in direction f_{IC} is 0.41 g.

Other examples of a similar nature and cases of wedge sliding are analyzed in the books by Hoek and Bray (1977) and Goodman (1976).[3] In the general case

---

[3] See also references by John, Londe et al., Hoek and Bray, and Hendron et al. The analyses of rock blocks using stereographic projection was pioneered by K. John. Wittke and Londe et al. proposed other approaches. Hendron, Cording, and Aiyer summarized this area in U.S. Army, Corps of Engineers, Nuclear Cratering Group Report 36, "Analytical and graphical methods for the analyses of slopes in rock masses" (1971).

of a block with more than one plane of possible sliding and one or more lines of intersection vectors, the friction circle is replaced by a more complex figure composed of small circles and great circles. In all cases, however, the whole sphere is divided into *safe* and *unsafe* regions. Forces are plotted and manipulated as in the example discussed here.

8.6 Analysis of Wedge Sliding Using the Stereographic Projection

As an illustration of the general method for blocks with more than one potential sliding surface, let us consider the case of a tetrahedral wedge with two free surfaces and two contacting planes (Figure 8.1*b*). The basic steps in the construction of a "generalized friction circle" will be sketched here. For a fuller explanation and exploration of these relationships, see Goodman (1976).

A wedge with two faces contacting rock presents three fundamental sliding modes: two modes of sliding on either plane alone or one mode sliding down the line of intersection. We cannot allow the block to slide down the dip vector of either plane, however, since it would then try to close the other plane. Thus, a set of directions is safe from sliding by virtue of kinematics.

To plot the safe zone corresponding to a wedge resting on planes 1 and 2, with line of intersection $\hat{I}_{12}$, perform the following steps.

1. Plot the normal vectors $\hat{n}_1$, $\hat{n}_2$ pointed out of the block, into the supporting planes. (In some cases, one or both of these normal vectors will be in the upper hemisphere, in which cases two separate hemisphere tracings can be used, or the projection can be done on one tracing extending the region of projection outside the horizontal circle. See Goodman (1976).)
2. Plot the line of intersection vector $\hat{I}_{12}$ pointed into the free space. (In some cases, $\hat{I}_{12}$ is directed into the upper hemisphere.)
3. Trace the great circles common to $\hat{n}_1$ and $\hat{I}_{12}$ (plane $\hat{n}_1\hat{I}_{12}$ on Figure 8.16) and common to $\hat{n}_2$ and $\hat{I}_{12}$ (plane $\hat{n}_2\hat{I}_{12}$ on Figure 8.16).
4. Along $\hat{n}_1\hat{I}_{12}$, mark points $\hat{p}$ and $\hat{q}$ at a distance ϕ_1 from $\hat{n}_1$ (Figure 8.16) where ϕ_1 is the friction angle for plane 1.
5. Along $\hat{n}_2\hat{I}_{12}$ mark points $\hat{s}$ and $\hat{t}$ at distance ϕ_2 from $\hat{n}_2$, where ϕ_2 is the friction angle for plane 2.
6. Draw great circles: through $\hat{p}$ and $\hat{s}$ and through $\hat{q}$ and $\hat{t}$.
7. Construct a friction circle of radius ϕ_1 about $\hat{n}_1$ and ϕ_2 about $\hat{n}_2$. Use only the portions of these friction circles shown on Figure 8.16, the remainders being kinematically inadmissible.

The generalized friction circle for the wedge is the ruled area in Figure 8.16. The three sliding modes belong to the three labeled sectors of this area. As in

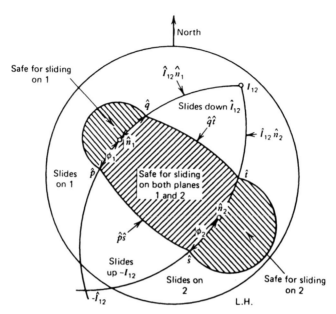

Figure 8.16 Example of analysis for a rock wedge: n_1 plunges 38° to N 60°W; $\phi_1 = 25°$; n_2 plunges 32° to S 52°E; $\phi_2 = 30°$.

application with the simple friction circle of Figure 8.15, a resultant is acceptable if it plots within the ruled area, and unacceptable if it plots outside the ruled area. In the two zones representing sliding on either plane alone, factor of safety can be defined and applied as before (Equation 8.11). However, in the mode of sliding on the line of intersection, there is an infinite combination of values of ϕ_1 and ϕ_2 that can shrink the safe zone down so as to pass through a given point inside the ruled area. Thus, there are an infinite number of factors of safety. A plot of the value of $\phi_{1,required}$ versus $\phi_{2,required}$ will reveal the sensitivity of the stability of the block to changes in either of these parameters. In view of the uncertainty with which friction angles are assigned in practice, it is more useful to express the degree of stability in terms of such a sensitivity study than to force it to respect the factor of safety concept.

8.7 Application of Block Theory to Rock Slopes

The principles of block theory introduced in Section 7.8 can be applied to rock slopes. Figure 8.17 shows a compound rock slope in two dimensions, contain-

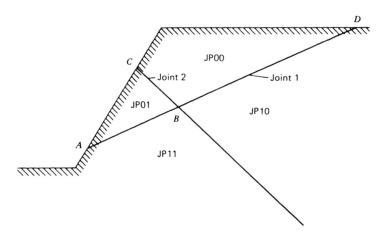

Figure 8.17 A two-dimensional rock slope, with two joints creating four joint pyramids.

ing a rock mass cut by two joint sets. The two joint sets create four joint pyramids (JPs): 00, 01, 10, and 11. By applying Shi's theorem to this simple case, we can determine that JP 01 can yield removable blocks in the steep face and JP 00 can yield removable blocks in either the face or upper surface of the excavation, or both simultaneously. It is also possible to produce a removable block with joint 1 alone, that is, with JP 20 where the symbol 2 represents the omission of a joint. The other JPs cannot produce removable blocks.

The first requirement for the removability of a block is that it be finite, that is, completely isolated from the mass by a continuous series of faces. Since the simplest such solid is a tetrahedron, the least number of faces that can isolate a block in three dimensions is four. The number of joint faces that will serve to isolate a tetrahedral block is four or less, depending on the number of faces that are formed by excavation planes.

The purpose of this discussion is to establish criteria for key blocks—blocks that tend to move as soon as freedom to do so is created by cutting the excavation slopes. Key blocks rarely contain parallel faces, that is, opposite faces produced by pairs of joints from the same set; such blocks tend to lock in place by virtue of the difficulty of overcoming joint roughness, as explored in Problem 7.19. Furthermore, key blocks cannot be created by excavating through already existing joint blocks because such blocks will prove to be tapered and therefore unremovable. Key blocks are removable blocks with nonparallel faces that tend to slide into the excavated space.

Figure 8.18 shows four types of tetrahedral key blocks, formed upon excavating slopes in a rock mass with three joint sets. Block 1 is removable in the steep excavated face alone, so that it has three faces created by joint planes. It

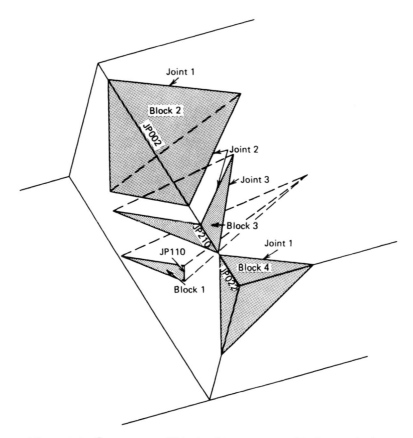

Figure 8.18 Some types of blocks that are removable in a rock slope.

is formed by JP 110, that is, by the intersection of the lower half spaces of joint planes 1 and 2 and the upper half space of joint plane 3. Block 2, an example of a common, important class of rock slope blocks, intersects both the steep excavation face and the upper free face, and therefore requires only two joint faces. The joint pyramid creating this block is 002, which intersects the upper half spaces of joint planes 1 and 2 and does not involve joint plane 3. Block 3 also intersects both the upper free surface and the free face and therefore needs only two joint planes; it differs from block 2 in that it intersects the lower half space of one joint and the upper half space of a second (JP 210). Finally, Block 4 intersects three excavation surfaces and therefore requires only one joint plane. It is rare to have a key block formed with only one joint plane; this happens when two excavations intersect, as depicted, or when a rock cut intersects a tributary valley.

Block theory allows the analysis of removability to be conducted in three dimensions, as discussed previously, using Shi's theorem and the stereographic projection. Recall that the joint pyramids project as spherical polygons, as in Figure 7.20. A removable block must be formed with a JP that has no intersection with the excavation pyramid (EP). For a simple, planar rock cut formed by one excavation plane, like face 1 in Figure 8.19a, the excavation pyramid is the rock surface *below* the excavation plane. The EP for this case is therefore the region *outside* the great circle corresponding to the dip and dip direction of the face. (This rule assumes that the projection is made with a lower focal point, so

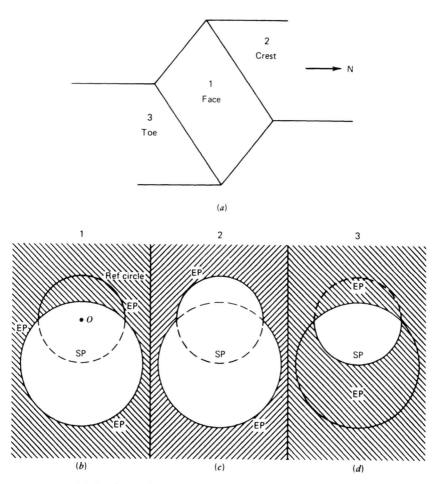

(a)

(b)   (c)   (d)

Figure 8.19 (a) Regions of a compound rock slope; (b) EP and SP for the face, (c) for the crest, (d) for the toe.

that the upper hemisphere is inside the reference circle. If an upper focal point projection is used, which places the lower hemisphere inside the reference circle, the EP will be inside the great circle for the excavated face.) The reference circle and the great circle for the rock cut are drawn in Figure 8.19*b*. The EP is the ruled area outside the latter. It is convenient to define a "space pyramid" (SP) as the complement of the EP. The SP is then the unruled region inside the great circle for the cut. A JP is removable if and only if it projects entirely within the SP.

Figure 8.19 also shows how to form the EP and SP for a compound excavation. The crest of the slope involves the face and the upper free surface, which is horizontal in this illustration. The excavation pyramid for the crest is the convex region formed by the intersection of the lower half spaces of each of these planes; as shown in Figure 8.19*c*, it is projected by the regions that are simultaneously outside both great circles—the smaller one for the upper free face and the larger one for the steeply inclined free face. The space pyramid is then the nonconvex region simultaneously above both planes, which projects as the region either in one circle or in the other. This larger SP provides a longer list of removable JPs. Conversely, the toe region creates an EP that is the nonconvex region below both the steep face and the lower free surface. It projects as the shaded region outside of either circle. The SP is the unshaded region simultaneously inside each circle. Since the SP for the toe of the slope is so much smaller than the SP for the face, there are fewer JPs that will be removable in the toe. The principles stated here can be applied to an inside or outside edge of any two nonparallel intersecting excavation faces, so that the removable blocks can be found for excavations of any shape in plan or section.

Every JP that has a removable block deserves further analysis. If the resultant force on a block is due only to gravity, we can disregard any block whose removability requires lifting or upward sliding. For example, even though block 2 of Figure 8.18 is removable in the upper free surface, it cannot fail that way under gravity alone (although it could during an earthquake or blast or in response to water forces or the pull of a cable). A complete analysis of permissible modes for all JPs, termed a "mode analysis," will be discussed later. When there are four or more joint sets, it will be found that certain JPs lack any permissible mode. If the orientation of the excavation can be established such that only blocks lacking a sliding mode are removable, the excavation will then be completely safe without any support.

Blocks that are removable, and do have a mode of sliding consistent with the direction of the resultant force must be subjected to limit equilibrium analysis, to determine whether they require support or will be restrained safely by friction. We have already discussed methods for limit equilibrium analysis of blocks in rock slopes. The contribution of block theory is that it lets us determine immediately the controlling directions of normals and lines of intersection to be used in such analyses.

LIMIT EQUILIBRIUM ANALYSIS OF JPS

Each JP has a specific equilibrium analysis. It makes no sense to consider the results of such an analysis if the blocks formed by the JP are not demonstrably removable. Therefore, we must first determine which blocks are removable and then select the critical JPs for stability analysis.

For example, consider again the rock mass having the system of joints projected in Figure 7.20; the dip/dip direction for the three joint sets are 30°/90°, 60°/45°, and 70°/330°. We will consider a possible surface excavation making a rock cut dipping 70° to azimuth 300° (70/300). Figure 8.20 is a lower focal point stereographic projection of the joints with the JPs identified. The great circle for the rock cut is the dashed circle. The EP is the region outside this circle and the SP is the region inside it. The removable blocks are those formed of JPs that plot entirely inside the SP. There is only one—JP 100.

Let us review the concepts presented earlier in our discussion of the friction circle (Section 8.5) by means of a simple two-dimensional example. In Figure 8.21 we examine the stability of blocks formed with JP 01, which is the angle above joint plane 1 and below joint plane 2. If the resultant force is directed from O with its tip anywhere in this angle, any block created from JP 01 must lift off both joints. We term this mode of failure "mode O." Now we establish normals to each joint plane *pointed out of the JP*. If the resultant force is directed from O with its tip in the angle between these normals, the blocks

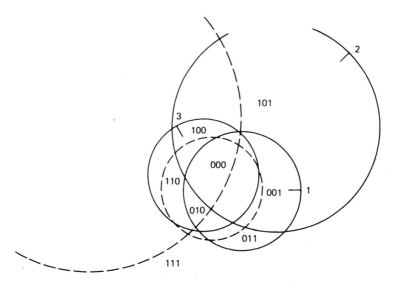

Figure 8.20 Stereographic projection (lower focal point) for the rock mass of Fig. 7.20, with the addition of an inclined rock cut with dip/dip-direction 70/300 (dashed). The JPs are labelled.

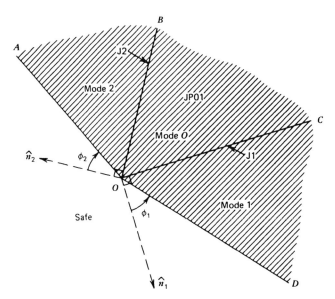

Figure 8.21 Graphical representation of the limit equilibrium analysis in two dimensions. The safe zone is unshaded.

formed of JP 01 will be absolutely stable, regardless of the friction angles. For a specific pair of friction angles on joints 1 and 2, a limit to the safe orientations of the resultant force can be established by laying off the friction angles from the normal to each plane toward its trace in the figure as shown. The whole set of all orientations for the resultant force is now divided into a safe subset, represented by directions in the unshaded region, and an unsafe subset, represented by directions in the shaded region.

The same principles apply in three dimensions. In place of an angle, a JP becomes a pyramid with apex at the origin and no base. The stereographic projection of the pyramid is a curved polygon bounded by the great circles of the appropriate joint planes. The corners of the curved polygons are the edges of the JPs, which are the lines of intersection of joint planes. Figure 8.22a shows the stability analysis for one JP (000) conveniently located in the center of the upper hemisphere.

Lifting of a block off all planes, in mode O, occurs if and only if the resultant force, with its tail at the origin, is directed inside the joint pyramid. The region inside JP 000 has accordingly been labeled "O."

Sliding on one plane necessitates a slip direction in that plane that is intermediate between its bounding edges, as depicted in Figure 8.23. Denoting the sliding plane as plane i, and the adjacent planes as j and k, the slip direction $\hat{s}$ must therefore be determined by the sum of positive components in directions $\hat{I}_{ij}$ and $\hat{I}_{ik}$. The resultant force is uniquely composed of a shear force parallel to $\hat{s}$

and a normal force directed along the *outward* normal $\hat{n}_i$ (the normal to plane i that lies along a trajectory away from the center of the JP). Combining these statements requires that the resultant force lie inside the spherical triangle whose corners are $\hat{I}_{ij}$, $\hat{n}_i$, $\hat{I}_{ik}$. This triangle has as its corners two corners of the JP and the outward normal to plane i. The three modes of sliding on one plane, modes 1, 2, and 3, are thus found by plotting each of the outward normals and constructing great circles to the corners of the JP as shown in Figure 8.22a.

For sliding on two planes, say i and j, the direction of sliding must be parallel to their line of intersection $\hat{I}_{ij}$; then there is only one direction of sliding $\hat{s}$. Consequently, the resultant force is derived uniquely from positive components parallel to the outward normals $\hat{n}_i$ and $\hat{n}_j$ and the direction of sliding $\hat{I}_{ij}$. On the stereographic projection, the two-plane sliding mode ij will lie inside the spherical triangle whose corners are $\hat{n}_i$, $\hat{I}_{ij}$, and n_j. Figure 8.22 labels (for JP 000) the three modes of this type: 12, 23, and 31.

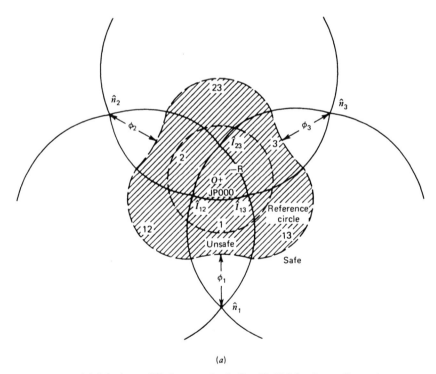

(a)

Figure 8.22 (a) Limit equilibrium analysis for JP 000 in three dimensions (lower focal point stereographic projection). The dip/dip-direction of the joints are: 45/0; 45/120; and 45/240. All joints have a friction angle of 25°. The safe zone is unshaded.

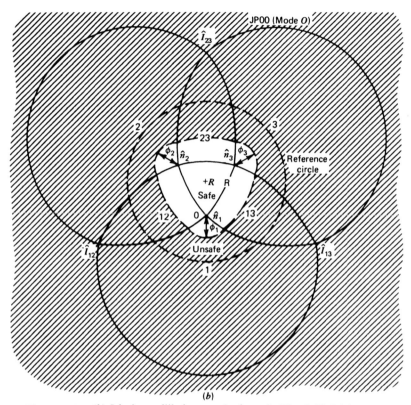

Figure 8.22 (*b*) Limit equilibrium analysis as in Fig. 8.22 (*a*) but projected from an upper focal point.

The limit of inclination of the resultant force from the direction $\hat{n}_i$ is determined by the friction angle ϕ_i. The friction angles determine partial friction circles in each of the regions corresponding to sliding on one plane. The three partial friction circles are shown in Figure 8.22a; their intersections with the great circles bounding the modes ij determine points like $\hat{p}$, $\hat{q}$, and t of Figure 8.16. Great circles constructed between these points, through the respective regions ij, complete the construction of the generalized friction circle for the JP, subdividing the whole sphere of directions into a subset of *safe* orientations and a subset of *unsafe* directions of force. Equilibrium can be maintained without acceleration of the block only when the resultant force plots inside the safe region.

For a stability analysis under gravity as the main force, most of the region of interest is in the lower hemisphere. Therefore, it is useful to prepare a stereographic projection of the stability regions with an upper focal point. Figure 8.22b shows the stability analysis for JP 000 projected from an upper

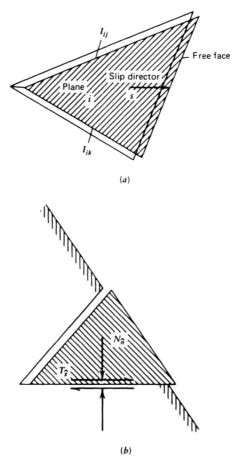

Figure 8.23 Sliding on plane i: (a) seen
normal to the plane; (b) seen in a section
parallel to the sliding direction s.

focal point, producing the lower hemisphere inside the reference circle. Now
the generalized friction circle can be seen as a locus surrounding the spherical
triangle defined by great circles connecting each outward normal. When the
resultant force plots inside the spherical triangle of the normals, the block is
absolutely safe even in the absence of any friction.

The preceding example was developed for JP 000 simply to center the JP in
the lower focal point projection. However, JP 000 can never slide under gravity
alone so it will not yield key blocks of a rock slope. In Figure 8.20, we determined that JP 100 was the sole removable block in a cut dipping 70/300 for a
rock mass with three sets of joints: 30/90; 60/45; 20/330; Figure 8.24a shows the
stability analysis for JP 100, with a lower focal point stereographic projection.

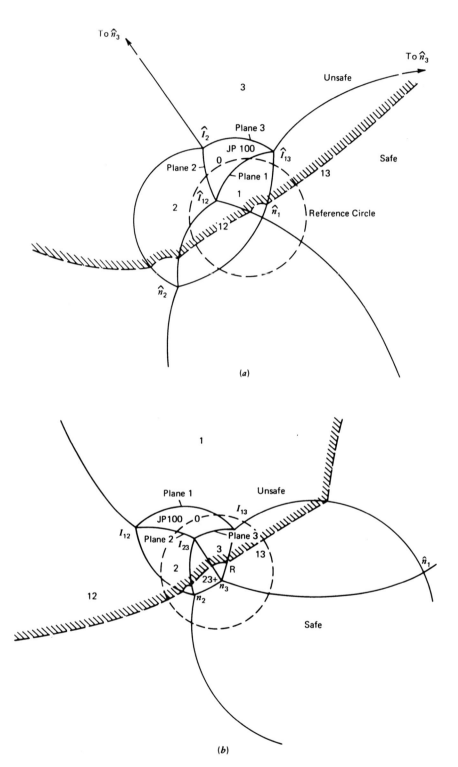

(a)

(b)

The outward normals to planes 1 and 2 plot inside the figure but $\hat{n}_3$ is too far away to be plotted at the selected scale. The stability regions have been identified, with one or two digits according to the number of sliding planes. The friction angles 25°, 16°, and 35° (as in Figure 7.20) have been laid off from the normals and the generalized friction circle constructed, as shown. An upper focal point projection of the stability analysis for JP 100 is presented in Figure 8.24b. Also shown is the direction of gravity, denoted by a cross at the center of the reference circle, from which we see that JP 100 is safe and its sliding tendency is in mode 3 (sliding on plane 3 alone). A very small force toward the west, as, for example, a water pressure on plane 1 (directed into the block and therefore opposite to $\hat{n}_i$) would drive the block into mode 23 (see Problem 17).

MODE ANALYSIS

The application of block theory to slope design has made use of a "removability analysis" and a limit equilibrium analysis. The former determines which JPs can define removable blocks in a specific excavation. The latter examines the statics of blocks corresponding to a given JP. The limit equilibrium analysis plotted on the stereographic projection may be considered as a map showing the influence of changing the direction of the resultant force, everything else being held constant. It is also useful to study the influence of changing the JP code while the direction of the resultant force is held constant. Such a graph is termed a "mode analysis."

Goodman and Shi (1985) show how to perform the mode analysis for a general direction of the resultant force. Here we examine the simple special case corresponding to the resultant force being identical to the direction of gravity. The object is to determine the mode of behavior for each JP when the rock mass is subjected to gravity alone. The mode can be realized, of course, only if the block is removable, and this is not represented in the output, but must be determined independently.

Each JP has one of the following possibilities: mode O, corresponding to opening of each joint; mode i corresponding to sliding on plane i and opening from all the other planes; mode ij corresponding to sliding on planes i and j simultaneously (with movement along their line of intersection) and opening from all other joint planes; and no mode, corresponding to safety even with

---

Figure 8.24 (*a*) Stability analysis for JP 100 in the rock mass of Figure 8.20; a lower focal point projection (placing the upper hemisphere inside the reference circle).

Figure 8.24 (*b*) The same as Fig. 8.24 (*a*) but projected from an upper focal point, placing the lower hemisphere inside the reference circle.

zero friction on every joint plane. Each mode can be established by observing the following rules.

• *Mode O* belongs to the JP that contains the downward direction; in an upper focal point projection, this is the JP that contains the center of the reference circle.

• *Mode i* is the JP inside the segment of the great circle for plane i that contains the dip vector of plane i (in a lower focal point projection). In the lower focal point projection, the dip vector will be the farther intersection of a radius from the center of the reference circle with the great circle for plane i.

• *Mode ij* belongs to a JP that has the lower hemisphere intersection I_{ij} as one of its corners. There are several of these but only one obeys both of the following rules. If the dip vector of plane i is *inside* the circle for plane j, then the JP with mode ij is *outside* the circle for plane j, and vice versa; also, if the dip vector of plane j is *inside* the circle for plane i, then the JP with mode ij is *outside* the circle for plane i, and vice versa. (These rules derive from two inequalities, established by Goodman and Shi (1985).)

• JPs that have none of the modes have no mode.

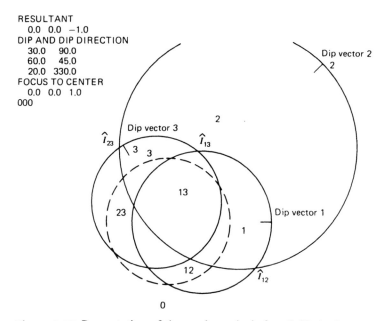

Figure 8.25 Presentation of the mode analysis for all JPs in the rock mass of Figure 8.20; lower-focal-point stereographic projection.

Figure 8.25 shows the results of a mode analysis for the rock mass of Figure 8.20, with three sets of joints. Note that JP 100 (compare with Figure 8.20) has mode 3, which is the result found in limit equilibrium analysis of this JP. There is one JP (010) without a mode.

8.8 Analysis of Slides Composed of Two Blocks

It was suggested that a useful method for back calculating an actual failure experience is to solve Equation 8.2 for S_j, the "cohesion" along the sliding surface. In Chapter 5, however, we observed that discontinuities in hard rock do not exhibit cohension at low pressures but rather acquire an enlarged friction angle from asperities along the shear surface. The normal pressures on slides beneath surface excavations and natural slopes are often quite low in comparison with the shear strength of intact rock so that if "cohesion" is calculated from analysis of an actual failure, it may really reflect some other mechanism.

One such mechanism is depicted in Figure 8.26 where a sliding surface that does not meet the kinematic condition for "daylighting" is joined by a second, flatter surface through the toe of the slope. The reserve of strength in the toe (the "passive region"), which rests on a relatively flat sliding surface, is over-

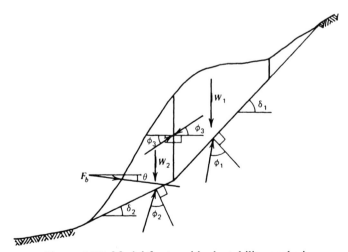

Figure 8.26 Model for two-block stability analysis.

come by excess force transmitted from the upper region (the "active block"), which cannot remain at rest by friction along its basal surface alone. A requirement for this mode of failure is that the upper surface be inclined steeper that ϕ_j while the lower surface be inclined flatter than ϕ_j. Assuming that the boundary between the active and passive blocks is vertical, analysis of the system of forces shown in Figure 8.26 yields

$$F_b = \frac{W_1 \sin(\delta_1 - \phi_1)\cos(\delta_2 - \phi_2 - \phi_3) + W_2 \sin(\delta_2 - \phi_2)\cos(\delta_1 - \phi_1 - \phi_3)}{\cos(\delta_2 - \phi_2 + \theta)\cos(\delta_1 - \phi_1 - \phi_3)}$$

$$(8.13)$$

where F_b is the support force in direction θ below horizontal required in the passive block to achieve limiting equilibrium with the friction angles input in the equation

ϕ_1, ϕ_2, ϕ_3 are the friction angles applicable to sliding along the upper, lower, and vertical slide surfaces, respectively

δ_1 and δ_2 are the inclinations of the upper and lower slide surfaces, respectively

W_1 and W_2 are the weights of the active and passive blocks per unit of slide width

For simplicity, assume that all the friction angles are identical.

The factor of safety of the slope can then be determined, given values for F_b, W_1, W_2, δ_1, and δ_2, by the following procedure. Calculate the value of the friction angle that makes Equation 8.12 true by solving for the root of this equation. This defines the required angle of friction to assure equilibrium (i.e., the value of $\phi_{required}$). For a given value of $\phi_{available}$, the factor of safety can now be calculated from Equation 8.12.

In an actual case of an incipient slide, survey data will define the magnitudes and directions of the resultant displacements at the top and toe of the slope, as well as elsewhere. If the magnitudes of the resultant displacement are constant throughout the slope, and the directions of displacement are outward and downward, a rigid, sliding type of motion is possible. The directions of the resultant displacement vectors can then be used to determine the values of δ_1 and δ_2 and the location of the tension crack will permit graphical determination of W_1 and W_2. Assuming that the factor of safety is unity, the value of $\delta_{available}$ may be calculated as the root of Equation 8.12. The increase in factor of safety achieved by a given quantity of excavation in the active region, or fill in the passive region, or anchoring in the passive region can be evaluated by inputting this value of $\phi_{available}$ and determining $\phi_{required}$ for the new conditions, as explained above. Problem 9 offers an example.

References

Goodman, R. E. (1976) *Methods of Geological Engineering in Discontinuous Rock*, Chapters 3 and 6, West, St. Paul, MN.

Goodman, R. E. and Bray, J. W. (1977) Toppling of rock slopes, *Proceedings, Speciality Conference on Rock Engineering for Foundations and Slopes*, ASCE (Boulder, Colorado), Vol. 2. pp. 201–234.

Goodman, R. E. and Shi, G. H. (1985) op. cit., Chapter 1.

Heuze, F. E., and Goodman, R. E. (1972) Three dimensional approach for design of cuts in jointed rock, *Proceedings, 13th Symposium on Rock Mechanics* (ASCE), p. 347.

Hoek, E. and Bray, J. W. (1974, 1977) *Rock Slope Engineering*, Institute of Mining and Metallurgy, London.

John, K. W. (1968) Graphical stability analyses of slopes in jointed rock, *J. Soil Mech. Found. Div.* (ASCE) **94** (SM2): 497–526.

Londe, P., Vigier, G., and Vormeringer, R. (1969) Stability of rock slopes, a three dimensional study, *J. Soil Mech. Found. Div.* (ASCE) **95** SM1): 235–262.

Londe, P., Vigier, G., and Vormeringer, R. (1970) Stability of rock slopes, graphical methods, *J. Soil Mech. Found. Div.* (ASCE) **96** (SM4): 1411–1434.

Pentz, D. T. (1976) Geotechnical factors in open pit mine design, *Proceedings, 17th Symposium on Rock Mechanics* (University of Utah), paper No. 2B1.

Schuster, R. L. and Krizek, R. J. (Eds.) (1978) *Landslides—Analysis and Control*, Trans. Res. Board Special Report 176 (NAS), including Chapter 9, Engineering of rock slopes, by D. R. Piteau and F. L. Peckover, and Chapter 2, Slope movement types and processes, by D. J. Varnes.

Wittke, W. (1965) Methods to analyze the stability of rock slopes with and without additional loading (in German), *Rock Mechanics and Engineering Geology, Supplement II*, p. 52.

Problems

1. A rock mass to be excavated in an open cut has the following recurrent discontinuities:

   > Set 1 (Bedding) strikes N 32° E, dipping 75° N 58 W.
   > Set 2 (Jointing) strikes N S and dips 65° E.
   > Set 3 (Jointing) is horizontal.

   Plot all the dip vectors, lines of intersection, and normal vectors (poles) on a lower hemisphere stereographic projection.

2. Assuming $\phi_j = 25°$ for each of the discontinuity surfaces, prepare a table giving the steepest safe slopes every 15° around a circular open cut in the

rock of Problem 1, respecting all modes of failure. What would be the best orientation for a highway cut through a ridge in this rock?

3. We are given a plane P daylighting into a cut and having attitude as follows: strike N 30° W, dip 50° NE. The weight of a potentially sliding mass on plane P is 400 tons on an area of 200 m² (metric tons). The friction angle is believed to be 30°.
   (a) Find the direction and magnitude of the minimum rock bolt force to achieve a factor of safety of 1.0, and a factor of safety of 1.5.
   (b) What water pressure acting on plane P could cause failure after rock bolts are installed for a safety factor of 1.5?
   (c) Is the direction of rock bolting for minimum required support force necessarily the best direction in which to install the rock bolts in this problem?

4. A block weighing 200 MN rests on a plane striking north and dipping 60° W The available friction angle is believed to be 33°.
   (a) Find the minimum force for stabilizing the block with a factor of safety of 2.0 using rock bolts.
   (b) Find the force for stabilizing the block with a factor safety of 2.0 if the bolts are installed 10° below horizontal to the N 76° E.
   (c) What seismic coefficient K initiates slip of the block if the inertia force acts horizontally to the north? (Assume the bolts were installed as in case (b) before the acceleration.)

5. (a) Using a kinematic analysis find the maximum safe angles for cuts on both sides of a highway oriented N 60° E, through a granitic rock mass with the following sets of discontinuities.
   (1) Strike N 80° E, dip 40° N.
   (2) Strike N 10° E, dip 50° E.
   (3) Strike N 50° W, dip 60° NE.
   Assume $\phi_j = 35°$.
   (b) Considering rock cut stability alone, what is the best direction for the cut?

6. The block of rock shown in the following diagram rests on a plane inclined δ with the horizontal. The angle of friction is ϕ_j. At what value of δ will the block:
   (a) Begin to slide?
   (b) Begin to overturn?

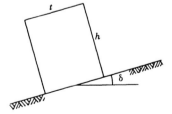

7. The two blocks of rock shown in following diagram rest on an inclined plane. Assume the friction angle is the same for sliding between the blocks and for sliding of a block along the surface. Discuss the conditions for equilibrium of the two block system.

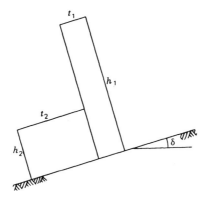

8. (a) The following diagram shows a mode of failure in high dip slopes. In coal mines, this behavior is sometimes called a "footwall failure." If the failure initiates by buckling as shown, derive a formula expressing the maximum length of slope l above the top of the buckling section such that failure does not occur. Assume the toe of the slope is fully drained and neglect the weight of the buckled column. The rock has modulus of elasticity E, unit weight γ, and interlayer friction angle ϕ_j.

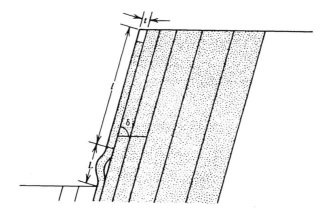

(b) Evaluate the formula to determine the critical slope length for $L = 40$ m, $\phi_j = 10°$, $\delta = 80°$, $t = 0.5$ m, $\gamma = 0.027$ MN/m³, and $E = 3 \times 10^3$ MPa.

9. A creeping rock slide above a road has moved a total of 3 m. The direction of resultant displacement at the top of the slope is 60° below the horizontal

while the direction of the resultant displacement in the lower portion of the slope is 25° below the horizontal. A cross section of the slope with these angles and the known position of the tension crack above the slide determines the area of the active block to be 10,000 m³ and the area of the passive block to be 14,000 m³. The rock weighs 0.027 MN/m³. There are no rock anchors in the slope.

(a) Assuming that the factor of safety of the slide is now 1.0 and that all the friction angles are the same (i.e., $\phi_1 = \phi_2 = \phi_3$), calculate the available friction angle.

(b) Calculate the increase in the factor of safety if 4000 m³ are excavated from the active block and removed from the slide.

(c) Calculate the horizontal anchorage force required per unit of slide width to achieve the same factor of safety as the excavation in (b). Roughly how many anchors do you think this requires?

10. Discuss how one could report a "factor of safety" for a case like that of Problem 9 where ϕ_1 and ϕ_2 are not constrained to be equal to each other. (Assume ϕ_3 is a fixed value.)

11. Derive an equation corresponding to Equation 8.6 for a potential slide in which the tension crack intercepts the face of the slope.

12. A prismatic block between two parallel vertical joints, J_1 and J_2, tends to slide on a fault $(P3)$ dipping $\delta°$, as shown in the figure. Assume the joints are smooth and have identical friction angles ϕ_j.

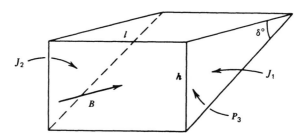

(a) A horizontal support force B is applied to the block. Derive a formula for the support force B corresponding to limiting equilibrium as a function of the stress σ_j normal to the joints. Express the answer in terms of the ratio B/W.

(b) If $\delta = 60°$, $\phi_3 = \phi_j = 30°$, find the σ_j required for limiting equilibrium without support $(B = 0)$ for blocks of widths $l = 1, 5, 10$, and 20 m. (Assume $\gamma = 27$ kN/m³.) Give your answer in kPa and psi.

13. (a) In Problem 12, the friction angle of the side joints is assumed to be given. Suppose there is no initial stress σ_j on these joints but that their

roughness angle i causes a stress to build up as the block displaces by amount u downslope. Assume that the rock to either side is rigid, and the sliding block has modulus of elasticity E. Determine the normal stress and normalized required support force B/W as a function of block displacement. (*Hint:* Assume that no dilatancy is allowed on the side joints.)

(b) Find the displacement u required for limit equilibrium in the case of Problem 12b if the initial normal stress is zero, the joints have dilatancy angle $i = 10°$, and the rock has modulus of elasticity $E = 2 \times 10^4$ *MPa*.

14. A block is formed from the upper half spaces of two joint planes and the lower half spaces of two excavation planes. Block theory analysis demonstrates that the block is removable in the excavation and we now want to assess its stability. The dip and dip direction values for the two bounding joint planes are as follows:

Joint plane	Dip	Dip Direction
1	40°	100°
2	50°	220°

The block is estimated to weight 100 metric tons.

(a) Assuming the friction angles are 20° on plane 1 and 15° on plane 2, compute the support force required to achieve a limiting equilibrium condition if the bolts are installed 10° below horizontal to the N 10°E (azimuth 10°). Use the stereographic projection, supplemented by vector calculations.

(b) If the friction angle is actually 30° on plane 2, and the bolts are installed as calculated in part (a), what angle of friction on plane 1 would be required to produce a limiting equilibrium condition for the block?

15. Label all the JPs of Problem 25, Chapter 7, with the appropriate mode for the resultant force of gravity alone.

16. For the conditions of Figure 8.24, calculate the water pressure on face 1 that will fail a block of 50 metric tons weight with JP 100. The area of the face with plane 1 is 7.5 square meters. (*Hint:* The water force on plane 1 acts in the direction $-\hat{n}_1$ opposite to the outward normal $\hat{n}_1$.)

17. A slope will be cut with dip and dip direction 50°/30° in the rock mass of Problem 25 in chapter 7, with joint sets 30°/70°, 50°/140°, and 60°/270°.

(a) Determine the JP that generates removable blocks in this rock cut.

(b) Anchors are installed horizontally to the south. Find the anchor force required to achieve a factor of safety of 2.0 in each relevant joint plane for a block weighing 90 tons. The friction angle believed to belong to each joint plane is 35°.

Chapter 9

Applications of Rock Mechanics to Foundation Engineering

9.1 Rock Foundations

This chapter concerns the behavior of rock as a structural foundation. Compared to soils, most rocks are strong and stiff and carrying a structural load down to rock usually assures a satisfactory bearing. However, large loads, as, for example, from a skyscraper or bridge pier, can by design cause pressures approaching the bearing capacity of even moderately strong rocks. If the rock is defective, this may provoke relatively large deformations, particularly when the rock is inherently weak, like some chalks, clay shales, friable sandstones, tuffs, or very porous limestones, or when the rock is weathered, cavernous, or highly fractured. Sowers (1977) reported, for example, that settlements of up to 8 in. occurred under loaded areas on weathered, porous limestone with foundation pressures less than 10 kPa. Thus there are numerous instances where the rock has to be evaluated carefully in foundation engineering.

Figure 9.1a shows the ideal condition for making use of rock as a foundation in preference to direct bearing on the soil. The rock is strong and relatively free from fractures and the bedrock surface is smooth, horizontal, and sharply defined. In weathered rock, by contrast (Figure 9.1b), the bedrock surface may be indefinable and the rock properties may vary widely over short distances vertically or horizontally, confusing those responsible for predicting the foundation elevation and allowable bearing values. Karstic limestones, depicted in Figure 9.1c, possess a highly sculptured, uneven bedrock surface, with cliffs, slopes, and variable and unknown soil depths, and irregular groundwater levels, as well as hidden caverns, clay seams, and rock of unpredictable quality.

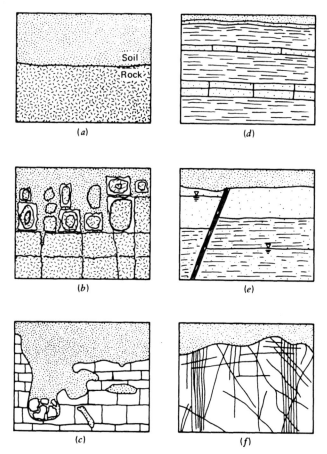

Figure 9.1 Types of bedrock surfaces. (*a*) Glacial till over bedrock. (*b*) Decomposed granite. (*c*) Karstic limestone. (*d*) Weathered rock and residual soil over sandstone and shale. (*e*) Soil over a fault in sedimentary rocks. (*f*) Transported soil over fractured rock.

Karst terrain can consequently create treacherous subsurface conditions. Figure 9.1*d* portrays rock having rhythmically changing properties due to interbedding of hard (cemented sandstone) and soft (claystone) layers. The net properties may be more troublesome than either alone since the strong layers may lack the flexural rigidity and strength to resist the bearing forces yet complicate the driving of piles or drilling of piers. Faults can cause additional foundation problems, by virtue of compressible gouge (Figure 9.1*e*), altered wall rock, and offset groundwater levels; faults also tend to complicate interpretation of depth to load-bearing strata. Highly fractured rock, depicted in Figure 9.1*f*, may also cause a marked reduction in safe-bearing pressures, as otherwise satisfactory rocks undergo appreciable deformations connected with closing and sliding of joints. Furthermore, when buildings are located near cliffs, throughgoing fractures "daylighting" in the cliff face can undermine their stability.

Another class of foundation problems occurs in rock with expansive or unstable minerals such as some montmorillonitic clay shales, nontronitic basalts, and rocks mineralized with pyrrhotite, marcasite, or certain other sulfides. Sulfuric acid released in the oxidation of the latter may attack concrete. Highly soluble rocks like gypsum and salt will require special attention in foundations of water-impounding structures, or structures located adjacent to operating wells or drains. Serious rock foundation problems also arise in formations underlain by completely or partially mined-out coal, sulfur, salt, or other mineral commodities. Assuring support to structures on the surface overlying abandoned mine workings requires special investigations and sometimes expensive treatment.

Engineering works present a wide variety of rock foundation problems. Homes, warehouses, and other light structures rarely create loads that test even weak rocks, but may require rock investigations in connection with cavernous, or mined-out substrata, or in areas of expansive rocks. Large public buildings like hospitals, office buildings, and airport terminals may have very large and rather modest loads acting near each other; because such facilities frequently cover a relatively large area, they may encompass varying foundation conditions and engineering solutions. Some industrial structures like turbines, boilers, reactors, and accelerators make stringent demands for precise and continued alignment that necessitate detailed investigations of foundation behavior even when dealing with good rock. Towers and very high buildings may generate large vertical and horizontal loads in response to wind or seismic forces. Bridges not only require foundations to be constructed through water and soil to bedrock but also place piers on steep valley sides where rock slope stability analysis becomes part of the foundation engineering work (Figure 9.2a). This is also true of dams, which can create relatively large inclined loads at their base and in their valley side abutments. Concrete arch dams transfer some of the reservoir and structural load to the abutment rock (Figure 9.2b) while concrete gravity and concrete buttress dams direct the load primarily into the foundation rock. Earth and rock-fill dams create smaller, usually tolerable stresses and deformations in rock foundations. All types of dams may suffer problems due to seepage in fractured or karstic foundations and all can be adversely affected by rock slides in the abutments, whether due to seepage forces, structural loads, or other causes.

To support building loads with tolerable deflections, it is possible to use several types of foundations. We will concern ourselves only with those intended to transfer some or all of the load to rock. Figure 9.3a shows a common solution where a modest excavation through the soil permits a *footing* to bear

Figure 9.2 Foundations of a bridge and a dam in very steep terrain. (*a*) Footings for the Glen Canyon Bridge, built by the U. S. Bureau of Reclamation across a precipitous canyon in Navajo sandstone. The small, dark squares on the rock are rock bolts. (*b*) The other abutment of the bridge and the left side of Glen Canyon arch dam.

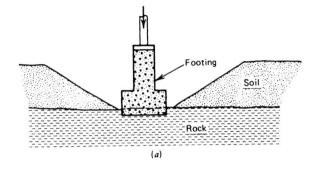

(a)

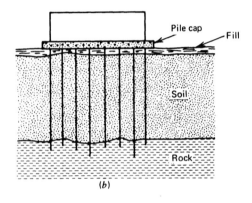

(b)

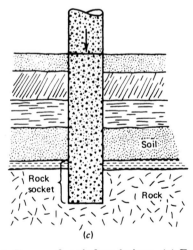

(c)

Figure 9.3 Types of rock foundations. (a) Footing on rock. (b) End-bearing piles on rock. (c) A pier socketed into rock.

directly against a prepared rock surface. Depending on the nature of the work and the magnitude of the load, the rock may be either simply inspected, drilled, and tested, or proof loaded prior to confirming the foundation grade. Setting foundation grade is often left to be determined during construction and may depend mainly on judgment or on rock classification (e.g., using the geomechanics classification discussed in Chapter 2). During construction the stability of the slope cut through the soil and through the weathered rock needs to be assured, the excavation must be drained and cleared of debris so that a good contact with concrete can develop, and the load-bearing surface must be kept from deteriorating in the interval before concreting. For footings carrying only moderate load, design may be dictated to meet special requirements of the structural engineer and architect without any reference to rock-bearing capacity or rock settlement. But large loads or rock marginal in quality may demand rational evaluation of the allowable loads through calculations or tests. Procedures for doing this will be discussed in subsequent sections.

Piles (Figure 9.3*b*) are driven to carry loads down to a satisfactory bearing layer. They may be driven from the ground surface or cast in drill holes. If the overburden is soft or if the piles are fairly short, most of the reaction comes from the pile tip; in this case, the pile is usually driven as much as a meter, occasionally more, into the rock until a specified number of blows is required to penetrate a fixed distance. Piles can be driven in this manner into weak rocks like chalk, tuff, claystone, and weathered rocks of many varieties, but they cannot be driven more than a few centimeters into fresh, hard rocks like limestone or sandstone unless equipped with hardened steel-driving points. It is difficult to guarantee pile seating in the case of an irregular or inclined bedrock surface. In fact, steel piles driven at small angles against a limestone surface have been destroyed by bending as they skidded off the rock. Piles cast in boreholes may develop significant side resistance in bond against weathered rock and overburden, then behaving like "friction piles" that are driven into clays. Cast-in-place piles may be "socketed" into rock by drilling some distance beyond the bedrock surface, in which case both bond along the sides and end resistance may be mobilized. Piles bearing on weak strata and soils are sometimes constructed with an enlarged base formed by reaming the bottom of the drill hole. This spreads the load to achieve restricted bearing pressures. As discussed later, the bearing capacity of most rocks is sufficiently high that enlarged bases are rarely necessary, the maximum loads being dictated by the concrete rather than by the rock strength.

Very heavy loads can be carried to bearing on rock through the use of *piers in drilled shafts* (Figure 9.3*c*). Large-diameter bucket augers, or spiral augers often mounted on cranes, enable drilling through overburden, weak and even moderately strong rocks like claystones, friable sandstones, chalk, weathered rocks, and evaporite deposits. The drilled shafts are then cleaned out and filled with concrete; if water conditions will not permit pouring concrete in the dry,

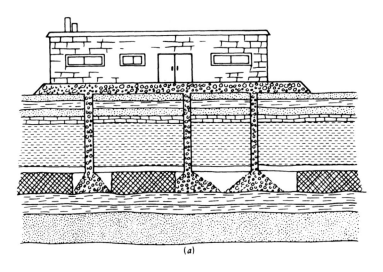

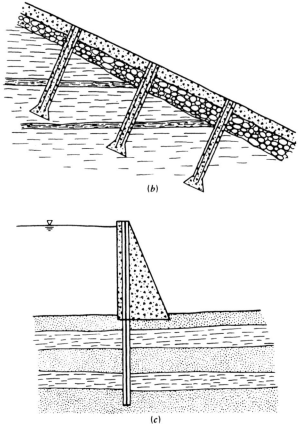

Figure 9.4 Special foundations. (*a*) Grout columns, for construction over old mines. (*b*) Hold-down piers, for swelling rock. (*c*) Deeply anchored cables, to increase the effective weight of a gravity structure.

tremie construction is used. To obtain satisfactory contact and bearing in good rock, it is common practice to drill the shafts several meters or more into the rock to form a "rock socket." In this case, the load is carried by a combination of end bearing and peripheral shear (bond or friction), as discussed later. Drilled piers with very large vertical loads (e.g., 10 MN) are economical if they do not require casing, work stoppage for handling flowing ground or drilling out hard rock blocks, or other special construction procedures. Inspections and tests to evaluate the bearing capacity and deformability of the rock can be conducted in the rock socket because the diameters are usually large enough to admit an engineer or geologist. This is an advantage over pile foundations where the bearing is remote and inaccessible. However, rocks that cannot be drilled due to hardness, pinnacles in the surface, or flowing water conditions can idle expensive equipment, squandering any cost savings.

Other types of foundations in rock are sometimes called for. Mass concrete structures like gravity dams, bridge piers, and powerhouses are sometimes founded on *caissons* sunk through overburden and water. Buildings over abandoned mine openings may be supported on grouted columns of crushed rock (*grout columns*) bearing on the floor of the old mine opening (Figure 9.4a). Structures placed in rock excavations like spillway gates and spillway slabs may require *hold-down piers* (Figure 9.4b) or tensioned *rock anchors* to reduce heave due to rock swelling. High-capacity, tensioned anchors are used to increase foundation compression in opposition to hydraulic uplift, for example, below buttresses of a dam on layered rock (Figure 9.4c).

9.2 Allowable Bearing Pressures in Codes: Behavior Modes

The design of a foundation requires that the bearing pressure and bond (adhesion) allowable in each geological unit be established for the base and sides of the foundation member. The values selected must have a margin of safety against loss of load-carrying capacity (bearing "failure") and must work without large deflections. In routine work, these values are usually taken from building codes, which provide conservative safe pressures and reflect regional experience. The more useful codes of practice reflect engineering geology history and incorporate local formation names as well as rock indexes. For example, Rochester, New York, specifies the bearing pressures for each of the local rock formations and defines defects that are unacceptable in the foundation, as summarized in Table 9.1. Table 9.2 cites allowable bearing pressures from a sampling of building codes—the stipulated pressures being intended to satisfy both bearing capacity and settlement limitations and to provide a factor of

Table 9.1 *Provisions of the Building Code for Rochester, New York (Dates Given in Parentheses)*

Rock is classified as:

Soft rock: Clinton and Queenston shale
Medium rock: Rochester shale
Hard rock: Lockport dolomite and Medina sandstone

If a hole below the bearing surface passes through at least 5 ft of rock, the bearing capacity shall be: 15 tons/ft^2 (1.4 MPa) in soft rock; 25 tons/ft^2 (2.4 MPa) in medium rock; and 50 tons/ft^2 (4.8 MPa) in hard rock (providing that all 5 ft are in the same kind of rock). (10/13/33)

For buildings less than six stories or 75 ft high, the Director of Buildings may reduce the number of drill holes required to be as few as, but not less than, one-fifth of the number of bearing areas, if in his or her opinion the nature and condition of the rock justify such omission. (1/11/66)

Seamy Rock: (11/29/60)

If seams of soil or soft rock having little or no bearing value occur within the 5-ft depth below a bearing area:

1. Seams less than $\frac{1}{4}$ in. thick (6 mm) may be ignored.
2. Seams $\frac{1}{4}$ to $\frac{1}{2}$ in. thick (6 to 13 mm) occurring deeper than 3 ft may be ignored.
3. Seams thicker than $\frac{1}{2}$ in. (13 mm) and deeper than 5 ft may be ignored depending upon the discretion of the building inspector.
4. Seams more than $\frac{1}{2}$ in. (13 mm) thick occurring within a depth of 5 ft, or more than $\frac{1}{4}$ in. (6 mm) thick occurring within the first 3 ft of depth are unsatisfactory. The bearing surface is to be lowered below the bottom of the lowest known seam of thickness greater than $\frac{1}{2}$ in. and further as required to meet these provisions. A new boring or borings shall then be required and any seam occurring in the new borings will be examined as above.
5. The Building Director may order pressure grouting of seams and tests to establish bearing values of grouted foundations.

safety. When there is little to be gained by deviating from local building codes or when it is not feasible to reach an independent assessment of bearing capacity and deformability, applicable codes should be followed. However, most codes do allow for variance if the request is supported by an engineering report and it will be economical to follow this course in many cases since the codes of practice tend to be so very conservative.

Since "rocks" embrace many kinds of materials, rock foundations behave in a number of modes. Unless the rock is known to be weakest in shear like some weathered clay shales and weathered volcanics, it is not obvious that the results of bearing capacity research in soil mechanics is applicable. Failures in

Table 9.2 *Allowable Bearing Pressures for Fresh Rocks of Various Types. According to Typical Building Codes; Reduce Values Accordingly to Account for Weathering, or Unrepresentative Fracturing.[a,b]*

Rock Type	Age	Location	Allow. Bear. Press. (MPa)
Massively bedded limestone[c]		U.K.[d]	3.8
Dolomite	L. Paleoz.	Chicago	4.8
Dolomite	L. Paleoz.	Detroit	1.0–9.6
Limestone	U. Paleoz.	Kansas City	0.5–5.8
Limestone	U. Paleoz.	St. Louis	2.4–4.8
Mica schist	Pre. Camb.	Washington	0.5–1.9
Mica schist	Pre. Camb.	Philadelphia	2.9–3.8
Manhattan schist[e]	Pre. Camb.	New York	5.8
Fordham gneiss[e]	Pre. Camb.	New York	5.8
Schist and slate		U.K.[d]	0.5–1.2
Argillite	Pre. Camb.	Cambridge, MA	0.5–1.2
Newark shale	Triassic	Philadelphia	0.5–1.2
Hard, cemented shale		U.K.[d]	1.9
Eagleford shale	Cretaceous	Dallas	0.6–1.9
Clay shale		U.K.[d]	1.0
Pierre Shale	Cretaceous	Denver	1.0–2.9
Fox Hills sandstone	Tertiary	Denver	1.0–2.9
Solid chalk	Cretaceous	U.K.[d]	0.6
Austin chalk	Cretaceous	Dallas	1.4–4.8
Friable sandstone and claystone	Tertiary	Oakland	0.4–1.0
Friable sandstone (Pico formation)	Quaternary	Los Angeles	0.5–1.0

[a] Values from Thorburn (1966) and Woodward, Gardner, and Greer (1972).

[b] When a range is given, it relates to usual range in rock conditions.

[c] Thickness of beds greater than 1 m, joint spacing greater than 2 m; unconfined compressive strength greater than 7.7 MPa (for a 4-in. cube).

[d] Institution of Civil Engineers Code of Practice 4.

[e] Sound rock such that it rings when struck and does not disintegrate. Cracks are unweathered and open less than 1 cm.

clays follow rotation and shear displacements as depicted in Figure 9.5*e*. Intact rocks are weakest in tension and it is the propagation of extension fractures that permits the indentation of a loaded area on rock.

Figure 9.5 traces the development of penetration into a brittle, nonporous rock as described by Ladanyi (1972). Assuming the rock mass is relatively

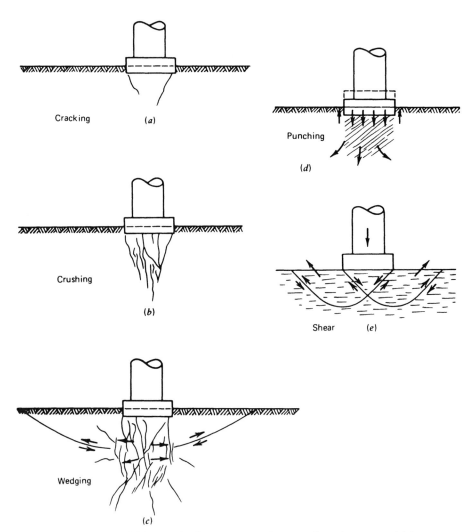

Figure 9.5 Modes of failure of a footing on rock. (*a–c*) Development of failure through crack propagation and crushing beneath the footing. (*d*) Punching through collapse of voids. (*e*) Shear failure.

unfractured, loading initially follows an elastic load-deflection relationship predictable by a formula like Equation 6.10, the precise form depending on the shape and deformability of the footing. After attaining a load such that cracks initiate, further loading extends cracks (Figure 9.5*a*) and at still higher loads they coalesce and interfere. Eventually, the cracks separate slivers and wedges that buckle and crush under additional increments of load (Figure 9.5*b*). Due to dilatancy, the bulb of cracked and crushed rock under the loaded area expands outward, eventually generating a radial network of cracks, one of which may finally propagate to the free surface as in Figure 9.5c. According to the load distribution on the footing and the properties of the rock in a cracked state, the maximum permissible deformations may be attained at any one of the stages depicted in Figure 9.5a–c.

In practice, rock masses undergo additional permanent deformation owing to closing of fissures, cracks, and pores. In open-jointed rock or rock with compressible seams, the deformations on closing or squeezing of fractures will most likely govern design even though the rock itself cannot be said to "fail." Highly porous rocks like some chalks, friable sandstones, and scoriaceous basalts may suffer destruction of the pore skeleton, as explored in Chapter 3 (Figure 3.6). In weakly cemented sedimentary rocks, irreversible settlements from this cause can occur at any level of stress without cracking and driving of wedges; this mode of "failure" is termed "punching" (Figure 9.5d). Fracturing, joint closing, and punching may occur simultaneously, or sequentially in any order; thus almost any load-deformation history is possible. Conversely, if the geotechnical examination of the foundation rock attempts to measure the openness of jointing, the strength of the pore skeleton, and the deformability and strength of seams, it may be possible to predict the load versus deformation response of the foundation under any prescribed intensity and character of foundation load. The allowable bearing pressures can then be selected with respect to the tolerance of the structure to deflections in its foundation.

9.3 Stresses and Deflections in Rock under Footings

When a rock foundation behaves elastically, the displacements and stresses in the neighborhood of a footing can be calculated using the theory of elasticity, either by reference to established results, for example, Equation 6.10, or through use of numerical modeling techniques, most noteably the finite element method. The stresses and displacements of footings loaded by any distribution of shear and pressure can also be obtained by superimposing solutions corresponding to a point load, generally inclined and acting on the surface of a half space. Poulos and Davis (1974) present results obtained in this manner for rigid and flexible footings of rectangular, circular, and other shapes.

Particular solutions using the finite element method may be required if the rock is heterogeneous or anisotropic (Figure 9.6). In this method described by Zienkiewicz (1971), the region of influence of the footing, generally at least six times its width in radial extent, is subdivided into elements, each of which is assigned a set of elastic properties. When the distribution of pressure and shear

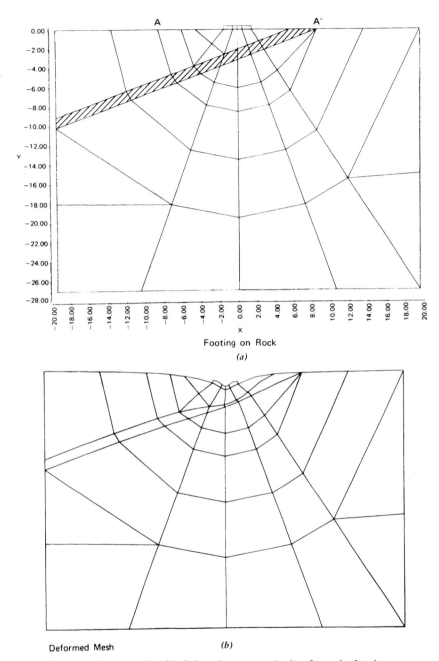

Footing on Rock

(a)

Deformed Mesh   *(b)*

Figure 9.6 Example of a finite element analysis of a strip footing under vertical load on a heterogeneous rock foundation. Analyzed by Victor Saouma, Cornell University. (*a*) Finite element mesh: the ruled elements have *E* equal to one-tenth that of the other elements. (*b*) Deformed mesh with greatly exaggerated displacements.

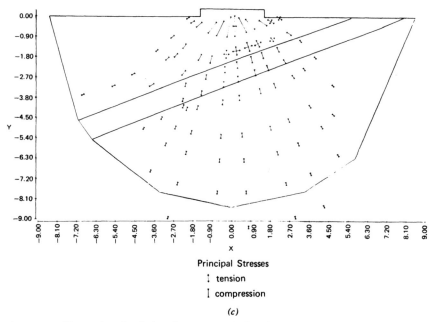

Principal Stresses

⁝ tension

I compression

(c)

Figure 9.6 Example of a finite element analysis of a strip footing under vertical load on a heterogeneous rock foundation. Analyzed by Victor Saouma, Cornell University. (*c*) Vectors showing directions and magnitudes of principal stresses in each element in the region enclosed within the locus *AA'* of the mesh.

on the footing are input, one obtains the stresses in all the elements and the displacements of a set of points throughout the medium; those of the footing itself and any instrumented points are the most interesting. Such programs are available in most engineering design offices. The representation of joints and seams of rock masses in finite element analysis is discussed by Goodman (1976) and special applications in geotechnical engineering are described in the book edited by Desai and Christian (1977).

Through the use of elastic solutions or special numerical models, it will be possible to find how a particular foundation responds to loads. It is not practical to explain such methods here but it is instructive to examine in particular how the load is transferred to the rock in the case of a general line load acting on rocks with various geological structures.

Consider a line load (force per unit length) acting normal to the surface of a semi-infinite, homogeneous, elastic, and isotropic medium as shown in Figure 9.7*a*. The problem depicted is one of plane strain, meaning that the load *P* continues indefinitely in the direction perpendicular to the paper. The principal stresses produced by *P* lie entirely along lines through the point of application of P (i.e., at a point in the medium located by polar coordinates r and θ [see Figure 9.7]), the normal stress acting along any radius (θ constant) is a principal stress and is equal to

$$\sigma_r = \frac{2P \cos \theta}{\pi r} \tag{9.1}$$

while the normal stress acting perpendicular to this direction and the shear stresses referred to these local axes are both zero,

$$\sigma_\theta = 0 \qquad \tau_{r\theta} = 0$$

The locus σ_r constant proves to be a circle tangent to the point of application of P and centered at depth $P/(\pi\sigma_r)$. A family of such circles, drawn for a set of

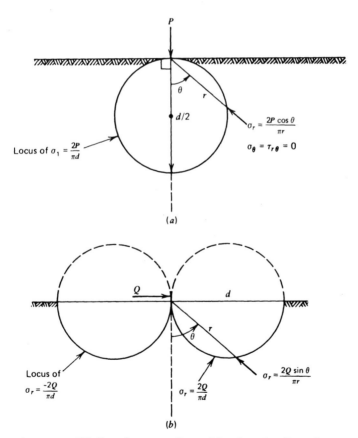

Figure 9.7 "Bulbs of pressure" resulting from loading of an elastic half plane by (a) a normal line load, (b) a shear line load.

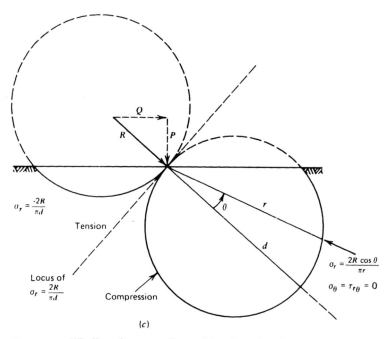

$\sigma_r = \frac{-2R}{\pi d}$

Tension

Locus of
$\sigma_r = \frac{2R}{\pi d}$

Compression

$\sigma_r = \frac{2R \cos \theta}{\pi r}$

$\sigma_\theta = \tau_{r\theta} = 0$

(c)

Figure 9.7 "Bulbs of pressure" resulting from loading of an elastic half plane by (c) an inclined line load.

values of σ_r are sometimes termed "bulbs of pressure." They show graphically how the applied load dissipates as it spreads out in the rock.

Similarly, when a line load acts in shear, the stress distribution is entirely radial (Figure 9.7b). At polar coordinates r, θ, the only nonzero stress is directed radially and has the value

$$\sigma_r = \frac{2Q \sin \theta}{\pi r} \tag{9.2}$$

The locus σ_r constant is represented by two circles tangent to each other and centered a distance $Q/(\pi\sigma_r)$ to the right and left along the surface from the point of application of Q. The left circle represents tensile stress while the right one represents compressive stress. Figures 9.7a and b can be combined into a single set of pressure bulbs centered along the line of action of R, the resultant of P and Q, as shown in Figure 9.7c. The upper circle now represents tensile stress while the lower represents compression. Near the ground surface, tensile stress is lost as the joints open; at greater depth the tensile stress increment adds to the initial horizontal compression, the net stress remaining compressive until a sufficient load is reached.

Another interpretation of the bulbs of pressure is possible. We have seen that the circle tangent at P or Q gives the locus of constant principal stress. It can also be viewed as the envelope to a bundle of vectors radiating from the point of application of P or Q and defining the radial pressure distribution on a circle centered about the point of load application. This is a useful image because it enables one to visualize how planes of limited friction like bedding, schistosity, faults, and joints must alter the contours of principal stress.

Figure 9.8 shows a halfspace in a regularly jointed rock loaded by inclined line load R. In isotropic rock, the pressure should distribute according to the dashed circle; but this cannot apply to the jointed rock mass because the

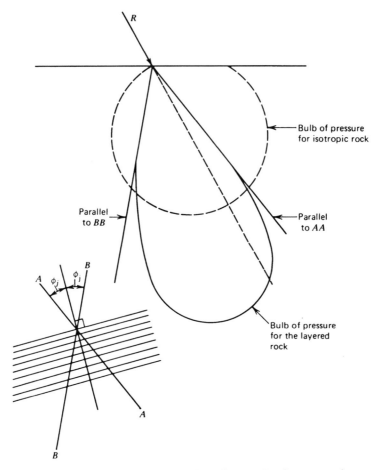

Figure 9.8 Narrowing and deepening of the bulb of pressure due to limited shear stress along discontinuities.

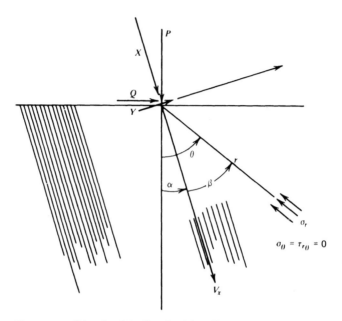

Figure 9.9 Line load inclined arbitrarily on a half space in transversely isotropic rock.

resultant stress cannot make all angles with the joint planes. According to the definition of interlayer friction, the absolute value of the angle between σ_r and the normal to the discontinuities must be equal to or less than ϕ_j. Therefore, the bulb of pressure cannot extend beyond lines *AA* and *BB*, drawn at an angle equal to ϕ_j with the normal to the layers (compare with Figures 7.6 and 7.7). Because the bulb of pressure is confined more narrowly than in isotropic rock, it must continue more deeply, meaning that the stresses are higher at a given depth below the load vector than they would be in rock without discontinuities. Depending on the orientation of the line load and the direction of the planes of discontinuity, some load can also flow into the rock parallel to the layers. In the particular case posed in Figure 9.8, any stress increment parallel to the layers would have to be tensile.

A more formal examination of the influence of discontinuities on the stress distribution beneath footings can be obtained by establishing an "equivalent" anisotropic medium for the rock mass as introduced in Chapter 6 (Equations 6.23 to 6.27). For the special case of a line load decomposed into components *X* and *Y* parallel and perpendicular to the planes of discontinuity (Figure 9.9), John Bray[1] showed that the stress distribution in the rock is still entirely radial

[1] Unpublished notes, 1977, Imperial College, London, Royal School of Mines. See also H. D. Conway (1955) Notes on the orthotropic half plane subjected to concentrated loads, *J. Appl. Mech.* **77**: 130.

with $\sigma_\theta = 0$, $\tau_{r\theta} = 0$, and

$$\sigma_r = \frac{h}{\pi r} \left(\frac{X \cos \beta + Yg \sin \beta}{(\cos^2 \beta - g \sin^2\beta)^2 + h^2 \sin^2 \beta \cos^2\beta} \right) \qquad (9.3)$$

where r is the distance from the point of load application and $\beta = \theta - \alpha$ as shown in Figure 9.9. β is the angle from the line of action of X to a radius through the point in question. Note that X is not normal to the surface but is parallel to the planes of discontinuity. The constants g and h are dimensionless

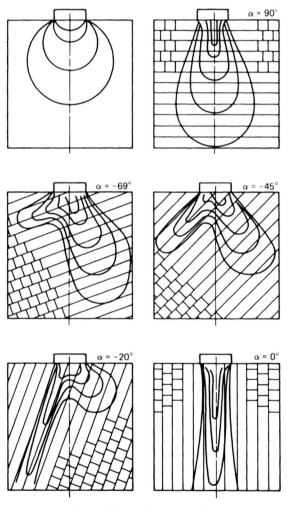

Figure 9.10 Lines of equal stress (bulbs of pressure) determined by Gaziev and Erlikhman (1971) from models. (α is defined in Figure 9.9.)

quantities describing the properties of a transversely isotropic medium "equivalent" to the discontinuous rock mass and are given as follows:

$$g = \sqrt{1 + \frac{E}{(1 - v^2)k_n S}} \tag{9.4}$$

$$h = \sqrt{\frac{E}{1 - v^2}\left(\frac{2(1 + v)}{E} + \frac{1}{k_s S}\right) + 2\left(g - \frac{v}{1 - v}\right)} \tag{9.5}$$

In the above expressions E and v are the elastic modulus and Poisson's ratio, respectively, of the rock itself, k_n and k_s are the normal and shear stiffnesses (FL^{-3}) of the discontinuities as discussed with respect to Equations 6.23 and 6.24, and S is the average spacing between discontinuities.

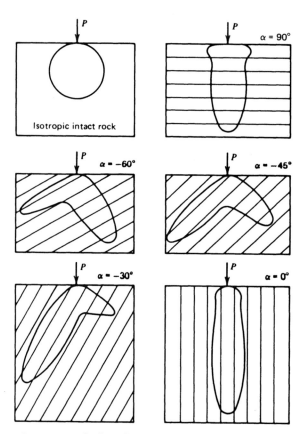

Figure 9.11 Bulbs of pressure under line loads calculated by John Bray using the method of Equation 9.3 to 9.5 (unpublished notes, 1977).

Using Equations 9.3 to 9.5, it is possible to calculate lines of equal radial stress under line loads with arbitrary inclination relative to the direction of layers. In all the equations concerning loci of constant radial stress, it can be noted that the inclination of the ground surface does not affect the answer except to establish which parts of the loci lie within the ground. John Bray compared the results of model studies published by Gaziev and Erlikhman (1971) with line-load solutions from Equation 9.3, calculated with joint properties such that closing of joints is equal in magnitude to the compression of the rock, that is, $E/(1 - \nu^2) = k_n S$; and the slip along joints is 5.63 times the shear displacement of the rock parallel to the joints, that is, $E/[2(1 + \nu)] = 5.63 k_s S$. With $\nu = 0.25$, this gives $g = \sqrt{2}$ and $h = 4.45$. The agreement in shapes between the principal stress contours found in model studies (Figure 9.10) and calculated with Equation 9.3 (Figure 9.11) shows that stresses caused by footings on layered, schistose, or regularly jointed rock can be predicted rationally.

9.4 Allowable Bearing Pressures on Footings on Rock

"Allowable pressure" on a footing is the maximum pressure against the rock surface consistent with both deflections, and limiting equilibrium (stability) as well as permissible stress values in the concrete; the latter may govern design with high loads or very good rock. Deflections are usually more limiting than stability when dealing with rock. An appropriate analysis of settlements and rotations under a footing on regularly bedded or fractured rock can be made by superposition and integration of Equation 9.3 using the stress-strain relations of Equation 6.9 with Equations 6.23 to 6.27. Kulhawy and Ingraffea (1978) and Kulhawy (1978) offered a simpler method to estimate settlement in fractured rocks under strip, circular, and rectangular footings. It is sometimes practical to conduct load tests on footings in the field, in which case safe pressures can be established directly without separately evaluating the structural and physical properties of the rock. However, such tests are expensive and can seldom encompass the whole range of rock and environmental conditions pertinent to a foundation. Finite element analysis offers another approach by which the variability of site conditions and rock properties can be studied to achieve an economical design.

The calculation of a *bearing capacity* according to limiting equilibrium calculations for a footing under load must respect the complexity and variety of the failure modes discussed earlier. Although we can give no universal formula for bearing capacity of rock, several simple results prove useful as tools to calculate the order of magnitude of a limiting safe pressure. Tests in isotropic rock have shown that this pressure often occurs at a settlement approximately equal to 4 to 6% of the footing width.

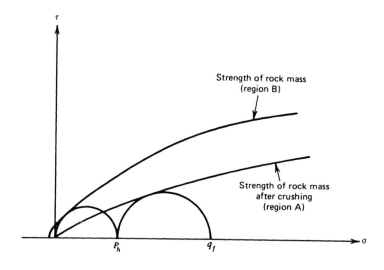

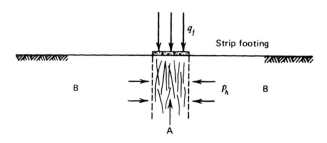

Figure 9.12 Analysis of bearing capacity on rock.

We consider now the mode of failure shown in Figures 9.5a–c, in which a laterally expanding zone of crushed rock under a strip footing induces radial cracking of the rock to either side. The strength of the crushed rock under the footing will be described by the lower failure envelope in Figure 9.12, while the strength of the less fractured, neighboring rock will be described by the upper curve in the same figure. The largest horizontal confining pressure that can be mobilized to support the rock beneath the footing (zone A in Figure 9.12) is p_h, determined as the unconfined compressive strength of the adjacent rock (zone B of Figure 9.12). This pressure determines the lower limit of Mohr's circle tangent to the strength envelope of the crushed rock under the footing.[2] Triaxial

---

[2] Suggested by Ladanyi (1972) who acknowledges R. T. Shield (1954), "Stress and velocity fields in soil mechanics" *J. Math. Phys.* **33:** 144–156.

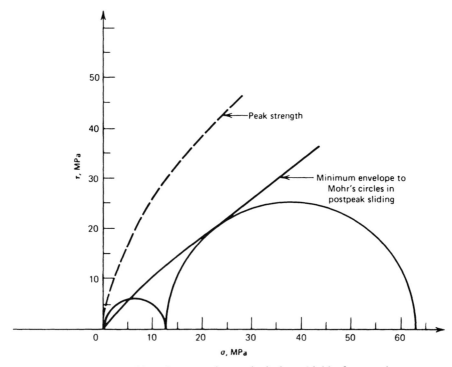

Figure 9.13 Example of bearing-capacity analysis for a highly fractured grey-wacke sandstone. (After Raphael and Goodman, 1979.)

compression tests on broken rock can define the latter strength envelope, and thus the bearing capacity can be found. Figure 9.13, for example, shows triaxial compression test results by Raphael and Goodman (1979) on intact and broken core samples from a foundation in highly fractured greywacke sandstone. The condition of the rock surrounding the footing can be represented by the envelope corresponding to the peak strengths of core samples in which all the fractures were carefully fitted together and held with tape before testing. The condition of the rock under the footing can be described by the envelope corresponding to residual strength of such specimens. With these strength properties determined and a factor of safety of 5, the bearing capacity is estimated as 12 MPa. For reference, the unconfined compressive strength of the intact rock is 180 MPa.

Examination of Figure 9.12 leads to the conclusion that the bearing capacity of a homogeneous, discontinuous rock mass cannot be less than the unconfined compressive strength of the rock mass around the footing, and this can be taken as a lower bound. If the rock *mass* has a constant angle of internal friction ϕ and unconfined compressive strength q_u (Mohr-Coulomb material), the method of Figure 9.12 establishes the bearing capacity as

$$q_f = q_u(N_\phi + 1) \tag{9.6}$$

where

$$N_\phi = \tan^2\left(45 + \frac{\phi}{2}\right) \tag{9.7}$$

Actual conditions may call for special analysis. Figure 9.14a, for example, shows a footing bearing on a thin, relatively rigid sandstone layer underlain by more flexible claystone. With sufficient load, the stiff layer will break in flexure, thereafter transferring a greater proportion of load to the clay shale. The deflections associated with the cracking of the upper layer will probably limit the design loads. Otherwise, the bearing capacity will be that calculated from the properties of the lower layer. The strength of the stiffer layer can be analyzed by considering it to be a thick beam.

Figure 9.14b depicts a footing resting on a portion of a single joint block created by orthogonal vertical joints each spaced distance S. Such a condition might arise, for example, in weathered granite. If the footing width B is equal to the joint spacing S, the rock foundation can be compared to a column whose strength under axial load should be approximately equal to the unconfined compressive strength q_u. If the footing contacts a smaller proportion of the joint block, the bearing capacity increases toward the maximum value consistent with the bearing capacity of homogeneous, discontinuous rock, obtained with the construction of Figure 9.12 or from Equation 9.6 as appropriate. This problem was studied by Bishnoi (1968), who assumed that some load is transferred laterally across the joints. Modifying this boundary condition for an open-jointed rock mass in which lateral stress transfer is nil yields

$$q_f = q_u\left\{\frac{1}{N_\phi - 1}\left[N_\phi\left(\frac{S}{B}\right)^{(N_\phi - 1)/N_\phi} - 1\right]\right\} \tag{9.8}$$

Comparing the results of computation with Equations 9.8 and 9.6 shows that open joints reduce the bearing capacity only when the ratio S/B is in the range from 1 to 5, the upper limit increasing with ϕ.

When determining the safe bearing pressures on a footing on rock, it is never permissible to use the bearing capacity as calculated, or even as measured by load tests in situ, without consideration of scale effects. There is an element of uncertainty associated with the variability of the rock and a significant size effect in strength under compressive loads. However, even with a factor of safety of 5, the allowable loads will tend to be higher than the code values sampled in Table 9.2, except when the foundation is on or near a rock slope.

Bearing capacity may be considerably reduced by proximity to a slope because modes of potential failure may exist in the region of the foundation

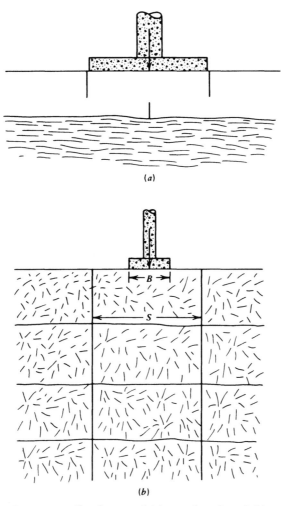

Figure 9.14 Footings on (*a*) layered rock and (*b*) rock with open, vertical joints.

with unsatisfactory degrees of safety even without added loads. The initiation of sliding could cause violent structural collapse for bridge piers, side-hill towers, and abutments of arch dams; thus the slopes must be explored and analyzed diligently. In such cases, special reinforcing structures may be needed. Figure 9.15*a* shows a concrete structure added downstream of the slender right abutment of the 151-m-high Canelles arch dam, Spain. By means of its own weight and the passive resistance of five tunnels filled with reinforced concrete (Figure 9.15*b*), the structure is supposed to increase the factor of safety against sliding on a daylighting system of vertical fractures in the Cretaceous lime-

Figure 9.15 Reinforcing structure for the abutment of Canelles arch dam, Spain. [Reproduced from Alvarez (1977) with permission.] (*a*) A view of the structure from downstream.

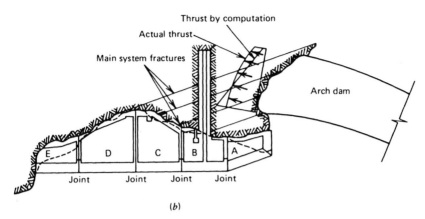

Figure 9.15 Reinforcing structure for the abutment of Canelles arch dam, Spain. (*b*) A horizontal section. (Reproduced from Alvarez (1977) with permission.)

stone. The fractures are filled with up to 25 cm of clay, and recur with average spacing of 5 m. The tunnels are intended to extend beyond the line of thrust of the arch, and can mobilize up to 5000 tons of tensile force.

Analysis of failure modes for foundations on rock slopes, assuming the geometry of failure to be determined by discontinuity planes, is an extension of methods discussed in Chapter 8. The addition of a force to the stereographic projection solution for plane and wedge slides was discussed in that chapter (e.g., Figure 8.12). The problems at the end of this chapter examine how the equations for stability under plane failure and for a slide composed of two planes can be modified to include one or more forces applied to the sliding mass.

Limestone is always suspect as a foundation rock for dams because past weathering may have opened up cavities that are not only capable of transmitting leakage but that may also reduce the bearing capacity of the foundation. This concern relates to earth and rock-fill dams as well as to concrete structures. Patoka Dam, Indiana, an earth and rock-fill embankment about 45 m high, illustrates foundation problems that can arise when dealing with limestone.[3] The dam was built over a series of upper Paleozoic sandstone, shale, and limestone formations. Solution cavities and solution-enlarged joints demanded considerable foundation treatment by the Corps of Engineers to provide bearing capacity and protection from erosion of the embankment material into the interstices of the rock mass. Concrete walls 30.5 cm thick were constructed against rock surfaces excavated by presplitting; these walls separate

---

[3] B. I. Kelly and S. D. Markwell (1978) Seepage control measures at Patoka Dam, Indiana, preprint, ASCE Annual Meeting, Chicago, October.

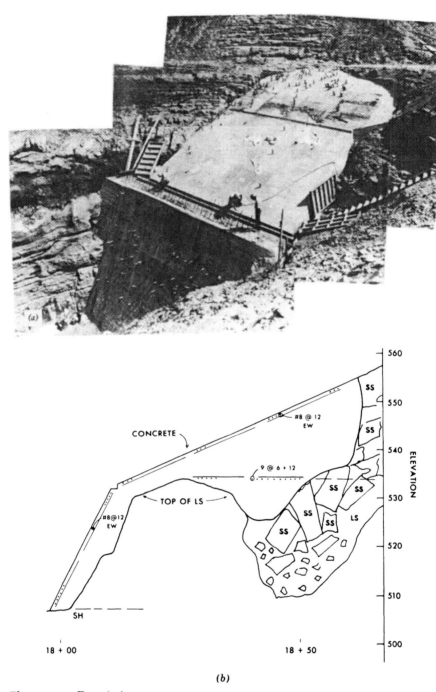

(b)

Figure 9.16 Foundation treatment required under a dike at Patoka Reservoir, Indiana. (*a*) View of a reinforced concrete plug to bridge across a rubble-filled cavity. (*b*) Vertical section through the structure. (Courtesy of Benjamin Kelly, Corps of Engineers, Louisville District.)

embankment materials from open-jointed limestone. The rock was grouted through the walls after they were constructed. Deep foundation grouting could not satisfactorily consolidate the rock and close seepage paths in the abutments due to excessive grout flow into open cavities, difficulty in drilling through collapsed, rubble-filled cavities, and hole alignment problems created by the irregular limestone surface. Instead, a cutoff trench averaging 8.5 m deep and about 1.7 m wide, and backfilled with lean concrete, was constructed along a side-hill length of 491 m in the right abutment to carry the foundation to the shale below the cavernous limestone. Roof collapses that had occurred under natural conditions left blocks of sandstone in clay as incomplete fillings of cavities reaching as much as 12 m above the top of the Mississippian limestone into the overlying Pennsylvanian sandstone. One large collapse feature under the abutment of a dike was bridged with a reinforced concrete plug and wall (Figure 9.16).

Although not nearly so unpredictable and treacherous as karstic limestones, decomposed granitic rocks may also require special foundations, particularly for large dams. Quite commonly, the degree of weathering of the rock forming a valley increases notably as the upper part of the valley is approached. Figure 9.17 shows a large gravity monolith that was required, for this reason, in the upper part of the abutment of an arch dam in Portugal.

Figure 9.17 Gravity block required for the upper part of the left abutment of Alto Rabagao Dam, Portugal. (Courtesy of Dr. Manual Rocha.)

9.5 Deep Foundations in Rock

When the allowable loads on the surface soils are low, it may be economical to carry the structural loads to rock by means of driven or cast-in-place piles, or piers cast in drilled shafts (Figure 9.3). When concrete is poured against drilled rock surfaces, it develops an adhesion ("bond"), which can carry shear stresses up to the shear strength of rock or of the concrete, whichever is less. To design the foundation, it is necessary to consider how the load will be distributed between *bond* on the sides of the pier or pile, and *bearing* resistance at its end. The length and diameter of the pier (or pile) can be selected to strike a balance between the two so that neither permissible bond stresses nor permissible bearing pressures are exceeded.

Bearing capacity increases when a footing is buried because it requires additional work to expand the failing region against an increased rock pressure. An exception to this rule is the case of failure by punching caused by the collapse of pore structure or the closing of joints. In cohesive soils, the bearing capacity beneath plates buried more than four diameters can be increased from the surface value of six times the undrained shear strength S_u to nine times the undrained shear strength (which corresponds to $4.5q_u$) (Woodward, Gardner, and Greer, 1972). Even this is conservative, as is shown in tests by Wilson (1977) on 900-mm-diameter cast-in-place concrete piles socketed into Cretaceous mudstone; the bearing strength was at least one-third greater than $9S_u$. The British code (Institution of Civil Engineers Code of Practice No. 4) permits a 20% increase in safe bearing capacity for each foot of depth up to a limit of twice the surface value.

The settlement of a rigid circular bearing plate on an isotropic, elastic half space was given in Equation 6.10. Following Poulos and Davis (1968), we introduce a depth factor n in that equation to express the settlement ω_{base} of the lower end of a pier or pile set in the base of a shaft below the bedrock surface (Figure 9.18a):

$$\omega_{base} = \frac{(\pi/2)p_{end}(1 - \nu_r^2)a}{E_r n} \tag{9.9}$$

where p_{end} is the normal pressure at the lower end of the pier or pile

ν_r and E_r are the Poisson's ratio and elastic modulus of the rock

a is the radius of the lower end of the pile or pier

n is a factor depending on relative depth and on ν_r as given in Table 9.3

If a pier is founded on top of the bedrock surface, it is prudent to neglect the adhesion along its sides in the soil layers and assume that the full pressure p_{total} acting on the top of the pier acts also on its base. However, when a pier is socketed in rock even several radii deep, a considerable proportion of the load is transferred to the perimeter and p_{end} is significantly less than p_{total}. As long as

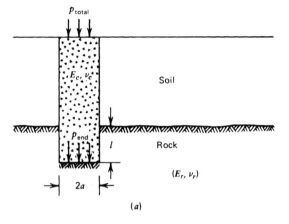

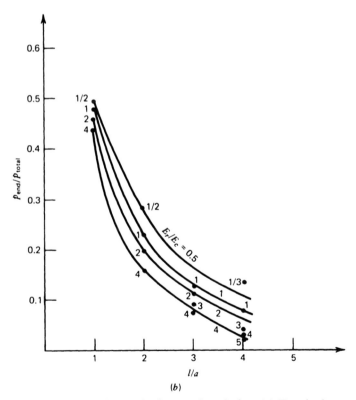

Figure 9.18 Load transfer in a socketed pier. (*a*) Terminology for the pier. (*b*) Data on load transfer calculated by Osterberg and Gill (1973) for indicated values of E_r/E_c—the curves were deduced by Ladanyi (1977).

Table 9.3 *Effect of Embedment Depth l on Displacements of a Rigid Plate According to Equation 9.9*

l/a	0	2	4	6	8	14
n: $v_r = 0$	1	1.4	2.1	2.2	2.3	2.4
n: $v_r = 0.3$	1	1.6	1.8	1.8	1.9	2.0
n: $v_r = 0.5$	1	1.4	1.6	1.6	1.7	1.8

the bond is maintained along the sides, analysis of the load transference corresponds to that of a cylindrical elastic inclusion "welded" to the surrounding medium. Finite element analysis of an elastic, axisymmetric system by Osterberg and Gill (1973) can therefore provide a useful starting point for analyzing load transfer in a pier socketed into rock, providing that the pier is not loaded beyond the limit of bond strength. Figure 8.18b presents a family of curves expressing the ratio p_{end}/p_{total} as deduced by Ladanyi (1977) from Osterberg and Gills's results. Notice that at embedment ratios l/a greater than 4, the end-bearing pressure under a pier on stiff rock is less than one-eighth of the pressure applied to the top of the pier.

When the rock is more compliant than the pier, as in chalk or compaction shale, or in the case of piles driven through rock to obtain a "set" in bedrock, the adhesion sustains a smaller proportion of the total load. This can be appreciated from the results of pile load tests like that presented in Figure 9.19 from Wilson (1977). His test was conducted by compressing a pile of 670 mm bottom diameter inside a socket augered at the base of an oversized hole; in this way, adhesion occurred only along a short section and the end-bearing capacity could be determined with minimal correction to the test data. The load was applied by jacking against a stiff steel girder held down by two piles cast against the rock over a predetermined length of 1 m. Monitoring the deflections of all three piles thus measures adhesion in the outer two piles simultaneously as the center pile is compressed. The adhesion measurement is conservative because pulling reduces the normal stress on the periphery of the pile, whereas the opposite is true in service under compression. After the right pile yielded, at 340 kN uplift load, the position of the jack was moved to the left end of the girder and the test was continued, eventually causing the left pile to yield at 520 kN.

Several principles are illustrated by these results. First, the adhesion is typically developed with a deflection of 10 mm or less, while mobilization of the full bearing capacity may require a settlement of 30 to 40 mm or more (typically 4 to 6% of the base diameter as noted previously). The curve of load versus deformation for the development of adhesion is steep with some loss of strength due to cracking in concrete or rock, or both, after the attainment of a

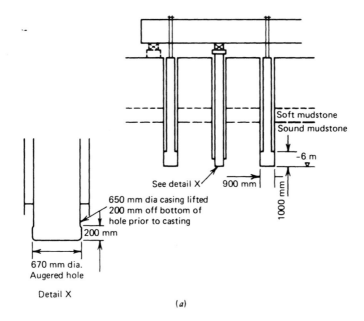

(a)

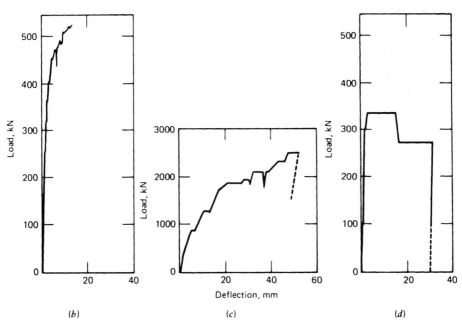

(b)   (c)   (d)

Figure 9.19 Pile load test, after Wilson (1977). (*a*) Test setup. (*b*) Pull test on left pile. (*c*) Results of compression test on center pile. (*d*) Results of pull test on right pile.

peak load. In contrast, the load-deformation curve in end bearing is curved downward almost from the beginning and may show strain hardening (i.e., upward curvature) after a plateau of strength is reached, although brittle behavior may occur with continued loading. A designer can choose to assign a large proportion of the pier load to perimeter bond only if he or she assures that deflections do not exceed those required for its full mobilization; but this is wasteful with respect to end bearing. With continued loading beyond the peak adhesion, the proportion of load carried by the base of the pier must increase.

In the limit where all bond is broken along the sides, it is useful to analyze the case posed by a pile or pier with frictional contact along its sides. The load transference corresponding to the elastic case charted in Figure 9.18b would change and approach the values corresponding to a frictional interface after the adhesion is broken by overloading or if a construction method is used in which bond is minimal (e.g., precast piles set into boreholes). Assuming the coefficient of side friction is zero between the pier or pile and the soil and is a constant value μ on the wall of the socket in the rock, it is shown in Appendix 4 that the vertical stress σ_y in the pier at depth y below the top of rock is

$$\sigma_y = p_{total}e^{-\{[2\nu_c\mu/(1 - \nu_c + (1 + \nu_r)E_c/E_r)](y/a)\}} \tag{9.10}$$

where the subscripts c and r denote concrete and rock, respectively, and p_{total} is the pressure applied to the top of the pier. If the depth l of the socket is input for y, σ_y calculated from the above equals the end-bearing pressure p_{end}. To approximate the results of the elastic analysis in which one assumes a welded contact between concrete and rock, large values of μ must be introduced into Equation 9.10, as examined in problem 7.

Bond strength is best determined by a field pullout test like the one described or by a compressive load test with a compressible filling placed beneath the end of the pile or pier to negate end bearing. In soft, clay-rich rocks like weathered clay shale, which tend to fail in shear rather than in compression, the bond strength is determined in relation to the undrained shear strength S_u:

$$\tau_{bond} = \alpha S_u \tag{9.11}$$

Recasting in terms of q_u and ϕ,

$$\tau_{bond} = q_u \frac{\alpha}{2\tan(45 + \phi/2)} \tag{9.12}$$

Typical values of α range from 0.3 to 0.9 but may be considerably greater if the surface is artificially roughened (Kenney, 1977). In hard rock, bond strength τ_{bond} reflects diagonal tension, and it may accordingly be approximated by the tensile strength of rock and concrete. A conservative value for bond strength in hard rocks is then

$$\tau_{bond} = \frac{q_u}{20} \tag{9.13}$$

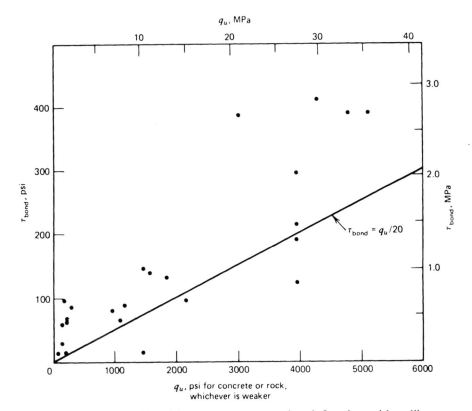

Figure 9.20 Strength of bond between concrete and rock for piers with radii greater than 200 mm. (Data from Horvath and Kenney (1979) based on load tests.)

in which q_u is the unconfined compressive strength of laboratory samples (see Figure 9.20). The allowable shear stress τ_{allow} must be less than τ_{bond}, in both the concrete and the rock.

Ladanyi (1977) proposed a method of design providing for full bond strength developed over a socket length sufficient to reduce the end-bearing pressures to acceptable values. The following iterative scheme will achieve this once the allowable bearing pressure and the allowable shear stress have been established.

Given the total vertical load F_{total} on the top of the pier:

1. Assume a value for the allowable bond stress τ_{allow} on the wall of the rock socket.
2. Select a radius a. This may be dictated by the allowable load in the concrete.

3. Neglect end bearing and calculate the maximum length l_{max} of the rock socket:

$$l_{max} = \frac{F_{total}}{2\pi a \tau_{allow}}$$

4. Choose a value l_1 less than l_{max} and corresponding to l_1/a determine p_{end}/p_{total} from Figure 9.18b. Alternatively, corresponding to a lower value of bond stress, choose a value for μ and calculate $p_{end}/p_{total} = \sigma_y/p_{total}$ from Equation 9.10 with $y = l_1$.

5. Calculate $p_{end} = (F_{total}/\pi a^2)(p_{end}/p_{total})$.

6. Compare p_{end} to the allowable bearing pressure q_{allow} appropriate for the material at depth l_1 with relative embedment ratio l_1/a (see Equation 9.9).

7. Calculate $\tau = (1 - p_{end}/p_{total})(F_{total}/2\pi a l_1)$.

8. Compare τ with τ_{allow}.

9. Repeat with l_2 and a until $\tau = \tau_{allow}$ and $p_{end} \leq q_{allow}$.

If a low factor of safety is used for bond strength, a higher factor of safety is required for bearing to assure that the displacements are compatible. Kenney (1977) suggested that bond and end resistance could be developed at compatible displacements by preloading the base using flat jacks or hydraulic cylinders between the pier base and the rock.

As shown in Figure 9.21, the settlement of a pier on rock can be calculated as the sum of three terms: (1) the settlement of the base (ω_{base}) under the action of p_{end}; (2) the shortening of the pile itself (ω_p) under a uniform compressive stress equal to p_{total}; and (3) a correction ($-\Delta\omega$) accounting for the transference of load through adhesion along the sides:

$$\omega = \omega_{base} + \omega_p - \Delta\omega \tag{9.14}$$

These terms are calculated as follows.

ω_{base} is calculated from Equation 9.9 for an isotropic material or using results of Kulhawy and Ingraffea for anisotropic materials:

$$\omega_p = \frac{p_{total}(l_0 + l)}{E_c}$$

where $l_0 + l$ is the total length of the pile and l is the length embedded in rock and

$$\Delta\omega = \frac{1}{E_c} \int_{l_0}^{l_0 + l} (p_{total} - \sigma_y)dy$$

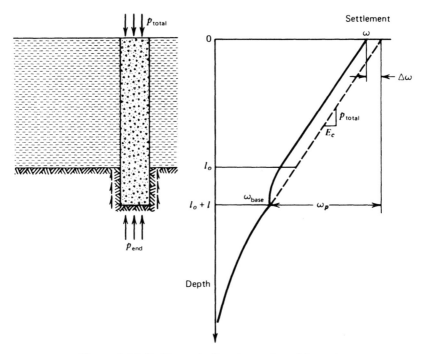

Figure 9.21 Settlement of a pier socketed in rock.

The last term is not important for socketed piers if most of the length of the pier is in soil.[4]

Shafts larger than about 1 m in diameter permit visual inspection and testing of the rock, subject to water conditions, wall stability, and air quality. Many types of tests have been tried to minimize the equipment "down time" yet assure satisfactory rock and accurate assertions concerning rock properties. Woodward, Gardner, and Greer (1972) recommend drilling inexpensive holes, without coring, in the base of the socket, then feeling the sides for open cracks and seams with a rod equipped with a sideward point. A borehole camera, television, periscope, or the Hinds impression packer can be used advantageously to inspect the rock. The latter device expands a packer in the hole to

[4] For the vertical stress distribution described by Equation 9.10,

$$\omega_p - \Delta\omega = \frac{p_{total} l_0}{E_c} + \frac{p_{total}}{E_c}\left(\frac{1}{\nu_c} - (1 + \nu_c)\right)\left(1 + \frac{E_c}{E_r}\frac{1 + \nu_r}{1 - \nu_c}\right)\frac{a}{2\mu}(1 - e^{-\delta})$$

where

$$\delta = \frac{2\nu_c\mu l/a}{1 - \nu_c + (1 + \nu_r)E_c/E_r}$$

squeeze a wax film against the wall of the borehole (Barr and Hocking, 1976; and Brown, Harper, and Hinds, 1979). Cracks, seams, and bedding can be seen clearly in the impression.

The depth of exploration necessary to assure satisfactory bearing under a pier depends on the depth of the rock socket and the shape and extent of the lines of equal principal stress. With vertical or horizontal strata having low interbed friction, the bulbs of pressure are narrow and deep as discussed previously. If the rock socket is short and the pressure bulbs are deep, stresses sufficiently large to cause appreciable settlement in a weak rock layer could occur more than 5 ft (depth of exploration required in the Rochester code, Table 9.1) below the base of the socket. In areas underlain by karstic limestone, it may be necessary to search below a shaft 10 m or more to find good rock, free of cavities continuously for at least 3 m.

Rock tests conducted on the walls of sockets or in the boreholes at the base of a socket can provide the data required for design. The borehole jack, which expands metal plates against opposite segments of a borehole, is well suited for this type of evaluation. (Borehole tests are discussed in Chapter 6.) In clay shales and other soft rocks free of hard concretions, the cone penetrometer has been used to evaluate the undrained shear strength below foundations (see Equation 9.11). The standard penetration test is also used in such rock. Wakeling (1970) correlated rock properties with standard penetration tests for chalk. When the rock has hard interbeds or concretionary lenses, standard penetration tests will be confusing. Rock mass classification by the geomechanics classification discussed in Chapter 2, together with the correlation of Figure 6.9, determines the modulus of elasticity of the foundation based on simple tests and observations.

9.6 Subsiding and Swelling Rocks

In previously mined regions, karst topography, highly soluble rocks, and rocks with swelling minerals, foundations may be displaced by rock movements quite apart from deflections caused by the foundation bearing pressures. In each case, the potential problems are best handled through judicious siting after thorough subsurface exploration. Locations and elevations of structures may need repeated shifting according to the results of core borings. In mined-out terrain, it may be possible to avoid the chance of subsidence by choosing a location underlain by barrier pillars between panels. In karst terrain, surprises can happen despite the most thorough exploration, or conditions can deteriorate after construction following a lowering of the groundwater table (Foose, 1968). Lowering the groundwater table increases effective stresses and brings additional load on existing cavities, while reducing capillarity of overlying soils that can then run into them (Sowers, 1976).

If a room and pillar mine occurs beneath a building, four possibilities must be recognized: (1) the mine is so deep that subsidence at the surface is extremely unlikely; (2) the mine is definitely caving with loss of support at the ground surface; (3) the mine openings are presently stable but could collapse in the future; or (4) the mine openings are stable and unlikely to deteriorate.

Mine openings more than 100 m deep rarely cave to the surface but it is not impossible for them to do so. The geological section will establish the presence or absence of thick, strong formations able to bridge a cave of given dimensions. Based on assumptions of the maximum size of opening that could occur at the base of a bridging formation, an analysis can be made to indicate the likelihood of roof destruction through flexure. High horizontal stresses tend to reinforce such bridging formations. When an opening of original height h stopes upward, broken roof rock tumbles down and eventually fills it; as the caving progresses, the former cavity in rock with density γ is replaced by a larger inclusion of crushed rock with density γ/B. The maximum possible height H of the inclusion above the previous roof is therefore

$$H = \frac{h}{B - 1} \tag{9.15}$$

Price et al. (1969) used this expression to establish the depth H to old mine workings such that surface subsidence is not likely. In highly fractured roof rock lacking appreciable horizontal stress, a cave may narrow upward but subsequently open upward reaching the surface through hundreds of meters. Thus, local experience in a mining district should be carefully considered.

In areas with active mining nearby, one may be able to acquire a mine map showing the plan and configuration of rock pillars at depth. If the accuracy of the plan can be determined, Equation 7.4 is applicable to calculating the safety of each pillar. Goodman et al. (1980) suggested that some pillar failure is acceptable if it can be shown that progressive failure is unlikely. Repeated pillar strength calculations with updated tributary areas reflecting reassignment of load from failed pillars will establish the maximum dimensions of potential caves. The capability of the roof rock to span such caves is then determined. If there is any doubt as to the safety of existing pillars, artificial support must be provided or the structure must be relocated.

Foundations for structures over old mines likely to collapse can be established safely in a number of ways as reviewed by Gray, Salver, and Gamble (1976). If the openings are at shallow depth, it may be cheapest to excavate the rock to a level below them and backfill or establish footings at that level. Deeper openings can be filled with grout or with low-strength soil cement (e.g., lime and fly ash). They can also be propped with grout columns (Figure 9.4a). Alternatively, drilled piers socketed below the floor of the openings or piles driven through drill holes into the floor of the mine openings can support the structure below the potentially caving levels. Deep foundations may be subjected to downdrag or to lateral loads from continued subsidence of the over-

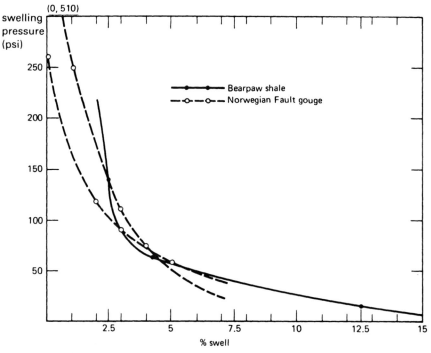

Figure 9.22 Swelling test data for Norwegian fault gouge from Brekke (1965), and for Bearpaw shale from Peterson and Peters (1963).

burden. Lightly loaded areas over sinkholes in karst terrain can be filled with crushed stone reinforced with wire mesh, and then tested with a compacting fill. Concrete fill is appropriate for small cavities beneath footings only if there is no risk of their enlargement; enlargement of a sinkhole filled with concrete can cause sudden, violent collapse.[5]

Swelling rock like montmorillonitic shales, weathered nontronite basalts, and some salts found in evaporite deposits can create uplift pressures on foundations. The expansion pressure is greatly reduced if some deflection is permitted; therefore, one should attempt to measure the relationship between swelling pressure and permitted expansion for representative core samples. Such data can be obtained in a consolidometer, bringing a dry specimen to an initial state of precompression and then monitoring the normal force and expansion as the rock is saturated. If a suitable consolidometer is not obtainable, one can place various dead weights on core samples and monitor the increase in length with time after saturation. Figure 9.22 shows data from expansion pressure measurements with a Norwegian fault gouge and with a Cretaceous shale.

[5] R. Foose, personal communication.

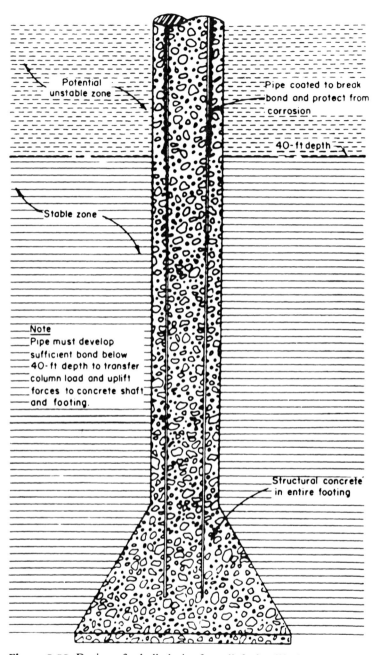

Within the figure:

Potential unstable zone

Pipe coated to break bond and protect from corrosion

40-ft depth

Stable zone

Note
Pipe must develop sufficient bond below 40-ft depth to transfer column load and uplift forces to concrete shaft and footing.

Structural concrete in entire footing

Figure 9.23 Design of a belled pier for relief of uplift due to expansion of the upper layer (dashed lines). The outer annulus of concrete is expected to break in tension near the bottom of the expansive layer; by Raba-Kistner Consultants, Inc. [Reproduced from Woodward, Gardner, and Greer (1972) with permission.]

The designer can either accept the pressures and deformations or place the foundations deep enough to inhibit access of water to the rock. Figure 9.23 from Woodward et al. (1972) shows a pier design used in Texas to accommodate uplift along the walls resulting from swe⁀ng soil and rock. A pipe coated with bond-breaking mastic on its outer surface separates the main load-carrying stem of the pier from the surrounding annulus, which breaks off in tension and moves up with the expanding soil. Anchor piles that reduced expansive heave of spillway slabs in Bearpaw shales are described by Jaspar and Shtenko (1969). In California, foundation redesign required on a housing project in expansive claystones could assure stable support only with piers at least 6 m deep (Meehan et al., 1975).

Fortunately, ground movements are rare in rock. However, the engineer must always be on guard for special problems, almost any of which can be solved economically if recognized in time.

References

Alvarez, A. (1977) Interpretation of measurements to determine the strength and deformability of an arch dam foundation, *Proceedings, International Symposium on Field Measurements in Rock Mechanics* (ISRM) (Balkema, Rotterdam), Vol. 2, pp. 825–836.

Ashton, W. D. and Schwartz, P. H. (1974) H bearing piles in limestone and clay shales, *J. Geotech. Eng. Div.,* (ASCE) **100** (GT7): 787–806.

Aurora, R. P. and Reese, L. C. (1977) Field tests of drilled shafts in clay shales, *Proceedings, 9th International Conference on Soil Mechanics and Foundation Engineering,* Vol. 1, pp. 371–377.

Barr, M. V. and Hocking, G. (1976) Borehole structural logging employing a pneumatically inflatable impression packer, *Proceedings, Symposium on Exploration for Rock Engineering,* Vol. 1, pp. 29–34.

Bell, F. G. (Ed.) (1978) *Foundation Engineering in Difficult Ground,* Newnes-Butterworths, London.

Bishnoi, B. W. (1968) Bearing capacity of jointed rock, Ph.D. thesis, Georgia Institute of Technology.

Brekke, T. L. (1965) On the measurement of the relative potential swellability of hydrothermal montmorillonite clay from joints and faults in PreCambrian and Paleozoic rocks, Norway, *Int. J Rock Mech. Min. Sci.* **2:** 155–165.

Brown, E. T., Harper, T. R., and Hinds, D. V. (1979) Discontinuity measurements with the borehole impression probe—a case study, *Proc. 4th Cong. ISRM* (Montreux), Vol. 2, pp. 57–62.

Carter, J. P. and Kulhawy, F. H. (1988) Analysis and design of drilled shaft foundations socketed into rock, Electric Power Research Institute, Report EL-5918.

Coates, D. F. (1967) *Rock Mechanics Principles,* op. cit. Chapter 1.

Conway, H. D. (1955) Note on the orthotropic half plane subjected to concentrated loads, *J. Appl. Mech.* **77:** 130.

David, D., Sroka, E., and Goldberger, M. (1977) Small diameter piles in karstic rock, *Proceedings, 9th International Conference on Soil Mechanics and Foundation Engineering*, Vol. 1, pp. 471–475.

Desai, C. S. and Christian, J. T. (Eds.) (1977) *Numerical Methods in Geotechnical Engineering*, McGraw-Hill, New York.

Dvorak, A. (1966) Tests of anisotropic shales for foundations of large bridges, *Proc. 1st Cong. ISRM* (Lisbon), Vol. 2, pp. 537–541.

Foose, R. M. (1968) Surface subsidence and collapse caused by ground water withdrawal in carbonate rock areas, *Proc. 23rd Int. Geol. Cong.* (Prague), Vol. 12, pp. 155–166.

Gaziev, E. and Erlikhman, S. (1971) Stresses and strains in anisotropic foundations, *Proceedings, Symposium on Rock Fracture, ISRM* (Nancy), Paper II-1.

Goodman R. E. (1976) *Methods of Geological Engineering in Discontinuous Rocks*, West, St. Paul, MN.

Goodman, R. E., Buchignani, A., and Korbay, S. (1980) Evaluation of collapse potential over abandoned room and pillar mines, *Bull. Assoc. Eng. Geol.* **18** (1).

Grattan-Bellew, P. E. and Eden, W. J. (1975) Concrete deterioration and floor heave due to biogeochemical weathering of underlying shale, *Can. Geot. J.* **12**: 373–378.

Gray, R. E., Salver, H. A., and Gamble, J. C. (1976) Subsidence control for structures above abandoned coal mines, *Trans. Res. Record 612 (TRB)*, pp. 17–24.

Harper, T. R. and Hinds, D. V. (1977) The impression packer: A tool for recovery of rock mass fracture geometry, *Proceedings, Conference on Storage in Evacuated Rock Caverns (ROCKSTORE)*, Vol. 2, pp. 259–266.

Horvath, R. G. (1978) Field load test data on concrete to rock "bond" strength for drilled pier foundations, University of Toronto, Department of Civil Engineering Publication 78–07.

Horvath, R. G. and Kenney, T. C. (1979) Shaft resistance of rock—socketed drilled piers, *Proceedings, Symposium on Deep Foundation Case Histories* (Atlanta). (ASCE). Preprint 3698.

Jackson, W. T., Perez, J. Y., and Lacroix, Y. (1974) Foundation construction and performance for a 34-story building in St. Louis, *Geotechnique* **24**: 69–90.

Jaspar, J. L. and Shtenko, V. W. (1969) Foundation anchor piles in clay shales, *Can. Geot. J.* **6**: 159.

Kenney, T. C. (1977) Factors to be considered in the design of piers socketed in rock, *Proceedings, Conference on Design and Construction of Deep Foundations* (Sudbury, Ont.), (Can. Soc. for C.E.).

Komornik, A. and David, D. (1969) Prediction of swelling pressures in clays, *Proc. ASCE, Soil Mech. Foundations Div.* **95** (SM1): 209–255.

Kulhawy, F. H. (1978) Geomechanical model for rock foundation settlement, *J. Geotech. Eng. Div.*, ASCE **104** (GT2): 211–227.

Kulhawy, F. H. and Ingraffea, A. (1978) Geomechanical model for settlement of long dams on discontinuous rock masses, *Proceedings, International Symposium on Rock Mechanics Related to Dam Foundations (ISRM)*, Rio de Janeiro, Vol. I theme III, p. 115–128.

Ladanyi, B. (1972) Rock failure under concentrated loading, *Proceedings, 10th Symposium on Rock Mechanics*, pp. 363–386.

Ladanyi, B. (1977) Discussion on "friction and end bearing tests on bedrock for high capacity socket design," *Can. Geot. J.* **14**: 153–156.

Londe, P. (1973) *Rock Mechanics and Dam Foundation Design,* International Commission on Large Dams (ICOLD).

Meehan, R. L., Dukes, M. T., and Shires, P. O. (1975) A case history of expansive claystone damage, *J. Geot. Div. (ASCE)* **101** (GT9): 933–948.

Meyerof, G. G. (1953) Bearing capacity of concrete and rock, *Magazine Concrete Res.,* No. 12, pp. 107–116.

Oberti, G., Bavestrello, F., Rossi, R. P., and Flamigni, F. (1986) Rock mechanics investigations, design, and construction of the Ridracoli Dam, *Rock Mech. Rock Eng.* **19**: 113–142.

Osterberg, J. O. and Gill, S. A. (1973) Load transfer mechanism for piers socketed in hard soils or rock, *Proceedings, 9th Canadian Symposium on Rock Mechanics* (Montreal), pp. 235–262.

Parkin, A. K. and Donald, I. B. (1975) Investigation for rock socketed piles in Melbourne mudstone, *Proceedings, 2nd Australia–New Zealand Conference on Geomechanics* (Brisbane), pp. 195–200.

Peck, R. B. (1977) Rock foundations for structures, *Rock Eng. Foundations Slopes* (ASCE) **2**: 1–21.

Peterson, R. and Peters, N. (1963) Heave of spillway structure on clay shale, *Can. Geot. J.* **1**: 5.

Poulos, H. G. and Davis, E. H. *Elastic Solutions for Soil and Rock Mechanics,* Wiley, New York.

Price, D. G., Malkin, A. B., and Knill, J. L. (1969) Foundations of multi-story blocks on the coal measures with special reference to old mine workings, *Q. J. Eng. Geol.* **1**: 271–322.

Raphael, J. and Goodman, R. E. (1979) Op. cit., Chapter 6.

Rosenberg, P. and Journeaux, N. L. (1976) Friction and end bearing tests on bedrock for high capacity socket design, *Can. Geot. J.* **13**: 324–333.

Sowers, G. B. and Sowers, G. F. (1970) *Introductory Soil Mechanics and Foundations,* 3d ed., Macmillan, New York.

Sowers, G. F. (1975) Failures in limestone in humid subtropics, *J. Geot. Div.,* ASCE **101** (GT8): 771–788.

Sowers, G. F. (1976) Mechanism of subsidence due to underground openings, *Trans. Res. Record 612 (TRB),* pp. 1–8.

Sowers, G. F. (1977) Foundation bearing in weathered rock, *Rock Eng. Foundations Slopes* (ASCE) **2**: 32–42.

Thorburn, S. H. (1966) Large diameter piles founded in bedrock, *Proceedings, Symposium on Large Bored Piles* (Institute for Civil Engineering, London), pp. 95–103.

Tomlinson, M. J. (Ed.) (1977) *Piles in Weak Rock,* Institute for Civil Engineering, London.

Underwood, L. B. and Dixon, N. A. (1977) Dams on rock foundations, *Rock Eng. Foundations Slopes* (ASCE) **2**: 125–146.

Wakeling, T. R. M. (1970) A comparison of the results of standard site investigation methods against the results of a detailed geotechnical investigation in Middle Chalk at Mundford, Norfolk, *Proceedings, Conference on In-Situ Investigation in Soils and Rocks, British Geotechnical Society* (London) pp. 17–22.

Webb, D. L. (1977) The behavior of bored piles in weathered diabase, in *Piles in Weak Rock,* Institution of Civil Engineering, London.

Wilson, L. C. (1972) Tests of bored and driven piles in Cretaceous mudstone at Port Elizabeth, South Africa, in *Piles in Weak Rock,* Institute of Civil Engineering, London.

Woodward, R. J., Gardner, W. S., and Greer, D. M. (1972) *Drilled Pier Foundations,* McGraw-Hill, New York.

Zienkiewicz, O. C. (1971) *The Finite Element Method in Engineering Science,* McGraw-Hill, New York.

Problems

1. Derive an analog to Equation 9.6 for the case where the strength envelope of the foundation rock has peak parameters ϕ_p, S_p and residual parameters ϕ_r, S_r.

2. Modify Equation 8.2 to include a structural load P oriented β degrees below horizontal toward the free surface and bearing on the top surface of the slide.

3. Discuss the stability of the block in the following sketch under its own weight W and the applied load P; α, b, and h are variables. In (*a*) P acts through the centroid; in (*b*) it acts at the upper right corner.

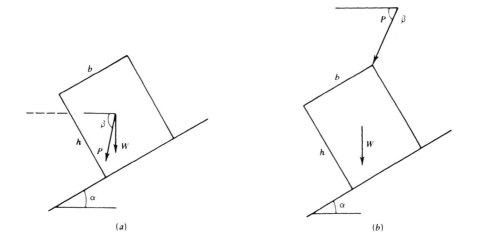

(*a*)   (*b*)

4. Modify Equation 8.12 to include a structural load P oriented as in Problems 2 and 3 and bearing on the surface of the upper block (the "active" block).

5. Compare p_{end}/p_{total} calculated with Equation 9.10 using $\mu = \tan 59°$, and as given by Osterberg and Gill's results (Figure 9.18b) for $E_c/E_r = 1/4$ and $\nu_c =$

$v_r = 0.26$. What value of μ seems appropriate for a pier loaded so that it settles 40 mm? Explain any discrepancy from the value of μ used in the first part to fit Osterberg and Gill's results.

6. Given a fractured cemented shale with unconfined compressive strength $q_u = 18$ MPa, obtained from tests with cylinders 4 in. long by 2 in. in diameter, estimate the allowable bearing pressure q_{allow} and the allowable bond stress τ_{allow} for design of a pier approximately 2 m in diameter. The rock in situ is fresh but is broken by three sets of joints spaced on the average 30 cm apart.

7. Discuss the design of a pier passing through soil and into the rock. The properties of the concrete and rock are as follows: $E_r/E_c = 0.5$; $v_r = v_c = 0.25$, $q_{allow} = 2$ MPa; $\tau_{allow} = 0.1$ MPa, and the maximum allowable compressive stress on the concrete is 10 MPa. The applied load at the surface of the pier is 20 MN (downward). Direct shear tests of concrete sliding on representative rock in a pier configuration gave a friction angle of 40°. Consider design for a bonded pier and for a pier unbonded to the wall.

8. Calculate the minimum width of a long cave necessary to fail the roof formed by a 200-ft-thick ledge of sandstone. The sandstone has $q_u = 20$ MPa and $T_0 = 2$ MPa.

9. Modify Equation 9.15 for the case of a triangular zone of caving above an opening of height h and width L (see the following diagram).

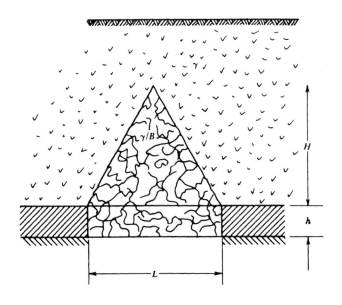

10. An approach like that used to derive Equation 9.10 can be used to derive the required support pressure p_b for a weightless beam in elastic rock under vertical pressure q. The figure shows a prismatic free body diagram from which the requirement for vertical equilibrium yields

$$s^2 d\sigma_v + 4\tau s \; dy = 0$$

in which s is the spacing between passive supports under the beam, placed on a rectangular pattern.

(a) Assume that the horizontal stress is $\sigma_h = k\sigma_v$ and that at limiting equilibrium $\tau = \sigma_h \tan \phi$. Solve the differential equation, with the boundary condition $\sigma_v = q$ when $y = 0$ to determine the support pressure $p_B = \sigma_v$ when $y = t$.

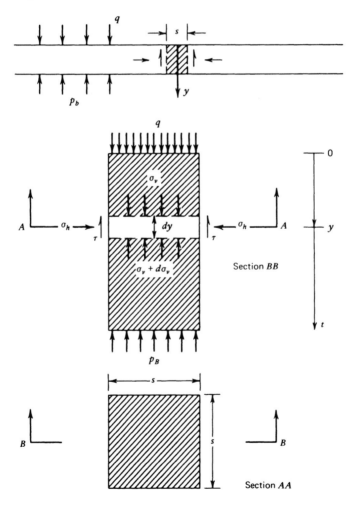

(b) Assume that the rock behaves like a cohesionless material with $\sigma_h = \sigma_v/\tan^2(45 + \phi/2)$. Calculate the force per support for a beam with $s = 1.5$ m, $\phi = 25°$, $t = 1$ m, $q = 21$ kPa, and a factor of safety of 1.0.

11. (a) Repeat Problem 10 for a beam with self weight.
    (b) Recalculate the answer in 10(b) when $\gamma = 27$ kN/m³.
    (c) How would the derivation in (a) differ if rock bolts, rather than passive roof supports were being installed?

Stresses

State of Stress at a Point—Two Dimensions

The "state of stress" at a point 0, Figure A1.1, is defined by the net forces per unit length (the "traction") across any two perpendicular lines, $0x$ and $0y$, through the point. If the stress varies in the body, the force per unit length is understood to apply only to the immediate neighborhood of 0. If the body is in equilibrium, the tractions are balanced by equal and opposite forces across the selected lines $0x$, $0y$. The state of stress is not altered by the choice of axes but its components are. The components of the traction on the x plane (perpendicular to $0x$) are σ_x normal to the x axis and τ_{xy} along it. If σ_x when compressive is pointed toward positive x, τ_{xy} is reckoned positive when pointed toward positive y, and vice versa. (Compression is being considered a positive stress here; tension is negative.) Rotational equilibrium of the small square of Figure A1.2

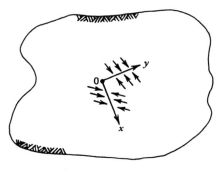

Figure A1.1

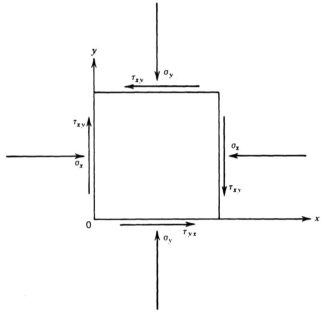

Figure A1.2

requires $\tau_{xy} = \tau_{yx}$; the state of stress is then defined by the values of the three components σ_x, σ_y, and τ_{xy}. We can denote this by

$$\{\sigma\}_{xy} = \begin{Bmatrix} \sigma_x \\ \sigma_y \\ \tau_{xy} \end{Bmatrix} \tag{A1.1}$$

Normal and Shear Stress Across a Given Plane through 0—Two Dimensions

Once $\{\sigma\}_{xy}$ is known, we can calculate the traction across a line of any other orientation through 0. Imagine the line as the trace of a plane parallel to z (so that we can refer to the stress across a "plane" rather than across a line) and consider a plane whose normal $0x'$ (Figure A1.3) is inclined α with $0x$. If AB has unit length, then $0A$ has length $\cos \alpha$ and $0B$ has length $\sin \alpha$. Let $S_{x'}$ be the net force perpendicular to AB (parallel to $0x'$). $S_{x'}$ is the vector sum of the component of forces produced by σ_x and τ_{xy} acting on $0A$, and τ_{yx} acting on $0B$.

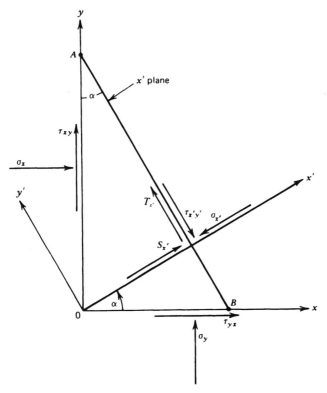

Figure A1.3

Therefore we can write

$$S_{x'} = (\sigma_x \cos \alpha)\cos \alpha + (\tau_{xy} \cos \alpha)\sin \alpha + (\sigma_y \sin \alpha)\sin \alpha$$
$$+ (\tau_{yx} \sin \alpha)\cos \alpha$$

and since $\tau_{xy} = \tau_{yx}$,

$$S_{x'} = \sigma_x \cos^2 \alpha + \sigma_y \sin^2 \alpha + \tau_{xy} 2 \sin \alpha \cos \alpha$$

Similarly, let $T_{x'}$ be the net force parallel to AB, directed parallel to $0y'$. Then, the vector resultants give

$$T_{x'} = -(\sigma_x \cos \alpha)\sin \alpha + (\tau_{xy} \cos \alpha)\cos \alpha + (\sigma_y \sin \alpha)\cos \alpha$$
$$+ (\tau_{yx} \sin \alpha)\sin \alpha$$

and with $\tau_{yx} = \tau_{xy}$, we can write

$$T_{x'} = \sigma_x(-\sin \alpha \cos \alpha) + \sigma_y(\sin \alpha \cos \alpha) + \tau_{xy}(\cos^2 \alpha - \sin^2 \alpha)$$

The normal and shear stress $\sigma_{x'}$ and $\tau_{x'y'}$ on the x' "plane" equilibrate $S_{x'}$ and $T_{x'}$ as shown in Figure A1.3. The normal and shear stress on any plane through 0 whose normal $0X'$ is inclined α with $0X$ are thus

$$\left\{\begin{matrix} \sigma_{x'} \\ \tau_{x'y'} \end{matrix}\right\} = \left(\begin{matrix} \cos^2 \alpha & \sin^2 \alpha & \sin 2\alpha \\ -\frac{1}{2} \sin 2\alpha & \frac{1}{2} \sin 2\alpha & \cos 2\alpha \end{matrix}\right)\left\{\begin{matrix} \sigma_x \\ \sigma_y \\ \tau_{xy} \end{matrix}\right\} \qquad (A1.2)$$

Positive values of $\sigma_{x'}$ and $\tau_{x'y'}$ are directed as shown in Figure A1.3.

Use of Mohr's Circles

This well-known graphical approach can be used to determine $\sigma_{x'}$ and $\tau_{x'y'}$ when given σ_x, σ_y and τ_{xy}. If the signs as well as the magnitudes of the shear stresses $\tau_{x'y'}$ are to be understood correctly, take care to follow these rules[1]:

• Draw Cartesian coordinates x, y, and take σ positive parallel to x, τ positive parallel to y (Figure A1.4).

• Plot point Q at coordinate (σ_x, τ_{xy}).

• Plot point P at coordinate $(\sigma_y, -\tau_{xy})$. P is termed the "pole."

• Mark point C along the σ axis at the midpoint of PQ.

• Draw a circle with center C and radius CP.

• Through P, draw a line parallel to x' intersecting the circle at L. The coordinates of point L are $(\sigma_{x'}, -\tau_{x'y'})$.

• Through P, draw a line parallel to y' intersecting the circle at M. The coordinates at point M are $(\sigma_{y'}, \tau_{x'y'})$.

For example, Figure A1.4 shows the Mohr circle construction for a state of stress given by $\sigma_x = 8$, $\sigma_y = 3$, and $\tau_{xy} = 2$. The magnitudes of the normal and shear stresses on a plane perpendicular to x' directed $10°$ from x are 8.5 and 1.0, respectively, in the directions shown.

Principal Stresses

For a certain value of α, $\tau_{x'y'}$ is zero and $\sigma_{x'}$ is maximum or minimum. The directions of x' and y' are called *principal directions* and the respective normal stresses are the major principal stress σ_1 and the minor principal stress σ_3. To

[1] Told to the author by Dr. John Bray of Imperial College.

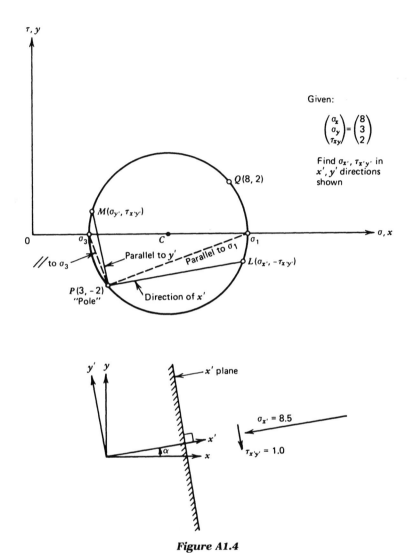

Given:

$$\begin{pmatrix} \sigma_x \\ \sigma_y \\ \tau_{xy} \end{pmatrix} = \begin{pmatrix} 8 \\ 3 \\ 2 \end{pmatrix}$$

Find $\sigma_{x'}$, $\tau_{x'y'}$ in x', y' directions shown

Figure A1.4

calculate the directions, set the second equation of (A1.2) equal to zero, giving

$$0 = \sin 2\alpha \, \frac{\sigma_y - \sigma_x}{2} + \cos 2\alpha \tau_{xy}$$

which yields

$$\tan 2\alpha = \frac{2\tau_{xy}}{\sigma_x - \sigma_y} \tag{A1.3}$$

The sign of α is determined as follows: Let δ ($-\pi/2 \le \delta \le \pi/2$) be the arctan of the term in parentheses in (A1.3). The major principal stress σ_1 acts in direction θ, measured counterclockwise from $0x$:

$$\theta = \frac{\delta}{2} \qquad \text{if } \sigma_x > \sigma_y$$

$$\theta = \frac{\delta}{2} + \frac{\pi}{2} \quad \text{if } \sigma_x < \sigma_y \quad \text{and} \quad \tau_{xy} > 0 \qquad \text{(A1.3a)}$$

and

$$\theta = \frac{\delta}{2} - \frac{\pi}{2} \quad \text{if } \sigma_x < \sigma_y \quad \text{and} \quad \tau_{xy} < 0$$

The two roots of (A1.3) define the principal directions and inserting them in the first row of (A1.2) gives the magnitudes of σ_1 and σ_2:

$$\sigma_1 = \tfrac{1}{2}(\sigma_x + \sigma_y) + [\tau_{xy}^2 + \tfrac{1}{4}(\sigma_x - \sigma_y)^2]^{1/2}$$
$$\sigma_2 = \tfrac{1}{2}(\sigma_x + \sigma_y) - [\tau_{xy}^2 + \tfrac{1}{4}(\sigma_x - \sigma_y)^2]^{1/2} \qquad \text{(A1.4)}$$

Mohr's circle can also yield the principal stresses and directions, as in Figure A1.4.

State of Stress at a
Point in Three Dimensions

In three dimensions, emulating the condition in two dimensions, the "state of stress" is defined by the net forces per unit area (tractions) across any three orthogonal planes through 0 (Figure A1.5). The state of stress is not altered by the choice of axes but the components are. The signs are defined as in the two-

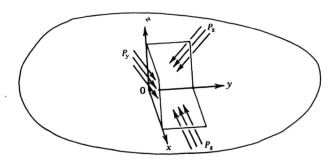

Figure A1.5

dimensional case: on each of the coordinate planes the traction is decomposed into one normal and two shear components; if a normal compression is directed parallel to a positive coordinate axis, the shear components are positive parallel to the other positive coordinate directions, and vice versa. On the plane perpendicular to x, for example, the positive shear stresses labeled τ_{yx} and τ_{yz} are directed as shown in Figure A1.6.

Rotational equilibrium of a small cube at 0 requires

$$\tau_{yz} = \tau_{zy}$$
$$\tau_{yx} = \tau_{xy} \qquad \qquad (A1.5)$$
$$\tau_{zx} = \tau_{xz}$$

Therefore the state of stress is completely defined by a symmetric matrix with six independent components:

$$(\sigma)_{xyz} = \begin{pmatrix} \sigma_x & \tau_{xy} & \tau_{xz} \\ \tau_{xy} & \sigma_y & \tau_{yz} \\ \tau_{xz} & \tau_{yz} & \sigma_z \end{pmatrix} \qquad \qquad (A1.6)$$

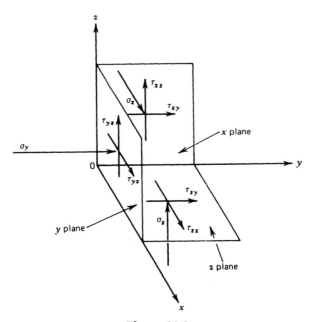

Figure A1.6

Normal and Shear Stresses across a Given Plane through 0—Three Dimensions

Consider the plane whose normal $0x'$ makes angles of $(x'x)$ with $0x$, $(x'y)$ with $0y$, and $(x'z)$ with $0z$ with direction cosines:

$$l_{x'} = \cos(x'x)$$
$$m_{x'} = \cos(x'y) \qquad \qquad \text{(A1.7)}$$
$$n_{x'} = \cos(x'z)$$

If ABC has unit area, then $0AC$ has area $l_{x'}$, OAB has area $m_{x'}$, and $0BC$ has area $n_{x'}$ (Figure A1.7).

Let $S_{x'}$ be the net force perpendicular to plane ABC (i.e., parallel to $0x'$). $S_{x'}$ is the vector sum of the components of forces produced by the tractions P_x, P_y, and P_z on the three coordinate planes of Figure A1.7, which have been decomposed into the nine stress components shown in Figure A1.8. The strategy will be to combine these components into forces $P_{x'x}$ parallel to x, $P_{x'y}$ parallel to y, and $P_{x'z}$ parallel to z and then to project each of these forces in turn in the direction of $S_{x'}$:

$$P_{x'x} = \sigma_x l_{x'} + \tau_{yx} m_{x'} + \tau_{zx} n_{x'}$$
$$P_{x'y} = \tau_{xy} l_{x'} + \sigma_y m_{x'} + \tau_{zy} n_{x'} \qquad \qquad \text{(A1.8)}$$
$$P_{x'z} = \tau_{xz} l_{x'} + \tau_{yz} m_{x'} + \sigma_z n_{x'}$$

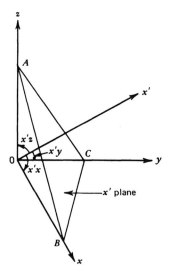

Figure A1.7

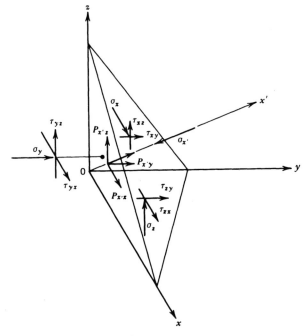

Figure A1.8

or

$$\langle P_{x'x} P_{x'y} P_{x'z} \rangle = \langle l_{x'} m_{x'} n_{x'} \rangle(\sigma)_{xyz} \tag{A1.9}$$

then

$$S_{x'} = P_{x'x} l_{x'} + P_{x'y} m_{x'} + P_{x'z} n_{x'} \tag{A1.10}$$

so that

$$S_{x'} = \langle l_{x'} m_{x'} n_{x'} \rangle(\sigma)_{xyz} \begin{Bmatrix} l_{x'} \\ m_{x'} \\ n_{x'} \end{Bmatrix} \tag{A1.11}$$

Similarly, let y' and z' be perpendicular axes in the x' plane and let $T_{x'y'}$ and $T_{x'z'}$ be the net forces on the x' plane parallel to y' and z' (Figure A1.9).

Let y' have direction cosines

$$\begin{aligned} l_{y'} &= \cos(y'x) \\ m_{y'} &= \cos(y'y) \\ n_{y'} &= \cos(y'z) \end{aligned} \tag{A1.12}$$

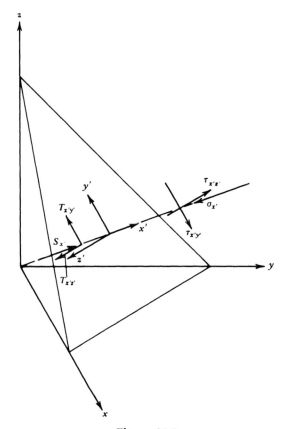

Figure A1.9

and let z' have direction cosines

$$
\begin{aligned}
l_{z'} &= \cos z'x \\
m_{z'} &= \cos z'y \\
n_{z'} &= \cos z'z
\end{aligned}
\tag{A1.13}
$$

where, as before, $y'x$ is the angle between the y' and x axes, etc.

To find $T_{x'y'}$, project $P_{x'x}$, $P_{x'y}$, and $P_{x'z}$ on y' and sum, giving

$$
T_{x'y'} = (P_{x'x}P_{x'y}P_{x'z})
\begin{Bmatrix}
l_{y'} \\
m_{y'} \\
n_{y'}
\end{Bmatrix}
\tag{A1.14}
$$

Similarly

$$
T_{x'z'} = (P_{x'x}P_{x'y}P_{x'z})
\begin{Bmatrix}
l_{z'} \\
m_{z'} \\
n_{z'}
\end{Bmatrix}
\tag{A1.15}
$$

Substituting (A1.9) in the above and combining with (A1.11) and observing that $S_{x'}$, $T_{x'y'}$, and $T_{x'z'}$ are equilibrated by stresses $\sigma_{x'}$, $\tau_{x'y'}$, and $\tau_{x'z'}$ acting over the x' plane (Figure A1.9), the latter can be calculated using the following compact formulas:

$$\langle \sigma_{x'} \tau_{x'y'} \tau_{x'z'} \rangle = (L_{x'})(\sigma)(L)^T \tag{A1.16}$$

where

$$(L_{x'}) = \langle l_{x'} m_{x'} n_{x'} \rangle \tag{A1.16a}$$

and

$$L = \begin{pmatrix} l_{x'} & m_{x'} & n_{x'} \\ l_{y'} & m_{y'} & n_{y'} \\ l_{z'} & m_{z'} & n_{z'} \end{pmatrix} \tag{A1.16b}$$

(The superscript T in A1.16) indicates the matrix transpose.)

With y' and z' axes directed as shown in Figure A1.9, the directions of positive shear stresses $\tau_{x'y'}$ and $\tau_{x'z'}$ are shown in the same figure. The shear stresses may be combined into a single resultant shear on the x' plane $\tau_{x',\max}$ whose magnitude is then

$$|\tau_{x',\max}| = \sqrt{\tau_{x'y'}^2 + \tau_{x'z'}^2} \tag{A1.17}$$

$\tau_{x',\max}$ makes a counterclockwise angle of θ with the negative direction of y' (Figure A1.10) where

$$\theta = \tan^{-1}\left(\frac{\tau_{x'z'}}{\tau_{x'y'}}\right) \tag{A1.18}$$

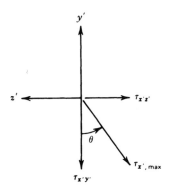

Figure A1.10

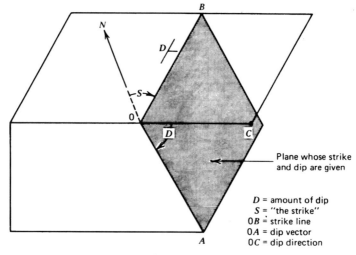

D = amount of dip
S = "the strike"
OB = strike line
OA = dip vector
OC = dip direction

Plane whose strike
and dip are given

Figure A1.11

Finding Direction Cosines for a Given Plane

Geological data will normally define the orientation of a plane by its "strike" and "dip" (Figure A1.11). The *strike* is the azimuth of a horizontal line in the plane. The *dip* is a vector pointed down the steepest slope of the plane and is defined by the azimuth of its horizontal projection (the *dip direction*), and the vertical angle between the horizontal projection and the dip vector (the *amount of dip*). For example, a bedding plane might be defined by strike N 40° E, dip

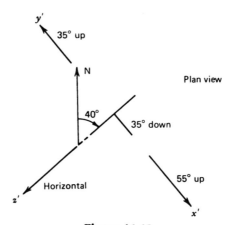

Figure A1.12

35° to the S 50° E. Coordinate axes will be taken as follows: x is horizontal to the east, y is horizontal north, and z is upward.

The bearing of the upward normal to the plane is the same as the bearing of its dip, and it rises at a vertical angle complement to the amount of dip. To apply Equations A1.16 to calculate the stresses across the given plane, given the local state of stress, we describe each of the axes x', y', z' by a horizontal angle β and a vertical angle δ as shown in Figure A1.13; β is the counterclockwise angle from x to the horizontal projection of the axis in question while δ is the vertical angle between the axis and its horizontal projection. The line $0x'$ is in the direction of the upward normal. It proves convenient to take y' positive when directed up the steepest slope of the plane, that is, opposite to the dip vector, and z' along the strike, as shown in Figure A1.9. For the indicated bedding plane then, the bearings and rise angles of the three axes are (Figure A1.12):

Line	Compass Bearing	β	Rise Angle δ (Positive if above Horizontal)
x'	S 50° E	−40°	55°
y'	N 50° W	140°	35°
z'	S 40° W	−130°	0

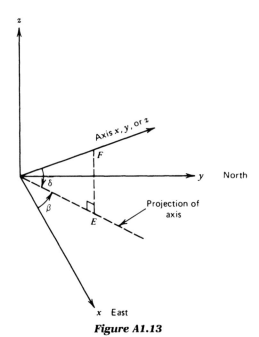

Figure A1.13

The direction cosines can be calculated from the known angles β and δ of each axis. From Figure A1.13:

$$l = \cos \delta \cos \beta$$
$$m = \cos \delta \sin \beta \qquad\qquad (A1.19)$$
$$n = \sin \delta$$

For the given bedding plane then

$$(L) = \begin{pmatrix} 0.44 & -0.37 & 0.82 \\ -0.63 & 0.53 & 0.57 \\ -0.64 & -0.77 & 0 \end{pmatrix}$$

Principal Stresses

No matter what the state of stress, it is possible to find three orthogonal planes along which the shear stresses vanish. The normal stresses on these three planes are called *principal stresses* σ_1, σ_2, σ_3. (We will adopt the convention $\sigma_1 > \sigma_2 > \sigma_3$ throughout.)

Consider a principal plane perpendicular to x^* with direction cosines l^*, m^*, n^* and normal stress (principal stress) σ^*. Since there is no shear component, σ^* is also the traction, that is, the resultant force is normal to the x^* plane. The x component of the traction (P_{x^*x}) must then equal the projection of σ^* along x, that is, $P_{x^*x} = \sigma^* l^*$ (Figure A1.14). Similarly, $P_{x^*y} = \sigma^* m^*$ and

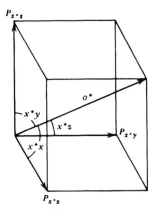

Figure A1.14

$P_{x^*z} = \sigma^* n^*$. Recalling (A1.9), these equations can be written

$$(l^* m^* n^*) \begin{pmatrix} \sigma_x & \tau_{xy} & \tau_{xz} \\ \tau_{xy} & \sigma_y & \tau_{yz} \\ \tau_{xz} & \tau_{yz} & \sigma_z \end{pmatrix} = (l^* m^* n^*) \begin{pmatrix} \sigma^* & 0 & 0 \\ 0 & \sigma^* & 0 \\ 0 & 0 & \sigma^* \end{pmatrix}$$

or

$$(l^* m^* n^*) \begin{pmatrix} \sigma_x - \sigma^* & \tau_{xy} & \tau_{xz} \\ \tau_{xy} & \sigma_y - \sigma^* & \tau_{yz} \\ \tau_{xz} & \tau_{yz} & \sigma_z - \sigma^* \end{pmatrix} = (0, 0, 0) \qquad \text{(A1.20)}$$

Equation A1.20 represents a homogeneous set of three simultaneous equations whose solution requires that the determinant of the square matrix be equal to zero.

Expanding this determinant, and setting it equal to zero, gives

$$\sigma^{*3} - I_1 \sigma^{*2} + I_2 \sigma^* - I_3 = 0 \qquad \text{(A1.21)}$$

where I_1 is the sum of the diagonal terms of $(\sigma)_{xyz}$, I_2 is the sum of the minors of each diagonal term of $(\sigma)_{xyz}$, and I_3 is the determinant of $(\sigma)_{xyz}$:

$$I_1 = \sigma_x + \sigma_y + \sigma_z$$
$$I_2 = (\sigma_y \sigma_z - \tau_{yz}^2) + (\sigma_x \sigma_z - \tau_{xz}^2) + (\sigma_x \sigma_y - \tau_{xy}^2) \qquad \text{(A1.21a)}$$

and

$$I_3 = \sigma_x(\sigma_y \sigma_z - \tau_{yz}^2) - \tau_{xy}(\tau_{xy}\sigma_z - \tau_{yz}\tau_{xz}) + \tau_{xz}(\tau_{xy}\tau_{yz} - \sigma_y\tau_{xz})$$

I_1, I_2, and I_3 are uniquely defined regardless of the choice of the x, y, z axes. They are thus spoken of as "invariants of stress." The three roots of this cubic equation are the three principal stresses (the eigenvalues); putting each root in turn in (A1.20) yields a homogeneous set of three equations. Discard any one and the remaining two equations may be solved for two of the direction cosines in terms of the third and since $l^{*2} + m^{*2} + n^{*2} = 1$, the direction cosines $l^* m^* n^*$ can be determined for two roots; the third direction is perpendicular to the two axes thus defined. These eigenvectors define the three principal stress directions.

Transformation to a New Coordinate System

Given the state of stress relative to x, y, z, Equation A1.16 shows how to find $\sigma_{x'}\tau_{x'y'}$ and $\tau_{x'z'}$, the components of stress on the x' plane. Repetition of this approach for the y' and z' planes would provide formulas for six additional stress components, completing the transformation of stresses to an entirely new set of coordinate axes. The results are

$$(\sigma)_{x'y'z'} = (L)(\sigma)_{xyz}(L)^T \tag{A1.22}$$

Since only six components are unique, it is possible to rewrite (σ) as a list of six numbers in a row or column, that is, to represent it by a six-component "vector." This proves useful in manipulating data from stress measurements in nonorthogonal directions as discussed in Chapter 4. Expanding (A1.22) and rearranging gives

$$
\begin{Bmatrix} \sigma_{x'} \\ \sigma_{y'} \\ \sigma_{z'} \\ \tau_{y'z'} \\ \tau_{z'x'} \\ \tau_{x'y'} \end{Bmatrix} =
$$

$$
\begin{bmatrix}
l_{x'}^2 & m_{x'}^2 & n_{x'}^2 & 2m_{x'}n_{x'} & 2n_{x'}l_{x'} & 2l_{x'}m_{x'} \\
l_{y'}^2 & m_{y'}^2 & n_{y'}^2 & 2m_{y'}n_{y'} & 2n_{y'}l_{y'} & 2l_{y'}m_{y'} \\
l_{z'}^2 & m_{z'}^2 & n_{z'}^2 & 2m_{z'}n_{z'} & 2n_{z'}l_{z'} & 2l_{z'}m_{z'} \\
l_{y'}l_{z'} & m_{y'}m_{z'} & n_{y'}n_{z'} & m_{y'}n_{z'} + m_{z'}n_{y'} & n_{y'}l_{z'} + n_{z'}l_{y'} & l_{y'}m_{z'} + l_{z'}m_{y'} \\
l_{z'}l_{x'} & m_{z'}m_{x'} & n_{z'}n_{x'} & m_{x'}n_{z'} + m_{z'}n_{x'} & n_{x'}l_{z'} + n_{z'}l_{x'} & l_{x'}m_{z'} + l_{z'}m_{x'} \\
l_{x'}l_{y'} & m_{x'}m_{y'} & n_{x'}n_{y'} & m_{x'}n_{y'} + m_{y'}n_{x'} & n_{x'}l_{y'} + n_{y'}l_{x'} & l_{x'}m_{y'} + l_{y'}m_{x'}
\end{bmatrix}
\begin{Bmatrix} \sigma_x \\ \sigma_y \\ \sigma_z \\ \tau_{yz} \\ \tau_{zx} \\ \tau_{xy} \end{Bmatrix}
$$

$$\tag{A1.23}$$

or

$$\{\sigma\}_{x'y'z'} = (T_\sigma)\{\sigma\}_{xyz} \tag{A1.24}$$

"Octahedral Stresses"

Theories of failure are often plotted in terms of principal stresses: $f(\sigma_1, \sigma_2, \sigma_3) = 0$. To achieve a two-dimensional representation of the surface represented by f, it proves helpful to view it in a series of sections looking down an axis (x') making equal angles with each of the principal stresses—the "octahedral axis." Let σ_1 correspond to z, σ_2 to x, and σ_3 to y (Figure A1.15); then the direction cosines of x' are

$$1_{x'} = m_{x'} = n_{x'} = \frac{1}{\sqrt{3}} \tag{A1.25}$$

As we did earlier, let y' be directed up the x' plane (the "octahedral plane") and let z' lie along the strike of the x' plane (Figure A1.15).

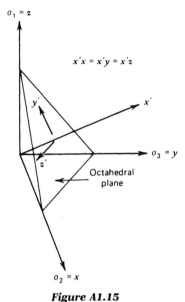

Figure A1.15

The matrix of direction cosines for transformation of axes is then

$$(L)_{\text{octahedral}} = \begin{bmatrix} \dfrac{1}{\sqrt{3}} & \dfrac{1}{\sqrt{3}} & \dfrac{1}{\sqrt{3}} \\[2mm] -\dfrac{1}{\sqrt{6}} & -\dfrac{1}{\sqrt{6}} & \dfrac{\sqrt{2}}{\sqrt{3}} \\[2mm] \dfrac{1}{\sqrt{2}} & -\dfrac{1}{\sqrt{2}} & 0 \end{bmatrix}$$

(A1.26)

whereupon the normal and shear stresses are

$$\sigma_{x'} = \frac{\sigma_1 + \sigma_2 + \sigma_3}{3}$$

(A1.27a)

$$\tau_{x'y'} = \frac{1}{3\sqrt{2}} (2\sigma_1 - \sigma_2 - \sigma_3)$$

(A1.27b)

$$\tau_{x'z'} = \frac{1}{\sqrt{6}} (\sigma_2 - \sigma_3)$$

(A1.27c)

The directions of the shear stresses in the octahedral plane (looking toward the origin down the octahedral axis, that is, down x') are as shown in Figure A1.16.

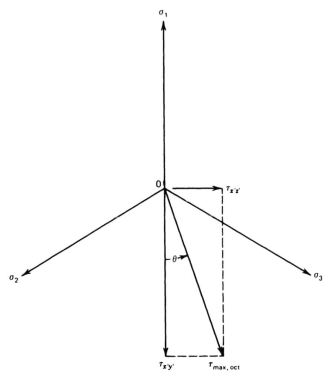

Figue A1.16

Combining Equations A1.27b and c into a resultant shear stress in the octahedral plane gives

$$\tau_{max,oct} = \frac{\sqrt{2}}{3} \sqrt{\sigma_1^2 + \sigma_2^2 + \sigma_3^2 - \sigma_1\sigma_2 - \sigma_1\sigma_3 - \sigma_2\sigma_3} \quad \text{(A1.28a)}$$

The direction of $\tau_{max,oct}$ is θ degrees counterclockwise from $\tau_{x'y'}$ (Figure A1.16) where

$$\theta = \tan^{-1} \frac{\sigma_2 - \sigma_3}{\sqrt{3}(2\sigma_1 - \sigma_2 - \sigma_3)} \quad \text{(A1.28b)}$$

Note that the normal stress on the octahedral plane is equal to the mean principal stress ($=I_1/3$). Also the maximum shear stress ($\tau_{max,oct}$) proves to be equal to $\sqrt{\frac{2}{3}J_2}$ where J_2 is the second invariant of the deviatoric stress matrix; the latter is formed from (σ) by subtracting $I_1/3$ from each diagonal term.

Problems

1. Using Equations A1.2, with $\{\sigma_{xy}\}$ as given, find the normal and shear stresses for the following planes and show the directions of the stresses by arrows. (Remember α is the angle from $0x$ to the *normal* to the plane.)
   (a) $\sigma_x = 50$, $\sigma_y = 30$, $\tau_{xy} = -20$
      1. $\alpha = 30°$
      2. $\alpha = 45°$
      3. $\alpha = 90°$
      4. $\alpha = 0$
   (b) $\sigma_x = 100$, $\sigma_y = 60$, $\tau_{xy} = 20$
      1. $\alpha = -75°$
      2. $\alpha = -60°$
      3. $\alpha = -30°$
      4. $\alpha = 22.5°$

2. Confirm the answers to each of the above by using Mohr's circle.

3. For the stresses of Problems 1a and 1b, find the directions and magnitudes of the principal stresses, (a) using Mohr's circles, and (b) using Equations A1.3 and A1.4.

4. Derive a formula for $\sigma_{y'}$ where $0y'$ is an axis through 0, perpendicular to $0x'$ as shown in Figure A1.2.

5. Use your answer to Problem 4 to show that $\sigma_x + \sigma_y = \sigma_{x'} + \sigma_{y'}$ for any value of α.

6. Find the direction of major principal stress analytically if
   (a) $\sigma_x = 60$, $\sigma_y = 100$, $\tau_{xy} = 20$
   (b) $\sigma_x = 60$, $\sigma_y = 100$, $\tau_{xy} = -20$
   (c) Verify (a) using Mohr's circle.
   (d) Verify (b) using Mohr's circle.

7. Given the following strikes and dips of planes, calculate (L), the matrix of direction cosines of x', y', and z'. (x', y', and z' are directed as in Figure A1.9. The x axis is horizontal to the east. The y axis is horizontal to the north. The z axis is upward.)
   (a) Strike N, dip 30° E.
   (b) Strike N 70° W, dip 70° S 20° W.
   (c) Strike N 45° E, dip vertical.
   (d) Horizontal (take y' north).

8. A point in a rock mass has stresses given by

$$(\sigma) = \begin{pmatrix} 100 & 50 & 50 \\ 50 & 200 & 0 \\ 50 & 0 & 700 \end{pmatrix}$$

If the x, y, and z axes are directed as in Problem 7, find the stress components referred to the x' plane for each of the planes (a)–(d) of Problem 7. Also calculate the maximum shear stress in the x' plane in magnitude and direction and show it in a diagram.

9. In each case of Problem 8, calculate the magnitude of the maximum shear stress by vector subtraction of $\sigma_{x'}$ from the resultant $(R_{x'})$ of $P_{x'x}$, $P_{x'y}$, and $P_{x'z}$. Compare your answers with the answers to Problems 8 and 9.

10. For Problem 8d, compare the answer as calculated with the stresses given in the x, y, z system.

11. For the stresses of Problem 8, calculate I_1, I_2, and I_3.

12. Map the states of stress below into the octahedral plane.
    (a) $\sigma_1 = 150$, $\sigma_2 = 0$, $\sigma_3 = 0$
    (b) $\sigma_1 = 100$, $\sigma_2 = 50$, $\sigma_3 = 0$
    (c) $\sigma_1 = 100$, $\sigma_2 = 25$, $\sigma_3 = 25$
    (d) $\sigma_1 = 50$, $\sigma_2 = 50$, $\sigma_3 = 50$
    (e) $\sigma_1 = 75$, $\sigma_2 = 75$, $\sigma_3 = 0$
    (f) $\sigma_1 = 200$, $\sigma_2 = 0$, $\sigma_3 = -50$

Strains and Strain Rosettes

"Strain" refers to the change in shape of a body as its points are displaced. A full discussion of strain theory is found in Chapter 2 of *Fundamentals of Rock Mechanics* by Jaeger and Cook (1976). Here we can list some of the basic relationships required to work with strain in rocks.

The state of strain in two dimensions is described by three components:

$$\{\varepsilon\}_{xy} = \begin{Bmatrix} \varepsilon_x \\ \varepsilon_y \\ \gamma_{xy} \end{Bmatrix} \tag{A2.1}$$

The "normal strain" ε_x is the shortening of a unit line originally parallel to the x axis while ε_y is the shortening of a unit line originally parallel to y. In Figure A2.1, points 0, P, and R before deformation have moved to $0'$, P', and R', respectively. The normal strains are approximately

$$\varepsilon_x = -\frac{0'P'' - 0P}{0P} = -\frac{\partial u}{\partial x}$$

and

$$\varepsilon_y = -\frac{0'R'' - 0R}{0R} = -\frac{\partial v}{\partial y}$$

where u and v are the x and y displacements of a point resulting from straining. The "shear strain" γ_{xy} is the sum of angles δ_1 and δ_2; it may be considered the loss in perpendicularity of the originally orthogonal axes $0P$ and $0R$. In Figure A2.1,

$$\gamma_{xy} = \delta_1 + \delta_2 = -\left(\frac{\partial u}{\partial y} + \frac{\partial v}{\partial x}\right)$$

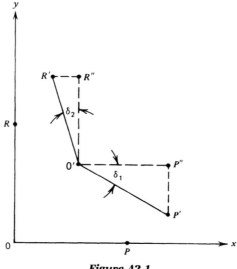

Figure A2.1

A positive shear strain is shown in the figure in which the angle $R0P$ has increased beyond 90° as a result of straining.

If the state of strain is known, the components of strain may be calculated for lines in all other directions, for example, parallel to axes x' and y' in Figure A2.2. Equations A1.2–A1.4 in Appendix 1, related to transformation of stresses, can be used directly if the following substitutions are made: (1) In place of any term named σ, introduce the corresponding ε term (e.g., ε_x for σ_x, ε_1 for σ_1). (2) In place of any term called τ, substitute $\gamma/2$ (e.g., $\frac{1}{2}\gamma_{x'y'}$ in place of $\tau_{x'y'}$).

To calculate ε_x, given $\{\varepsilon\}_{xy}$ make these substitutions in the first of Equations A1.2, giving

$$\varepsilon_{x'} = (\cos^2 \alpha \quad \sin^2 \alpha \quad \tfrac{1}{2} \sin 2\alpha) \begin{Bmatrix} \varepsilon_x \\ \varepsilon_y \\ \gamma_{xy} \end{Bmatrix} \tag{A2.2}$$

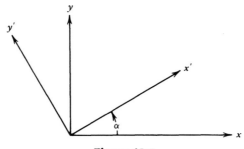

Figure A2.2

Use of Strain Gage Rosettes

A "rosette" is a set of three strain gages in different known orientations, α_A, α_B, α_C, (Figure A2.3).

When the state of strain is given with respect to the x, y coordinates, the normal strain along another direction inclined α with x (counterclockwise positive) is given by Equation A2.2. Applying this in turn to each of the rosette "arms" gives

$$\begin{Bmatrix} \varepsilon_A \\ \varepsilon_B \\ \varepsilon_C \end{Bmatrix} = \begin{bmatrix} \cos^2 \alpha_A & \sin^2 \alpha_A & \tfrac{1}{2}\sin 2\alpha_A \\ \cos^2 \alpha_B & \sin^2 \alpha_B & \tfrac{1}{2}\sin 2\alpha_B \\ \cos^2 \alpha_C & \sin^2 \alpha_C & \tfrac{1}{2}\sin 2\alpha_C \end{bmatrix} \begin{Bmatrix} \varepsilon_x \\ \varepsilon_y \\ \gamma_{xy} \end{Bmatrix} \tag{A2.3}$$

The magnitudes of the strain components may be found by inverting the above.

For a rosette with $\alpha_A = 0$, $\alpha_B = 45°$, and $\alpha_C = 90°$ (a "45° rosette") inversion of (A2.3) yields

$$\begin{Bmatrix} \varepsilon_x \\ \varepsilon_y \\ \gamma_{xy} \end{Bmatrix} = \begin{bmatrix} 1 & 0 & 0 \\ 0 & 0 & 1 \\ -1 & 2 & -1 \end{bmatrix} \begin{Bmatrix} \varepsilon_A \\ \varepsilon_B \\ \varepsilon_C \end{Bmatrix} \tag{A2.4}$$

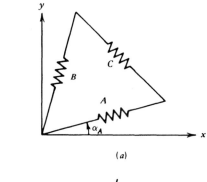

(a)

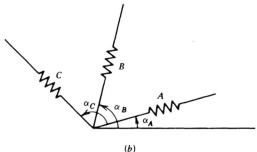

(b)

Figure A2.3

For rosette with $\alpha_A = 0$, $\alpha_B = 60°$, $\alpha_C = 120°$, (a 60° rosette), inversion of (A2.3) yields

$$\begin{Bmatrix} \varepsilon_x \\ \varepsilon_y \\ \gamma_{xy} \end{Bmatrix} = \begin{bmatrix} 1 & 0 & 0 \\ -\frac{1}{3} & \frac{2}{3} & \frac{2}{3} \\ 0 & 1.1547 & -1.1547 \end{bmatrix} \begin{Bmatrix} \varepsilon_A \\ \varepsilon_B \\ \varepsilon_C \end{Bmatrix} \tag{A2.5}$$

Principal Strains

The principal strains are the normal strains along a set of directions such that originally perpendicular lines remain perpendicular as a result of deformations. Principal strains ε_1, ε_2 ($\varepsilon_1 > \varepsilon_2$) may be found for a given state of strain by making the required substitutions in Equations A1.3 and A1.4 (Appendix A1). The principal strains are

$$\varepsilon_1 = \tfrac{1}{2}(\varepsilon_x + \varepsilon_y) + \tfrac{1}{2}[\gamma_{xy}^2 + (\varepsilon_x - \varepsilon_y)^2]^{1/2}$$

and (A2.6)

$$\varepsilon_2 = \tfrac{1}{2}(\varepsilon_x + \varepsilon_y) - \tfrac{1}{2}[\gamma_{xy}^2 + (\varepsilon_x - \varepsilon_y)^2]^{1/2}$$

and the direction of major principal strain is given by

$$\tan 2\alpha = \frac{\gamma_{xy}}{\varepsilon_x - \varepsilon_y} \tag{A2.7}$$

The sign of α is determined by application of the rules given after (A1.3).

Deviatoric and Nondeviatoric Strains

The 3×3 strain matrix describing a three-dimensional state of strain can be subdivided into "nondeviatoric" and "deviatoric" portions. The former describes the volume change; the latter describes the distortion:

$$\begin{pmatrix} \varepsilon_x & \gamma_{xy} & \gamma_{xz} \\ \gamma_{xy} & \varepsilon_y & \gamma_{yz} \\ \gamma_{xz} & \gamma_{yz} & \varepsilon_z \end{pmatrix} = \begin{pmatrix} \bar{\varepsilon} & 0 & 0 \\ 0 & \bar{\varepsilon} & 0 \\ 0 & 0 & \bar{\varepsilon} \end{pmatrix} + \begin{pmatrix} e_x & 2e_{xy} & 2e_{xz} \\ 2e_{xy} & e_y & 2e_{yz} \\ 2e_{xz} & 2e_{yz} & e_z \end{pmatrix} \tag{A2.8}$$

$$\text{Nondeviatoric} \qquad \text{Deviatoric}$$

where

$$\bar{\varepsilon} = \frac{\varepsilon_x + \varepsilon_y + \varepsilon_z}{3}$$

$$e_x = \varepsilon_x - \bar{\varepsilon}, \text{ etc.}$$

and

$$e_{xy} = \tfrac{1}{2}\gamma_{xy}, \text{ etc.}$$

Problems

1. Write formulas for the complete change of axes to express $\{\varepsilon\}_{x'y'}$ in terms of $\{\varepsilon\}_{xy}$.

2. Derive formulas for a rosette gage with arms at $\alpha_A = 0$, $\alpha_B = 60°$, and $\alpha_C = 90°$ expressing $\{\varepsilon\}_{xy}$ in terms of the three measured contractions ε_A, ε_B, and ε_C.

3. A strain rosette with arms at $\alpha_A = 0$, $\alpha_B = 60°$, and $\alpha_C = 120°$, gives the following readings:
   (a) $\varepsilon_A = 10^{-3}$, $\varepsilon_B = 0.5 \times 10^{-3}$, $\varepsilon_C = 0$
   (b) $\varepsilon_A = 10^{-2}$, $\varepsilon_B = 2 \times 10^{-2}$, $\varepsilon_C = 3 \times 10^{-2}$
   (c) $\varepsilon_A = 2 \times 10^{-4}$, $\varepsilon_B = 3.8 \times 10^{-4}$, $\varepsilon_C = 5.2 \times 10^{-4}$
   Calculate the state of strain $\{\varepsilon\}_{xy}$.

4. In regard to Problem 3, compute the magnitudes and directions of principal strains ε_1 and ε_2.

Identification of Rocks and Minerals

How Many Rocks and Minerals Must an Engineer Know?

Textbooks of mineralogy commonly list determinative properties for about 200 minerals. A good book on petrography will mention more than 1000 types of rocks. The subject is interesting and has many practical offshoots. Fortunately, however, the list of the most common rock-forming minerals is rather short— about 16—and many of the rock types fall naturally into groups with similar engineering attributes, so that only about 40 rock names will suffice to describe most of the individuals of real interest for civil engineering purposes. There are exceptional cases, however, when rather bizarre rock types cause unusual problems for excavations or rock materials. Rather than learn 1000 varieties to be equipped for the one special case, it is more efficient to enlist the aid of a petrologist when this happens. For the basic education of the geotechnical engineer, it will usually suffice to become familiar with the 16 minerals and 40 rocks discussed below; that is, to be able to identify them and to know something of their occurrence and properties.

Rock-Forming Minerals

The common rock-forming minerals are silicates, carbonates, and several salts (sulfates and chlorides). The silicate minerals are formed from the silica tetrahedra (SiO_4) linked together in "island structures," sheets, chains, and networks by iron, magnesium, calcium, potassium, and other ions. The island structures, such as olivine, are tetrahedra without shared corners—they are the highest temperature minerals of the silicate group (earliest formed when a melt cools) and they are generally the first to weather when exposed to the atmosphere. The sheet structures (e.g., mica) have easy parting (cleavage) in one direction and generally low shear strength along that direction (parallel to the sheets). Chains (e.g., pyroxenes and amphiboles) and networks like feldspars and quartz are usually very strong and durable.

The carbonates are weakly soluble in water but more highly soluble if water has been enriched in acid by percolation through soil or by industrial pollution. The carbonate minerals also have the characteristic of twinning readily by gliding on intracrystalline planes, so rocks composed of these minerals behave plastically at elevated pressures. Other salts (e.g., gypsum and halite) are readily soluble in water. The sulfide pyrite is present in small amounts in almost all rocks and occasionally occurs as a significant percentage of rocks.

The common rock-forming minerals that you should be able to identify are:

Silicates

Quartz, feldspar (orthoclase, and plagioclase), mica (biotite and muscovite), chlorite, amphibole, pyroxene, and olivine.

Carbonates

Calcite and dolomite.

Others

Gypsum, anhydrite, halite, pyrite, and graphite.

Table A3.1 will assist you to identify these minerals. Since minerals forming the rock fabric are usually found in fragments or crystals less than a centimeter in maximum dimension, it is necessary to view the rock using a hand lens, or even better, a binocular microscope. The minerals are divided into those that can be scratched by the fingernail, those that can be scratched by a knife blade but not by the fingernail, and those that cannot be scratched by a knife blade. On Moh's scale of relative hardness, the fingernail will usually have a hardness between 2 and $2\frac{1}{2}$, while the average knife will have a hardness of between 5 and $5\frac{1}{2}$. The presence or absence of cleavage is one of the easily noted diagnostic features of the minerals listed. Cleavage surfaces are smooth and uniform and reflect incident light uniformly at one orientation. The angles between the cleavages can be measured by rotating the specimen in the hand to move from the orientation of a reflection on one surface to the reflection orientation for the adjacent surface. As an illustration of how the table works, compare calcite, feldspar,

TABLE A3.1

Simplified Mineral Identification Flowchart: Common Rock-forming Minerals

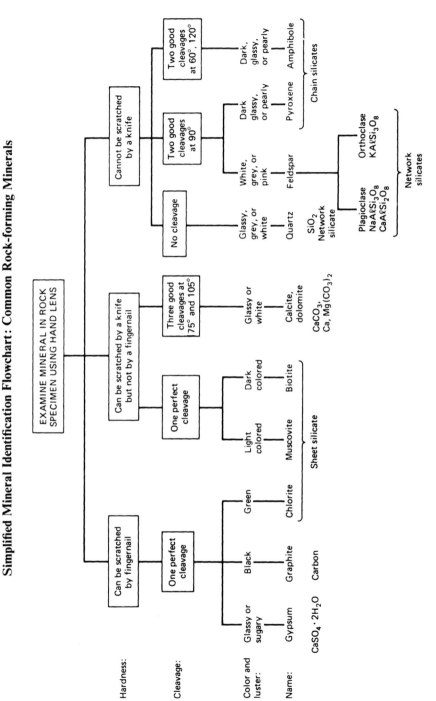

and quartz, three minerals that are frequently confused by engineers. Quartz presents no cleavage and will not be scratched by a knife. (It may display crystal faces; crystal faces are destroyed when the crystal is broken, whereas cleavage surfaces will be found on all the mineral chips after the crystal is broken.) Feldspar is harder than the knife and also presents two good directions of cleavage. Calcite also has good cleavage but it can be scratched. Moreover, the calcite presents rhombohedral angles between the cleavage surfaces (75° and 105°) whereas the feldspar cleavages have approximately 90° angles between them.

Other Important Minerals

A small number of minerals account for most of the special problems that sometimes occur with rocks. These special problems concern pollution, rapid weathering, swelling, chemical attack of neighboring rocks, deleterious behavior in concrete, and very low friction. Some of the minerals involved are difficult to identify in hand specimens but the engineer should recognize the names and look for them in geological reports. Academic geologists are not always aware of the influence some of these individuals can exert on engineering properties and behavior of rocks. A partial listing of potential problem minerals follow.

Soluble Minerals

Calcite, dolomite, gypsum, anhydrite, salt (halite), and zeolite.

Unstable Minerals

Marcasite and pyrrhotite.

Potentially Unstable Minerals

Nontronite (iron-rich montmorillonite), nepheline, leucite, micas rich in iron.

Minerals Whose Weathering Releases Sulfuric Acid

Pyrite, pyrrhotite, and other sulfides (ore minerals).

Minerals with Low Friction Coefficients

Clays (especially montmorillonites), talc, chlorite, serpentine, micas, graphite, and molybdenite.

Potentially Swelling Minerals

Montmorillonites, anhydrite, vermiculite.

Minerals That React or Interfere with Portland Cement

Opal, volcanic glass, some cherts, gypsum, zeolite, and mica.

Bergeforsen Dam affords an example of special problems due to mineralogy. A series of basic dikes under the concrete gravity structure decomposed in a few years after impoundment of water due to rapid decay of calcite. Solution of calcite was accelerated by carbon dioxide in vugs within the rock; the percolating water dissolved the carbon dioxide, thereby becoming enriched in carbonic acid. Originally hard rock was thus transformed to clay. The foundation was washed and pressure grouted. Also, lime-saturated water was circulated continuously through the rock at a pressure higher than that of the reservoir, thereby excluding the reservoir water from the foundation (Aastrup and Sallstrom, 1964; see references in Chap. 2).

Identification of Common Rocks

One cannot expect to be able to assign the correct geological name to all specimens found in an engineering project; sometimes it requires not only a thorough training in petrology but also petrographic examination of a thin section to determine the rock type. However, there is a system to rock identification and most engineers can manage to become fairly proficient at classifying rocks with a little guidance. It should be appreciated that the geological classification of rocks is not intended to group rocks with like engineering properties; in fact, its prime purpose is to group rocks of similar mode of origin. Nevertheless, a rock name with a short description of the nature and arrangement of the component particles or crystals often connotes much that is of practical value.

Table A3.2 presents a greatly simplified flow chart that will help you assign a name to an unknown specimen. In most cases using this chart, a rock group name can be assigned unambiguously after examination of a fresh surface of an unknown hand specimen. However, the chart is not infallible because the boundaries between grades are sometimes based upon subjective judgments, and the qualities being assessed are often gradational in character from one individual to the next. Of the many attributes presented by a rock specimen three are singled out dominantly in this chart—texture, hardness, and structure.

TABLE A3.2
Identification Scheme for Rocks

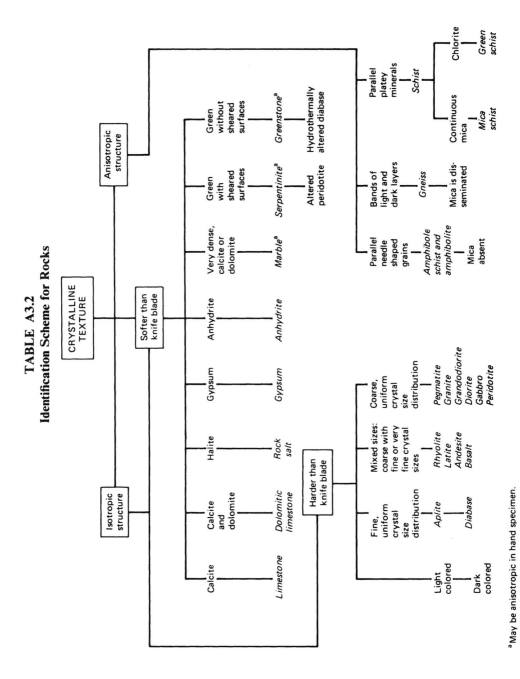

a May be anisotropic in hand specimen.

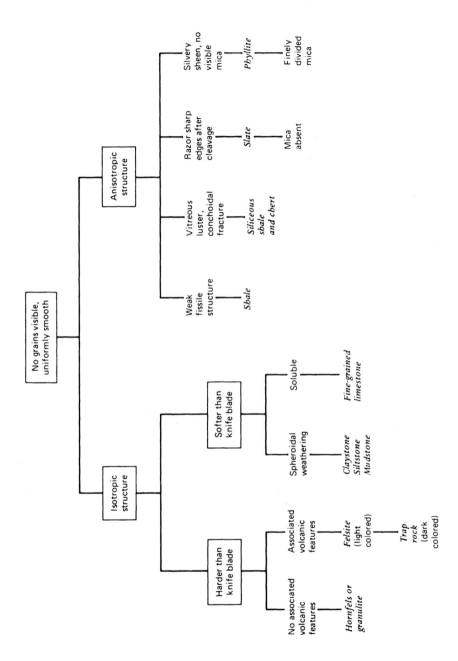

TABLE A3.2—*(continued)*

```
                        ┌─────────────┐
                        │   Clastic   │
                        │   texture   │
                        └──────┬──────┘
                        ┌──────┴──────┐
                        │ Isotropic or │
                        │ anisotropic  │
                        └──────┬──────┘
      ┌──────────┬──────────┬──┴───────┬──────────┐
```

Mainly volcanic pebbles and cobbles	Mainly nonvolcanic pebbles and cobbles	Angular blocks	Sand grains	Mainly volcanic sand (lapilli) and ash
Agglomerate	*Conglomerate*	*Breccia*	*Sandstone*	*Tuff*

		Quartzite	*Greywacke*	*Arkose*
		Uniform quartz grains	Dirty sand with rock grains	Quartz Feldspar (some mica and other minerals)

The major division is between *crystalline* and *clastic* textures. The crystalline rocks like granite, basalt, and marble possess an interlocking fabric of crystals with very little or no pore space. There may be grain boundary cracks and other fissures that can weaken the rock, and the crystals themselves may be deformable (e.g., calcite in marble) but the fabric is generally a strong one. In contrast, the clastic rocks consist of a collection of pieces of minerals and rocks with subspherical pore space more or less continuously connected through the rock. To the extent that this pore space is filled with durable cement, the rock will be strong and rigid. Some clastic rocks that are hard and rocklike in appearance contain only clay in the interparticle spaces and soften to soil-like consistency on soaking in water. Some rock specimens are so fine grained that the grains or crystals cannot be seen with the naked eye; in that case the rock can be classified with other tests.

The second classification index used in Table A3.2 is *hardness*. Although it is less well defined as a rock property than as a mineral property, the scratch hardness of a fresh rock surface gives a helpful index. By "fresh" we mean to exclude those specimens softened by weathering or localized alteration processes. Some rocks (e.g., greenstone) owe their characteristics to hydrothermal alteration, which occurred at considerable depth and uniformly altered a large rock volume. It is not our intention to exclude these rocks from the candidates for identification, but rather to exclude those weathered specimens that have sounder neighbors. The scratch hardness of a rock is not an infallible index, because some rocks lie close to the hardness of a knife and some straddle this boundary with a "scatter band" of variability. However, in certain cases the scratch test is useful, for example, in distinguishing aplite from marble, or hornfels from claystone. In micaceous rocks, what appears to be a scratch is revealed, under the microscope, actually to consist of a flaking off of cleavage fragments in front of the blade—a ploughing action. Scratch hardness is not usually useful as a classification index in the coarse clastic rocks.

A third division is made between *isotropic* and *anisotropic* structures. Metamorphic rocks (e.g., slate, schist, and gneiss) possess an incipient parting tendency parallel to one plane or one axis; consequently, these rocks present extreme anisotropy (i.e., directionality) in all physical properties. Some sedimentary rocks (e.g., shale, chert, and thin-bedded limestone) possess such closely spaced laminations that even the hand specimen shows strong directionality. In other rocks, the structure is massive on the hand specimen scale (e.g., thick-bedded sandstone and limestone, and basalt) so that the specimen appears to be isotropic. Some granites are isotropic even on the field scale. In coarse clastic rocks, although it is important as a physical attribute, the degree of isotropy is not useful as a classification index. The classification of these rocks is effected mainly in terms of grain size and fabric.

Some individual rock groups will now be considered. The hard, isotropic crystalline rocks present three series depending on the relative sizes of the crystals: the coarse-grained varieties are of *plutonic* igneous origin; those with relatively coarse crystals in a matrix of crystal sizes invisible to the naked eye (porphyritic texture) are *volcanic* in origin; rocks that are uniformly fine grained, or porphyritic with a fine-grained ground mass are usually of *dike* origin, having cooled at shallow or moderate depth. The many rock names within these groups reflect changes in mineralogic composition that are not always significant from an engineering point of view. For example, the difference between *granite* and *granodiorite* is mainly in the relative abundance of orthoclase and plagioclase minerals that are almost identical in physical properties. The darker varieties of these rocks, such as gabbro and peridotite, are composed of a relatively larger proportion of early formed, high-temperature pyroxene and olivine, which renders them more susceptible to weathering processes. Hard anisotropic crystalline rocks are usually rather strong (e.g., gneiss and amphibolite). Soft anisotropic crystalline rocks include schists, in which the softness may be due to a true indentation of chlorite or other soft minerals, or a ploughing of micas as previously noted.

Isotropic crystalline rocks that scratch readily include evaporite rocks—limestone, dolomite, gypsum, anhydrite, rock salt, etc.—and altered basic igneous rocks—serpentinite and greenstone. All these rocks can present undesirably weak and deformable properties to the civil engineer, serpentinites by virtue of internal surfaces of previous shear and associated weak minerals, and schist by virtue of the continuous bands of mica or chlorite or other minerals of low shear strength.

The rocks that are most difficult to identify are those without visible grains or crystals. Uniformly aphanitic basalts, cherts, shales, some slates, and some fine-grained limestones and dolomites can provide difficulty when the hardness and structure is overlooked. Associated rocks and structures that can be studied in the field usually make rock identification much easier in the field.

Table A3.3 lists the periods of geological history. Time names should be included with the petrologic rock name in engineering practice, particularly

Table A3.3 The Geologic Time Scale

Era	Period	Epoch	Age[a]
Cenozoic	Quaternary	Holocene	10,000 yr
		Pleistocene	2 m.y.
	Tertiary	Pliocene	
		Miocene	
		Oligocene	
		Eocene	65 m.y.
Mesozoic	Cretaceous		
	Jurassic		
	Triassic		225 m.y.
Paleozoic	Permian		
	Pennsylvanian		
	Mississippian		
	Devonian		
	Silurian		
	Ordovician		
	Cambrian		570 m.y.
Precambrian			

[a] m.y. = million years.

when dealing with sedimentary rocks. In a general way, the older rocks tend to be harder and more permanently cemented. There are, unfortunately, important and dramatic exceptions; for example, uncemented montmorillonite clays are found in rock units from the lower Paleozoic. To those conversant with engineering geology, however, rock age names do imply associated engineering attributes more effectively than does any single index property. Every worker in rock mechanics should know these names and use them routinely in rock descriptions.

Derivations of Equations

Equation 2.3

$$\gamma_{dry} = G\gamma_w(1 - n)$$

Assume a bulk rock of unit volume (i.e., $V_t = 1$). The volume of pores is then $V_p = nV_t = n$ and the volume of solids is $V_r = 1 - n$. The weight of pores, if they are dry, is zero; $W_p = 0$. The weight of solids with average specific gravity G is $W_r = (1 - n)G\gamma_w$. The dry density is W_r/V_t giving (2.3) directly.

Equation 2.4

$$\gamma_{dry} = \gamma_{wet}/(1 + w)$$

With the same assumptions as above, but with water filling the pores, $W_p = n\gamma_w$. The water content is $w = W_p/W_r = n\gamma_w/[(1 - n)G\gamma_w] = n\gamma_w/\gamma_{dry}$ or $w = n\gamma_w/\gamma_{dry}$ (1). The wet density $\gamma_{wet} = (W_p + W_r)/V_t = n\gamma_w + (1 - n)G\gamma_w$ (2). Substituting for $n\gamma_w$ from (1), and for $(1 - n)G\gamma_w$ from (2.3): $\gamma_{wet} = \gamma_{dry}(1 + w)$, which leads to (2.4).

Equation 2.5

$$n = \frac{wG}{1 + wG}$$

Assume unit volume of rock; the weight of rock W_r is then $G\gamma_w$. If the rock is saturated, the water content $w = W_w/W_r$. Thus $W_w = wG\gamma_w$ and the volume of water $V_w = wG$. The porosity $n = V_w/V_t = V_w/(1 + V_w)$ leading directly to (2.5).

Equation 2.6

$$n = \frac{w_{Hg}G/G_{Hg}}{1 + w_{Hg} \cdot G/G_{Hg}}$$

The derivation is the same as above except that the weight of mercury in the pores is $W_{Hg} = w_{Hg}G\gamma_w$ and the volume of mercury is therefore $V_{Hg} = w_{Hg}G\gamma_w/(G_{Hg}\gamma_w) = w_{Hg}G/G_{Hg}$.

Equation 2.9

For a cylinder between R_1 and R_2 with thickness L, the radial flow q_r across the circumference at radius r is, from Darcy's law

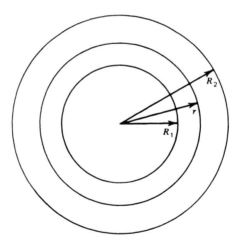

$$q = -k\frac{dh}{dr} 2\pi rL$$

or

$$\frac{dr}{r} = -\frac{k \cdot 2\pi L}{q} dh$$

Equation 2.10 **429**

Integrating between R_1 and R_2

$$\ln \frac{R_2}{R_1} = \frac{k2\pi L(h_1 - h_2)}{q}$$

giving

$$k = \frac{q \ln(R_2/R_1)}{2\pi L \Delta h}$$

Equation 2.10

Snow (1965, Chapter 2) shows that equations for laminar flow of an incompressible fluid between parallel plates with separation e lead to

$$q_x = \frac{\gamma}{12\mu} e^3 \frac{dh}{dx}$$

for a direction x parallel to the plates. For a regular system with interplate separation e and plate thickness S, the discharge can be written

$$q_x = \underbrace{\left(\frac{\gamma}{12\mu} \frac{e^3}{S}\right)}_{k} \underbrace{\left(\frac{dh}{dx}\right)(S)}_{A}$$

Consider three mutually perpendicular sets of fractures each spaced distance S and all having identical aperture e. The flow in a direction parallel to any one of the directions of fracture intersection is the sum of flow through two fracture sets. The flow in oblique directions can be described by considering its components in the fracture directions. However, this system is isotropic so the components must also form a resultant equal to twice q_x given above. Therefore, for any direction through the fracture system

$$q = \frac{\gamma}{6\mu} \frac{e^3}{S} \frac{dh}{dx} S$$

and the conductivity applicable to a unit area of the rock (however many fractures that embraces) is

$$k = \frac{\gamma}{6\mu} \frac{e^3}{S}$$

Equation 2.14

$$\frac{1}{V_l^*} = \sum_i \frac{C_i}{V_{l,i}}$$

The average velocity of a tightly packed collection of minerals with individual longitudinal wave velocities $V_{l,i}$ is computed by adding the travel times t_i for a wave to cross each mineral constituent in turn. If the rock has thickness L and mineral i has volume proportion C_i, the transit time for the wave across mineral i is $t_i = C_i L / V_{l,i}$ and the total time for the wave to cross the rock is $t = L \sum C_i / V_{l,i}$. The computed wave velocity $V_l^* = L/t$ giving Equation 2.13.

Equations 3.8, 3.9

$$\sigma_{1,p} = 2S_i \tan(45 + \phi/2) + \sigma_3 \tan^2(45 + \phi/2)$$

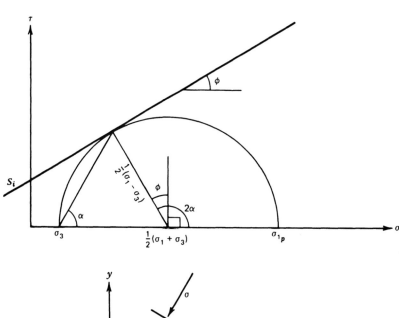

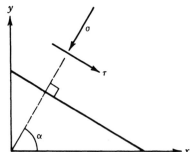

Equation 4.3 **431**

Taking σ_1 parallel to x and σ_3 parallel to y, the pole of the Mohr circle is at σ_3 (see Appendix 1) and the plane whose shear and normal stress are at the point of intersection A of the Coulomb failure line and the Mohr circle is described by α as shown. From the geometry of the upper diagram,

$$2\alpha = 90 + \phi \quad \text{and} \quad \alpha = 45 + \phi/2 \tag{1}$$

The equation of failure is

$$\tau_p = S_i + \sigma \tan \phi \tag{2}$$

From the upper figure,

$$\sigma = \tfrac{1}{2}(\sigma_1 + \sigma_3) + \tfrac{1}{2}(\sigma_1 - \sigma_3)\cos 2\alpha \tag{3}$$

and

$$\tau_p = \tfrac{1}{2}(\sigma_1 - \sigma_3)\sin 2\alpha \tag{4}$$

Inserting (3) and (4) in (2) with $\sigma_1 = \sigma_{1p}$ results in

$$(\sigma_{1P} - \sigma_3)\sin 2\alpha = 2S_i + (\sigma_{1p} + \sigma_3)\tan \phi + (\sigma_{1p} - \sigma_3)\cos 2\alpha \tan \phi \tag{5}$$

$$\sigma_{1p}(\sin 2\alpha - \cos 2\alpha \tan\phi - \tan \phi) = 2S_i + \sigma_3(\sin 2\alpha - \cos 2\alpha \tan \phi + \tan \phi) \tag{6}$$

$$\sin 2\alpha - \cos 2\alpha \tan \phi = \frac{1}{\cos \phi}$$

giving

$$\sigma_{1p}\left(\frac{1}{\cos \phi} - \tan \phi\right) = 2S_i + \sigma_3\left(\frac{1}{\cos \phi} + \tan \phi\right)$$

or

$$\sigma_{1p} = 2S_i \frac{\cos \phi}{1 - \sin \phi} + \sigma_3 \frac{1 + \sin \phi}{1 - \sin \phi}$$

Finally,

$$\frac{1 + \sin \phi}{1 - \sin \phi} = \tan^2\left(45 + \frac{\phi}{2}\right) \quad \text{and} \quad \frac{\cos \phi}{1 - \sin \phi} = \tan\left(45 + \frac{\phi}{2}\right)$$

Equation 4.3

$$K = \frac{\sigma_h}{\sigma_v} = \frac{K_0\gamma Z_0 - [v/(1 - v)]\,\gamma\Delta Z}{\gamma Z_0 - \gamma\Delta Z}$$

Let $Z_0 - \Delta Z = Z$; then $Z_0 = Z + \Delta Z$, giving

$$K(Z) = \frac{K_0(Z + \Delta Z) - [\nu/(1 - \nu)]\Delta Z}{Z}$$

or

$$K(Z) = K_0 + \left(K_0\Delta Z - \frac{\nu}{1 - \nu}\Delta Z\right)\frac{1}{Z}$$

Equation 4.4

For normal faulting, $\sigma_h = K_a\sigma_v = \sigma_3$ and $\sigma_v = \gamma Z = \sigma_1$. According to Coulomb's law, faulting occurs when σ_1 and σ_3 satisfy Equation 3.8. Combining these conditions yields

$$\gamma Z = q_u + K_a\gamma Z \tan^2\left(45 + \frac{\phi}{2}\right)$$

giving

$$K_a = \frac{\gamma Z - q_u}{\gamma Z \tan^2(45 + \phi/2)}$$

which simplifies to Equation 4.4

Equation 4.5

For reverse faulting $\sigma_h = K_p\sigma_v = \sigma_1$ and $\sigma_v = \gamma Z = \sigma_3$. Combining with Equation 3.8 gives

$$K_p\gamma Z = \gamma Z \tan^2\left(45 + \frac{\phi}{2}\right) + q_u$$

which, on solving for K_p, yields (4.5).

Equation 4.7

This result comes directly from Equation 7.1b. On the inner surface of a drill hole of radius a, $r = a$ and (7.1) gives

$$\sigma_\theta = (p_1 + p_2) - 2(p_1 - p_2)\cos 2\theta \tag{1}$$

Equation 4.11 and Discussion That Follows It   433

At points A and B in Figure 4.10, $\theta = 0$ and $\theta = 90°$, respectively. Let $p_1 = \sigma_{h,max}$ and $p_2 = \sigma_{h,min}$. Then at A, Equation 1 gives $\sigma_{\theta,A} = -\sigma_{h,max} + 3\sigma_{h,min}$.

Equation 4.10

On first pressuring, cracking and peak load are presumed to occur when Equation 4.8 is satisfied:

$$3\sigma_{h,min} - \sigma_{h,max} - p_{c1} = -T_0 \tag{1}$$

On repressuring, the peak load is p_{c2} and the tensile strength is zero since there is now a crack. Therefore,

$$3\sigma_{h,min} - \sigma_{h,max} - p_{c2} = 0 \tag{2}$$

Subtracting (1) from (2) gives

$$p_{c1} - p_{c2} = T_0$$

Equation 4.11 and the Discussion That Follows It

A vertical fracture forms when the tangential stress σ_θ on the wall of the hole equals

$$\sigma_\theta = 3\sigma_{h,min} - \sigma_{h,max} - p_{c1} = -T_0 \tag{1}$$

A horizontal fracture forms when the longitudinal stress σ_l on the wall of the hole equals

$$\sigma_l = \sigma_v - p_{c1} = -T_0 \tag{2}$$

A vertical fracture occurs first (assuming the tensile strength is isotropic) if $\sigma_\theta < \sigma_l$:

$$\sigma_v - p_{c1} > 3\sigma_{h,min} - \sigma_{h,max} - p_{c1} \tag{3}$$

Let

$$\frac{\sigma_{h,min}}{\sigma_{h,max}} = N$$

Then the vertical fracture is preferred if

$$\sigma_v > (3N - 1)\sigma_{h,max} \tag{4}$$

Adding and subtracting $4\sigma_{h,max}$ to (3) and regrouping terms gives

$$\sigma_v > 6 \frac{\sigma_{h,min} + \sigma_{h,max}}{2} - 4\sigma_{h,max} \tag{5}$$

or

$$\sigma_v > 6\overline{K}\sigma_v - 4\sigma_{h,max} \tag{6}$$

where $\overline{K} = \overline{\sigma}_h/\sigma_v$ is the ratio of mean horizontal stress to vertical stress. Dividing both sides of (6) by $\overline{K}\sigma_v$ gives

$$\frac{1}{\overline{K}} > 6 - \frac{4\sigma_{h,max}}{\overline{K}\sigma_v} \tag{7}$$

Now,

$$\frac{\sigma_{h,max}}{\overline{K}\sigma_v} = \frac{2\sigma_{h,max}}{\sigma_{h,min} + \sigma_{h,max}} = \frac{2}{N + 1} \tag{8}$$

Substituting (8) in (7) gives

$$\frac{1}{\overline{K}} > \frac{6N - 2}{N + 1} \tag{9}$$

which, on inverting, gives the reported condition for a vertical fracture:

$$\overline{K} < \frac{N + 1}{6N - 2} \left(N > \frac{1}{3}\right) \tag{10}$$

Equation 4.13

The stress concentrations -1 and 3 were calculated from the Kirsch solution (Equation 7.16) in the derivation to Equation 4.7. Substitute:

$\sigma_{\theta,w}$ for $\sigma_{\theta,A}$ $\sigma_{\theta,R}$ for $\sigma_{\theta,B}$ $\sigma_{h,max}$ for σ_{horiz} and $\sigma_{h,min}$ for σ_{vert}

Equation 4.14

$$\begin{pmatrix} \frac{1}{8} & \frac{3}{8} \\ \frac{3}{8} & \frac{1}{8} \end{pmatrix} \text{ is the inverse of } \begin{pmatrix} -1 & 3 \\ 3 & -1 \end{pmatrix}$$

Equation 4.15   435

PROOF

$$\begin{pmatrix} (\tfrac{1}{8})\cdot(-1) + (\tfrac{3}{8})\cdot(3) & (\tfrac{1}{8})\cdot(3) + (\tfrac{3}{8})\cdot(-1) \\ (\tfrac{3}{8})\cdot(-1) + (\tfrac{1}{8})\cdot(3) & (\tfrac{3}{8})\cdot(3) + (\tfrac{1}{8})\cdot(-1) \end{pmatrix} = \begin{pmatrix} 1 & 0 \\ 0 & 1 \end{pmatrix}$$

Equation 4.15

The derivation of this relation, for the case with $\tau_{xz} = 0$, is given by Jaeger and Cook (op. cit. Chapter 1), Section 10.4. The results for f_1, f_2, and f_3 can be found in their Equation 26, which is therefore equivalent to (4.15) with $\tau_{xz} = 0$.

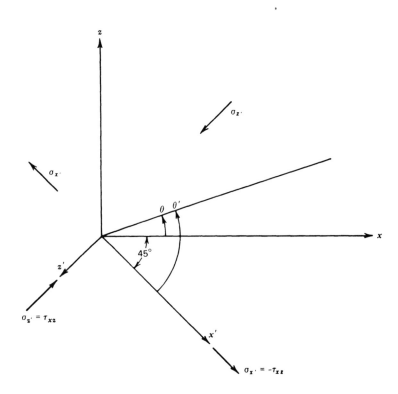

To find the displacement induced by an additional stress component τ_{xz}, proceed as follows (see diagram). Consider a set of principal stresses $\sigma_{x'} = -\tau_{xz}$, $\sigma_{y'} = 0$, and $\sigma_{z'} = +\tau_{xz}$ where $0x'$ is $45°$ clockwise from $0x$. With respect to the x' axis, the displacement Δd along a line inclined θ with $0x$ is found from (4.15) by substituting $-\tau_{xz}$ for σ_x, 0 for σ_y, τ_{xz} for σ_z, for τ_{xz}, and $\theta + 45°$ for θ.

Introducing these substitutions in Equation 4.15 gives

$$\Delta d = -\tau_{xz}\left(d[1 + 2\cos(90 + 2\theta)]\frac{1 - v^2}{E} + \frac{dv^2}{E}\right)$$

$$+\tau_{xz}\left(d[1 - 2\cos(90 + 2\theta)]\frac{1 - v^2}{E} + \frac{dv^2}{E}\right)$$

or

$$\Delta d = \tau_{xz}\frac{d(1 - v^2)}{E}\,4\sin 2\theta = \tau_{xz}f_4$$

Equation 4.17

The stress-strain relations for isotropic, linearly elastic bodies are stated in Equation 6.1. The hole being parallel to y, on the bottom surface of the drill hole $\sigma_y = \tau_{xy} = \tau_{yz} = 0$. Then, the independent variables in (6.1) reduce to 3 and the stress strain relations become

$$\varepsilon_x = \frac{1}{E}\sigma_x - \frac{v}{E}\sigma_z \tag{1}$$

$$\varepsilon_z = \frac{-v}{E}\sigma_x + \frac{1}{E}\sigma_z \tag{2}$$

$$\gamma_{zx} = \frac{2(1 + v)}{E}\tau_{zx} \tag{3}$$

Multiplying (2) by v, and adding (1) and (2) gives

$$\frac{1 - v^2}{E}\sigma_x = \varepsilon_x + v\varepsilon_z \tag{4}$$

Multiplying (1) by v, and adding (1) and (2) gives

$$\frac{1 - v^2}{E}\sigma_z = v\varepsilon_x + \varepsilon_z \tag{5}$$

Equations 3, 4, and 5 correspond to the three rows of (4.17) if the strains are zero before deepening the drill hole.

Equation 4.23   437

Equation 4.21

We need to derive a formula for the radial displacement of a point near a circular hole resulting from creating that hole in the surface of an initially stressed isotropic, elastic rock mass. If the rock surface is normal to y, $\sigma_y = \tau_{yz} = \tau_{yz} = 0$. The state of stress in the rock near the surface is, therefore, one of *plane stress*. Equation 7.2a gives the radial displacement of any point around a circular hole in an initially stressed rock mass corresponding to *plane strain*. As discussed with the derivation to Equation 7.2 later on in this appendix, there is a simple connection between plane stress and plane strain. To convert a formula derived for plane strain to one correct for plane stress, substitute $(1 - v^2)E$ in place of E, and $v/(1 + v)$ in place of v. $G = E/[2(1 + v)]$ is unaffected. Making this substitution in (7.2a) gives

$$u_r = \frac{a^2}{r} \frac{2(1 + v)}{4E} \left\{ (p_1 + p_2) + (p_1 - p_2) \left[4\left(1 - \frac{v}{1 + v}\right) - \frac{a^2}{r^2} \right] \cos 2\theta \right\} \quad (1)$$

Simplifying and substituting σ_x in place of p_1, and σ_z in place of p_2 yields

$$u_r = \frac{1}{2E} \frac{a^2}{r} [(1 + v)(\sigma_x + \sigma_x) + H(\sigma_x - \sigma_z)\cos 2\theta] \quad (2)$$

where

$$H = 4 - (1 + v) \frac{a^2}{r^2}$$

Then using the same procedure as in the derivation to Equation 4.15, we find the influence of a shear stress:

$$u_r = \frac{1}{E} \frac{a^2}{r} H\tau_{xz} \sin 2\theta \quad (3)$$

Arranging (2) and (3) in the form of Equation 4.21 determines f_1, f_2, and f_3 as given.

Equation 4.23

Leeman (1971, Chapter 4) presents complete formulas for the stresses around a long circular hole of radius a bored in an isotropic, elastic medium with an initial state of stress. The initial stress components will be represented here by

x', y', z' subscripts with y' parallel to the axis of the bore as shown in Figure 4.16. For a point on the wall of the bore ($r = a$) and located by angle θ counterclockwise from x' as shown, Leeman's equations reduce to

$$\sigma_r = 0 \tag{1}$$

$$\sigma_\theta = (\sigma_{x'} + \sigma_{z'}) - 2(\sigma_{x'} - \sigma_{z'})\cos 2\theta - 4\tau_{x'z'} \sin 2\theta \tag{2}$$

$$\sigma_x = -\nu[2(\sigma_{x'} - \sigma_{z'})\cos 2\theta + 4\tau_{x'z'} \sin 2\theta] + \sigma_{y'} \tag{3}$$

$$\tau_{r\theta} = 0 \tag{4}$$

$$\tau_{rx} = 0 \tag{5}$$

$$\tau_{x\theta} = -2\tau_{y'x'} \sin \theta + 2\tau_{y'z'} \cos \theta \tag{6}$$

In our coordinate system, at any measuring station j the direction radial to the surface is denoted by y_j; and the direction tangential to the surface is denoted by z_j. Then σ_θ of Equation 2 should be renamed σ_z, and $\tau_{x\theta}$ of Equation 6 should be renamed τ_{xz}. Grouping the terms and arranging in matrix fashion defines all the terms of Equation 4.23.

Equations 4.24, 4.25

See Appendix 1.

Equation 5.9

Since

$$A = (\sigma_1 - \sigma_3)\sin \psi \tag{1}$$

and

$$\sigma = \sigma_3 + A \sin \psi$$

then

$$\sigma = \sigma_3 + (\sigma_1 - \sigma_3)\sin^2 \psi \tag{2}$$

also

$$\tau = A \cos \psi$$

or with (1)

$$\tau = (\sigma_1 - \sigma_3)\sin \psi \cos \psi \tag{3}$$

Equation 6.3 **439**

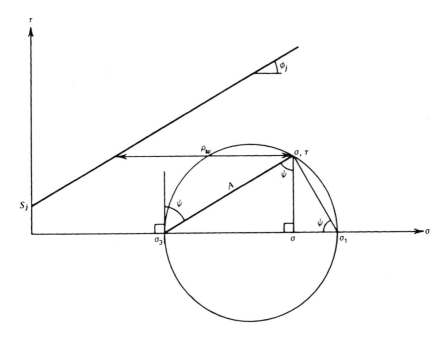

For slip, the water pressure must reach a value p_w such that

$$S_j + (\sigma - p_w)\tan \phi_j = \tau \tag{4}$$

Inserting (2) and (3) in (4) gives

$$S_j + [\sigma_3 + (\sigma_1 - \sigma_3)\sin^2 \psi - p_w]\tan \phi_j = (\sigma_1 - \sigma_3)\sin \psi \cos \psi \tag{5}$$

Solving for p_w results in

$$p_w \tan \phi_j = S_j + \sigma_3(\tan \phi_j) + (\sigma_1 - \sigma_3)(\sin^2 \psi \tan \phi_j - \sin \psi \cos \psi) \tag{6}$$

which leads directly to Equation 5.9.

Equation 6.3

Shear stress and shear strain can be related by considering the shear strain produced by a condition of pure shear. The latter can be achieved by superimposing the effects of principal normal stresses equal to $-\tau$ and τ along lines at $\theta = -45°$, $\theta = +45°$, respectively (directions x' and y') (see the derivation of Equation 4.15). For this state of stress, Equation 6.1 gives

$$\varepsilon_{x'} = \frac{1}{E}[-\tau - (\nu)(\tau)] = \frac{-\tau}{E}(1 + \nu) \tag{1}$$

and

$$\varepsilon_{y'} = \frac{1}{E}\left[-(\nu)(-\tau) + \tau\right] = \frac{\tau}{E}(1 + \nu) \tag{2}$$

Also

$$\gamma_{x'y'} = 0 \tag{3}$$

Introducing the discussion above Equation A2.2 (Appendix 2) in the second of Equations A1.2 (Appendix 1) gives an equation for shear strain γ_{xy} in terms of the strain components referred to x', y' axes:

$$\gamma_{xy} = -\sin 2\alpha\varepsilon_{x'} + \sin 2\alpha\varepsilon_{y'} + \cos 2\alpha\gamma_{x,y'}$$

For $0x$, $\alpha = 45°$ from $0x'$. Introducing this with (1) to (3) gives

$$\gamma_{xy} = \frac{2}{E}(1 + \nu)\tau = \frac{\tau}{G}$$

Equation 6.4

The upper left quarter of Equation 6.1 is

$$\begin{pmatrix} \varepsilon_x \\ \varepsilon_y \\ \varepsilon_z \end{pmatrix} = \frac{1}{E}\begin{pmatrix} 1 & -\nu & -\nu \\ -\nu & 1 & -\nu \\ -\nu & -\nu & 1 \end{pmatrix}\begin{pmatrix} \sigma_x \\ \sigma_y \\ \sigma_z \end{pmatrix} \tag{1}$$

Then, using Cramer's rule to invert the 3×3 matrix, we have

$$\begin{pmatrix} \sigma_x \\ \sigma_y \\ \sigma_z \end{pmatrix} = \frac{E}{\text{Det}}\begin{pmatrix} 1 - \nu^2 & \nu(1 + \nu) & \nu(1 + \nu) \\ \nu(1 + \nu) & 1 - \nu^2 & \nu(1 + \nu) \\ \nu(1 + \nu) & \nu(1 + \nu) & 1 - \nu^2 \end{pmatrix}\begin{pmatrix} \varepsilon_x \\ \varepsilon_y \\ \gamma_{xy} \end{pmatrix} \tag{2}$$

where Det is the determinant of the 3×3 matrix in (1). Expanding down the first column of the latter, gives

$$\text{Det} = (1 - \nu^2) - (\nu^2 + \nu^3) + (-\nu^2 - \nu^3) = 1 - 3\nu^2 - 2\nu^3$$

or

$$\text{Det} = (1 + \nu)^2(1 - 2\nu) \tag{3}$$

Substituting (3) in (2) gives

$$\begin{pmatrix} \sigma_x \\ \sigma_y \\ \sigma_z \end{pmatrix} = \frac{E}{(1 + \nu)(1 - 2\nu)}\begin{pmatrix} 1 - \nu & \nu & \nu \\ \nu & 1 - \nu & \nu \\ \nu & \nu & 1 - \nu \end{pmatrix}\begin{pmatrix} \varepsilon_x \\ \varepsilon_y \\ \varepsilon_z \end{pmatrix} \tag{4}$$

Equation 6.13 **441**

Let

$$\lambda = \frac{E\nu}{(1 + \nu)(1 - 2\nu)} \tag{5}$$

Since

$$G = \frac{E}{2(1 + \nu)} \tag{6}$$

then

$$\lambda + 2G = \frac{E(1 - \nu)}{(1 + \nu)(1 - 2\nu)} \tag{7}$$

Since with (5), (6), and (7), Equation 4 is equivalent to Equation 6.2, we have established the validity of (5), which is the same as (6.4).

Equation 6.6

The bulk modulus K can be expressed in terms of E and ν by calculating the volumetric strain $\Delta V/V$ caused by all-round pressure $\sigma_x = \sigma_y = \sigma_z = p$. For this case, Equation 6.1 gives

$$\varepsilon_x = \varepsilon_y = \varepsilon_z = \frac{1}{E} p - \frac{2\nu}{E} p \tag{1}$$

As shown in Problem 3.10, for small strains,

$$\Delta V/V = \varepsilon_x + \varepsilon_y + \varepsilon_z \tag{2}$$

Combining (2) and (1) yields

$$\frac{\Delta V}{V} = \frac{3(1 - 2\nu)}{E} p \tag{3}$$

Equation 6.13

This important result derives from the theory of an axisymmetrically loaded thick-walled cylinder in an isotropic, linearly elastic, homogeneous and continuous medium. See Jaeger and Cook (op. cit. Chapter 1). Inserting the strain-displacement equations in the equation of equilibrium in polar coordinates yields a differential equation (Euler's equation) in one unknown—the radial displacement. This equation is solvable by a simple substitution. The constants of integration are evaluated with the conditions that displacement approaches zero as r grows without bound.

Equations 6.15, 6.16

Consider a bar of small-constant cross-sectional area A in which a stress wave travels to the right.

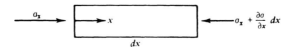

By Newton's second law, as a compressive stress wave travels down the bar:

$$A(\sigma_x) - \left(\frac{\partial \sigma_x}{\partial x} \, dx + \sigma_x\right)A = (A \, dx)\rho \, \frac{\partial^2 u}{\partial t^2} \tag{1}$$

where ρ is the mass per unit volume of the bar. The stress-strain relation is

$$\sigma_x = E\varepsilon_x = E \, \frac{-du}{dx} \tag{2}$$

Combining (2) and (1) and simplifying yields

$$E \, \frac{\partial^2 u}{\partial x^2} = \rho \, \frac{\partial^2 u}{\partial t^2} \tag{3}$$

The one-dimensional wave equation is

$$V_l^2 \, \frac{\partial^2 u}{\partial x^2} = \frac{\partial^2 u}{\partial t^2} \tag{4}$$

where V_l^2 is the phase velocity of the wave traveling in the bar. Therefore, the wave velocity is $V_l = (E/\rho)^{1/2}$. The derivation is identical in the case of a shear wave, except that G replaces E. Then $V_t = (G/\rho)^{1/2}$.

Equation 6.17

$$\frac{V_l^2}{V_t^2} = \frac{E/\rho}{G/\rho} = \frac{E}{E/2(1 + \nu)} = 2(1 + \nu)$$

Solving for ν yields Equation 6.17.

Equations 6.18, 6.19

Consider a stress pulse traveling through a three-dimensional elastic space, due to an unbalance in σ_x and with constraint such that there is no strain produced in the plane perpendicular to x.

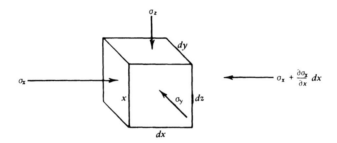

Equilibrium in the x direction gives

$$\sigma_x dy\, dz - \left(\sigma_x + \frac{\partial \sigma_x}{\partial x} dx\right) dy\, dz = \rho \frac{\partial^2 u}{\partial t^2} dx\, dy\, dz \tag{1}$$

where u is the displacement parallel to x. By the stress-strain relation (6.2), with $\varepsilon_y = \varepsilon_z = 0$

$$\sigma_x = (\lambda + 2G)\varepsilon_x = (\lambda + 2G)\frac{-\partial u}{\partial x} \tag{2}$$

Combining (2) with (1) and simplifying gives

$$(\lambda + 2G)\frac{\partial^2 u}{\partial x^2} = \rho \frac{\partial^2 u}{\partial t^2} \tag{3}$$

Let the pulse velocity be denoted by V_p. Comparing (3) with the wave equation (Equation 4 of the derivation to Equations 6.15 and 6.16) gives

$$V_p^2 = \frac{\lambda + 2G}{\rho}$$

The shear pulse in a three-dimensional physical space is mathematically one dimensional since the stress-strain relation between τ and γ is one dimensional. Therefore $V_s^2 = V_t^2 = G/\rho$ as in the case of a shear wave traveling in a slender bar.

Equation 6.20

$$\frac{V_p^2}{V_s^2} = \frac{(\lambda + 2G)/\rho}{G/\rho} \tag{1}$$

In the derivation to Equation 6.4, we observed that

$$\lambda + 2G = \frac{E(1 - \nu)}{(1 + \nu)(1 - 2\nu)} \tag{2}$$

Recall that

$$G = \frac{E}{2(1 + \nu)} \tag{3}$$

Then combining (2) and (3) with (1) yields

$$\frac{V_p^2}{V_t^2} = \frac{(1 - \nu)(2)(1 + \nu)}{(1 + \nu)(1 - 2\nu)} = \frac{2 - 2\nu}{1 - 2\nu} \tag{4}$$

Solving for ν yields (6.20).

Equation 6.22

$$\rho V_p^2 = \lambda + 2G$$

Introducing (2) from the preceding section gives

$$\rho V_p^2 = \frac{E(1 - \nu)}{(1 + \nu)(1 - 2\nu)} \tag{2}$$

Solving for E gives (6.22).

Equation 6.24

For a homogeneous anisotropic material loaded in direction n

$$\varepsilon_n = \frac{1}{E_n} \sigma_n \tag{1}$$

Equation 6.29 **445**

But

$$\varepsilon_n = \Delta V_{\text{rock}} + \Delta V_{\text{joint}} = \frac{\sigma_n}{E} + \frac{\sigma_n}{k_n S} \tag{2}$$

where E is the Young's modulus for solid (unjointed) rock, k_n is the joint normal stiffness per unit area, and S is the spacing between joints. Equating (2) and (1) yields (6.24).

Equation 6.29

A Burgers body, shown in Figure 6.17e, consists of a Kelvin solid (Figure 6.17b) and a Maxwell body (Figure 6.17a) connected in series. If we examine two springs in series subjected to an increment of load, we will discover that the displacement of the system is the sum of the displacements in each spring (as in the derivation to Equation 6.24). Similarly the displacement of a Burgers body subjected to an increment of shear stress is the sum of the shear strains in each of its components. Therefore, we can deduce the strain of a Burgers body by adding the strains of a Kelvin solid and a Maxwell body.

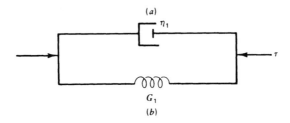

(a)

η_1

τ

G_1

(b)

Consider a Kelvin solid with viscosity η_1 in element (a) and shear modulus G_1 in element (b) (see diagram) and subjected to a constant shear stress increment τ. In this parallel arrangement,

$$\tau = \tau_a + \tau_b \tag{1}$$

and

$$\gamma = \gamma_a = \gamma_b \tag{2}$$

Shear stress and shear strain in component (a) are related by Equation 6.28:

$$\tau_a = \eta_1 \dot{\gamma}_a \tag{3}$$

(the dot signifies differentiation with respect to time); and in component (b) they are related by

$$\tau_b = G_1 \gamma_b \tag{4}$$

Combining Equations 1–4 yields

$$\tau = \eta_1 \dot{\gamma} + G_1 \gamma \tag{5}$$

or

$$\tau = \left(\eta_1 \frac{d}{dt} + G_1\right)\gamma \tag{6}$$

The behavior of the Kelvin solid in the creep test can be solved with the correspondence principle discussed by Flugge (1975) and Jaeger and Cook (1976). According to this principle, a solution to a given problem with an isotropic elastic material having elastic constants G and K offers corresponding solutions for the same problem posed in a linear viscoelastic material. For simplicity, we assume that the material in question is elastic with respect to compressibility (i.e., that only G is time dependent). In a linear viscoelastic material, shear stress and shear strain are related by

$$F_1(t)\tau = F_2(t)\gamma \tag{7}$$

According to the correspondence principle, the solution to the corresponding problem in elasticity is obtained by replacing G by $F_2(t)/F_1(t)$.

For the Kelvin solid, comparing (6) and (7) shows that

$$F_1(t) = 1$$

and

$$F_2(t) = \eta_1 \frac{d}{dt} + G_1$$

Therefore to convert an elastic solution into a corresponding Kelvin solid solution, we must substitute in place of G

$$\frac{F_2(t)}{F_1(t)_{\text{Kelvin}}} = \eta_1 \frac{d}{dt} + G_1 \tag{8}$$

As shown in Problem 6.1, the shear modulus governs the relation not only between shear stress and shear strain but between deviatoric stress and deviatoric strain. In particular,

$$\sigma_{1,\text{dev}} = 2G\varepsilon_{1,\text{dev}} \tag{9}$$

(Deviatoric stress and deviatoric strain are discussed in Appendices 1 and 2, respectively, and Equation 9 is developed in the answer to Problem 6.1.)

Also, since $\varepsilon_{\text{mean}}$ equals $\frac{1}{3}(\Delta V/V)$, then

$$\sigma_{\text{mean}} = 3K\varepsilon_{\text{mean}} \tag{10}$$

Equation 6.29   **447**

since

$$\varepsilon_{1,dev} = \varepsilon_1 - \varepsilon_{mean} \qquad (11)$$

$$\varepsilon_1 = \varepsilon_{1,dev} + \varepsilon_{mean} \qquad (12)$$

Introducing (9) and (10) in (12) gives

$$\varepsilon_1 = \frac{\sigma_{1,dev}}{2G} + \frac{1}{3}\frac{\sigma_{mean}}{K} \qquad (13)$$

For uniaxial compression with axial stress $= \sigma_1$ and $\sigma_2 = \sigma_3 = 0$,

$$\sigma_{mean} = \tfrac{1}{3}\sigma_1 \qquad (14)$$

and

$$\sigma_{1,dev} = \tfrac{2}{3}\sigma_1 \qquad (15)$$

so that (13) becomes

$$\varepsilon_1 = \frac{1}{3G}\sigma_1 + \frac{1}{9K}\sigma_1 \qquad (16)$$

Equation 16 expresses the strain corresponding to an increment of axial stress in an isotropic elastic material described by constants G and K. For a rock that behaves as a Kelvin solid in shear and an elastic material in compression, we replace G by $F_2(t)/F_1(t)$ as given by (8). This gives the differential equation

$$\varepsilon_1 = \sigma_1\left(\frac{1}{3[\eta_1(d/dt) + G_1]} + \frac{1}{9K}\right) \qquad (17)$$

the solution to which depends on the initial conditions. For a creep test, with σ_1 applied as an increment at time 0 and maintained constant thereafter, its solution[1] is

$$\varepsilon_1(t) = \frac{\sigma_1}{9K} + \frac{\sigma_1}{3G_1}(1 - e^{-(G_1 t/\eta_1)}) \qquad (18)$$

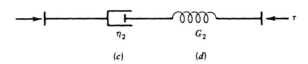

η_2 $\qquad$ G_2

(c) $\qquad$ (d)

Now consider a Maxwell body with constants η_2 in element (c) and G_2 in element (d) (see diagram) subjected to shear stress τ. Since the elements are in

[1] Solutions to differential equations of this sort are readily obtained using the Laplace transform. See Murray Spiegel, *Laplace Transforms* Schaum's Outline Series, McGraw-Hill, New York, 1965.

series

$$\tau = \tau_c = \tau_d \tag{19}$$

and

$$\gamma = \gamma_c + \gamma_d \tag{20}$$

Differentiating (20) yields

$$\dot{\gamma} = \dot{\gamma}_c + \dot{\gamma}_d \tag{21}$$

Introducing $\tau_c = \eta_2 \dot{\gamma}_c$ and $\tau_d = G_2 \gamma_d$ and combining with (19) to (21) gives

$$\dot{\gamma} = \frac{\tau}{\eta_2} + \frac{\dot{\tau}}{G_2} \tag{22}$$

or

$$\left(\frac{1}{\eta_2} + \frac{1}{G_2}\frac{d}{dt}\right)\tau = \frac{d}{dt}\gamma \tag{23}$$

Comparing (23) with (7) shows that for a Maxwell liquid

$$F_1(t) = \frac{1}{\eta_2} + \frac{1}{G_2}\frac{d}{dt}$$

and

$$F_2(t) = \frac{d}{dt}$$

so the correspondence principle requires substituting in place of G:

$$\frac{F_2(t)}{F_1(t)_{\text{Maxwell}}} = \frac{d/dt}{1/\eta_2 + [(1/G_2)(d/dt)]} \tag{24}$$

Making this substitution in Equation 16 gives

$$\varepsilon_1 = \sigma_1 \left(\frac{1/\eta_2 + [(1/G_2)(d/dt)]}{3(d/dt)} + \frac{1}{9K}\right) \tag{25}$$

For σ_1 applied as an increment at time zero, the solution to (25) is

$$\varepsilon_1(t) = \frac{\sigma_1 t}{3\eta_2} + \frac{\sigma_1}{3G_2} + \frac{\sigma_1}{9K} \tag{26}$$

Equations 18 and 26 give the creep of a cylinder subjected to constant axial stress σ_1. The term in K in each comes from the nature of uniaxial compression tests in which both the mean stress and deviatoric stress are increased when axial load is applied. The creep strain of the Burgers body is obtained by adding the right side of (26) to the right side of (18). This summation gives Equation 6.29.

Equation 6.34   449

Equation 6.33

Equation 6.13 gives the modulus of elasticity for a borehole dilatometer test in terms of the radial displacement Δu_r, the radius a, the internal pressure Δp, and Poisson's ratio v as

$$E = (1 + v)\Delta p \frac{a}{\Delta u} \tag{1}$$

Substituting u_r for Δu, p for Δp, and G for $E/[2(1 + v)]$ gives

$$u_r = \frac{pa}{2G} \tag{2}$$

There is no term in K because the mean stress remains constant as the pressure is raised inside a borehole under plane strain conditions.

In the previous section we considered the viscoelastic solution (Equation 6.29) corresponding to a compression test whose elastic solution was

$$\varepsilon_1 = \left(\frac{1}{3G} + \frac{1}{9K}\right)\sigma_1 \quad \text{(Equation 16 of the previous section)} \tag{3}$$

By analogy to the creep test we can write the solution for creep in a dilatometer test inside a Burgers material by comparing Equations 2 and 3. For the dilatometer creep test in a Burger's body,

• There must be no term in K.

• The other four terms on the right of Equation 6.29 all apply but in place of (σ_1/σ_3) substitute $(pa/2)$.

This deletion and substitution in Equation 6.29 gives Equation 6.33.

Equation 6.34

The average displacement of a flexible plate of radius a on an elastic solid with plate pressure p is given by Equation 6.10 as

$$\bar{\omega} = \frac{1.7pa(1 - v^2)}{E} \tag{1}$$

To use the correspondence principle, we recast this in terms of K and G by substituting:

$$E = \frac{9KG}{3K + G} \tag{2}$$

which follows from Equation 16 in the derivation of Equation 6.29 and

$$\nu = \frac{3K - 2G}{6K + 2G} \tag{3}$$

Making these substitutions in (1) gives

$$\bar{\omega} = \frac{1.7pa}{4} \frac{3K + 4G}{G(3K + G)} \tag{4}$$

For a material that behaves as a Burger's body in distortion and an elastic material in all-around compression, take the sum of the solutions applying to first the Kelvin solid and second the Maxwell body.

For a Kelvin solid in distortion, replace G in (4) by F_2/F_1 given by (8) in the derivation to (6.29). Then, corresponding to Equation 4, we obtain for a sustained increment of pressure

$$\eta_1^2 \frac{d^2\bar{\omega}}{dt^2} + (2G_1 + 3K)\eta_1 \frac{d\bar{\omega}}{dt} + G_1(G_1 + 3K)\bar{\omega}$$

$$= \frac{1.7a}{4} \left((4G_1 + 3K)p + 4\eta_1 \frac{dp}{dt} \right) \tag{5}$$

$\bar{\omega}$ at the time $t = 0$ is zero. The solution to (5) is then

$$\bar{\omega}_{Kelvin} = \frac{1.7pa}{4} \left(\frac{1}{G_1}(1 - e^{-G_1 t/\eta_1}) + \frac{3}{3K + G_1}(1 - e^{-(3K+G_1)t/\eta_1}) \right) \tag{6}$$

For the Maxwell body in distortion, in place of G in (4), substitute F_2/F_1 given by (24) in the derivation to 6.29. This gives, corresponding to (4),

$$\frac{d^2\bar{\omega}}{dt^2} \left(1 + \frac{3K}{G_2} \right) + \frac{d\bar{\omega}}{dt} \frac{3K}{\eta_2} = \frac{1.7a}{4} \left[\frac{3Kp}{\eta_2^2} + \frac{dp}{dt} \left(\frac{6K}{\eta_2 G_2} + \frac{4}{\eta_2} \right) + \frac{d^2p}{dt^2} \left(\frac{3K}{G_2} + \frac{4}{G_2} \right) \right] \tag{7}$$

The initial conditions are that $\bar{\omega} = d\bar{\omega}/dt = 0$. Solution of (7) with these conditions gives

$$\bar{\omega}_{Maxwell} = \frac{1.70pa}{4} \left(\frac{1}{G_2} + \frac{t}{\eta_2} + \frac{1}{K} - \frac{G_2}{K(3K + G_2)} e^{-3KG_2 t/[\eta_2(3K+G_2)]} \right) \tag{8}$$

Equation 6.34 is obtained as the sum of $\bar{\omega}_{Kelvin}$, given by (6) and $\bar{\omega}_{Maxwell}$, given by (8).

Equation 6.35

The differential equation is derived exactly as for Equation 6.29. However, the initial condition is no longer that of a step increment of load applied at time 0.

Instead the stress σ_1 is increased from zero at a constant rate $\dot{\sigma}_1$. Then in each of Equations 17 and 25 of the derivation to (6.29) substitute ($\dot{\sigma}_1 t$) in place of σ_1. The solutions to these equations and their summation without duplication of the term in K, as discussed with regard to Equation 6.29, yields

$$\varepsilon_1(t) = \dot{\sigma}_1 \left(\frac{1}{3G_1} + \frac{1}{3G_2} + \frac{2}{9K}\right)t + \frac{1}{2}\frac{\dot{\sigma}_1}{3\eta_2}t^2 - \frac{\dot{\sigma}_1\eta_1}{3G_1^2}(1 - e^{-(G_1 t/\eta_1)}) \tag{1}$$

Since $\dot{\sigma}_1$ is constant,

$$\sigma_1 = \dot{\sigma}_1 t \quad \text{and} \quad t = \frac{\sigma_1}{\dot{\sigma}_1}$$

Introducing these in (1) gives

$$\varepsilon_1(t) = \sigma_1\left(\frac{1}{3G_1} + \frac{1}{3G_2} + \frac{2}{9K}\right) + \frac{1}{6}\frac{\sigma_1^2}{\eta_2\dot{\sigma}_1} - \frac{\dot{\sigma}_1\eta_1}{3G_1^2}(1 - e^{-G_1\sigma_1/(\eta_1\dot{\sigma}_1)}) \tag{2}$$

This is Equation 6.35.

Equations 7.1, 7.2

The stresses and displacements around a circular hole in a homogeneous plate subjected to tension—the Kirsch solution (1898)—are derived in Jaeger and Cook, *Fundamentals of Rock Mechanics*, 2d ed., 1976, pp. 249, 251 (See Ch. 1 references.) The problem is complicated by the fact that the space is axisymmetric but the initial stresses are not, being defined by maximum and minimum normal stresses p_1 and p_2 in the plane of the hole. Jaeger and Cook's derivation uses the method of complex stress functions. Another derivation, using real stress functions, is presented by Obert and Duvall, *Rock Mechanics and the Design of Structures in Rock*, pp. 98–108. (See Ch. 1 references.)

The displacement can be calculated by integrating the stress equations (7.1) after inserting the stress-strain relations (6.2) and the strain displacement relations in polar coordinates:

$$\varepsilon_r = -\frac{\partial u}{\partial r} \qquad \varepsilon_\theta = -\frac{u}{r} - \frac{1}{r}\frac{\partial v}{\partial \theta} \quad \text{and} \quad \gamma_{r\theta} = -\frac{1}{r}\frac{\partial u}{\partial \theta} - \frac{\partial v}{\partial r} + \frac{v}{r}$$

where u and v are displacements in the r and θ directions. However, this integration will include a term representing the installation of the initial stresses, whereas, in fact, the displacements of a tunnel or borehole are measured relative to an initial condition where p_1 and p_2 were applied previously.

The displacements given in Equations 7.2a and b result from subtracting these initial displacements from those deduced by integration of (7.1). These are also the correct displacements with reference to measurement of stresses

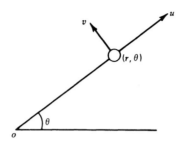

by undercoring (Equations 4.21). For overcoring, Equation 4.16, the displacements are calculated for a region containing a hole and acted on by applied tensions $-p_1$ and $-p_2$ at distances far from the central hole.

Equations 7.2 refer to a condition of *plane strain*, defined as a state of strain in which one normal strain and two shear strains are zero (e.g., $\varepsilon_z = 0$, $\gamma_{zx} = \gamma_{zy} = 0$). Physically, plane strain is associated with long structures or excavations with constant cross section and acted on by loads in the plane of the cross section. Solutions are also presented for conditions of *plane stress* where one normal stress and two shear stresses are zero (e.g., $\sigma_z = 0$, $\tau_{zx} = \tau_{zy} = 0$). The conditions of plane stress are encountered in thin plates loaded only in their plane. For plane stress, the isotropic stress strain relations (6.1) reduce to

$$\begin{pmatrix} \varepsilon_x \\ \varepsilon_y \\ \gamma_{xy} \end{pmatrix} = \frac{1}{E} \begin{pmatrix} 1 & -\nu & 0 \\ -\nu & 1 & 0 \\ 0 & 0 & 2(1 + \nu) \end{pmatrix} \begin{pmatrix} \sigma_x \\ \sigma_y \\ \tau_{xy} \end{pmatrix} \tag{1}$$

For conditions of plane strain with the long direction parallel to z,

$$\varepsilon_z = 0 = \frac{1}{E} \sigma_z - \frac{\nu}{E} \sigma_x - \frac{\nu}{E} \sigma_y$$

giving

$$\sigma_z = \nu(\sigma_x + \sigma_y) \tag{2}$$

and

$$\varepsilon_x = \frac{1}{E} \sigma_x - \frac{\nu}{E} \sigma_y - \frac{\nu}{E} \sigma_z \tag{3}$$

Substituting (2) in (3) gives

$$E\varepsilon_x = \sigma_x - \nu\sigma_y - \nu^2(\sigma_x + \sigma_y) \tag{4}$$

or

$$E\varepsilon_x = (1 - \nu^2)\sigma_x - \nu(1 + \nu)\sigma_y$$

or

$$\varepsilon_x = \frac{1}{E/(1 - \nu^2)} \sigma_x - \frac{\nu}{1 - \nu} \cdot \frac{1}{E/(1 - \nu^2)} \sigma_y \qquad (5)$$

A similar equation can be written in direction y. By comparing Equation 5 with the first row of (1), observe that a plane stress solution can be converted to a plane strain solution if the values of E and ν are modified to $E/(1 - \nu^2)$ and $\nu/(1 - \nu)$, respectively.

Equations 7.5, 7.6

For a beam long in direction x, with cross section in the xy plane and acted on by load q parallel to z, the deflection in the z direction obeys

$$EI_y \frac{d^4u}{dx^4} = q$$

Integrate four times and evaluate the four constants with these boundary conditions: $u = 0$ and $du/dx = 0$ at $x = L$ and $x = 0$. The deflection is then

$$u = \frac{qx^2}{24EI_y} (L - x)^2$$

The maximum deflection occurs at the center where $x = L/2$, giving

$$u_{max} = qL^4/(384EI_y)$$

For a rectangular beam, with unit width and thickness t, $I_y = \frac{1}{12}(t)^3$. Let $q = \gamma t$, Then

$$u_{max} = \frac{\gamma TL^4}{384E(t^3/12)} = \frac{\gamma L^4}{32Et^2}$$

The moment M obeys

$$M = -EI_y \frac{d^2u}{dx^2}$$

giving

$$M = \frac{-q}{12} (L^2 - 6Lx + 6x^2)$$

The maximum moment occurs at the ends $x = 0$, $x = L$:

$$M_{max} = \frac{-qL^2}{12} = \frac{-\gamma tL^2}{12} \qquad \text{and} \qquad \sigma_{max} = \frac{-M_{max}t/2}{\frac{1}{12}t^3} = \frac{\gamma L^2}{2t}$$

Equation 7.10

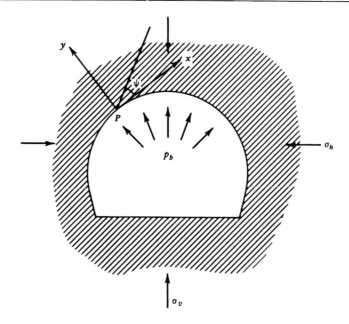

Consider a tunnel in laminated or schistose rock with initial stresses σ_v, σ_h. At a point P on the periphery, we set up coordinates x and y tangential and normal to the wall. The angle ψ is measured between the direction of the layers and x. Let p_b be the radial support pressure. The stresses at point P are

$$\tau_{xy} = 0 \qquad \sigma_x = N_h\sigma_h + N_v\sigma_v - Ap_b \tag{1}$$

and

$$\sigma_y = p_b \tag{2}$$

For a circular tunnel $A = 1$, while for a rock wall $A = 0$. We assume here that $A = 0$ (for simplicity).

Using Bray's formula, Problem 10 in Chapter 6,

$$\frac{\sigma_y}{\sigma_x} = \frac{\tan |\psi|}{\tan (|\psi| + \phi_j)} \tag{3}$$

Bray's formula is derived in the next section.

Equations 7.11 to 7.16

The theory of this section is based on the two papers by John Bray (1967) published in *Rock Mechanics and Engineering Geology*, (Springer, Berlin) Vol.

5, pp. 117–136 and pp. 197–216. Changes in notation and sign convention were made for consistency in this volume.

HYPOTHESIS

Assume the initial stress is $p_1 = p_2 = p$ in the plane perpendicular to the tunnel. This creates a stress difference near the tunnel that is sufficient to break the rock according to the Coulomb theory. The direction of the failure surface is fixed at $\psi_f = 45 - \phi/2$ with σ_θ, or $\delta = 45 + \phi/2$ with the direction of σ_r, where ϕ if the internal friction angle for rock. This means that the surfaces of failure are log spirals, since a log spiral is a locus making a constant angle (different from zero) with a bundle of radius vectors from a common point. Bray assumed two sets of log spirals making angles of $\delta = \pm 45 + \phi/2$ with r (see Figure 7.10).

We define the radius R as the outer limit of the zone of rock failure—the "plastic zone."

PRINCIPAL STRESSES IN THE PLASTIC ZONE, $r < R$

Let $K_f = \sigma_3/\sigma_1 = \sigma_r/\sigma_\theta$ for limiting equilibrium according to Coulomb's law for sliding on a joint inclined ψ with the direction of σ_1. If the friction angle is ϕ_j, then

$$2\sigma = \sigma_r + \sigma_\theta - (\sigma_\theta - \sigma_r)\cos 2\psi$$

and

$$2\tau_p = (\sigma_\theta - \sigma_r)\sin 2\psi$$

The condition for sliding is $|\tau_p| = \sigma \tan \phi_j$. Substituting for σ and τ_p gives

$$\sigma_r(\sin 2\psi + \tan \phi_j + \cos 2\psi \tan \phi_j) = \sigma_\theta(\sin 2\psi - \tan \phi_j + \cos 2\psi \tan \phi_j)$$

or

$$\sigma_r[\sin(2\psi + \phi_j) + \sin \phi_j] = \sigma_\theta[\sin(2\psi + \phi_j) - \sin \phi_j]$$

Using identities for $\sin A + \sin B$ and $\sin A - \sin B$, this reduces to

$$K_f = \frac{\sigma_r}{\sigma_\theta} = \frac{\cos(\psi + \phi_j)\sin \psi}{\sin(\psi + \phi_j)\cos \psi}$$

Therefore

$$K_f = \frac{\tan \psi}{\tan(\psi + \phi_j)}$$

Considering the absolute value sign on τ_p,

$$K_f = \frac{\tan |\psi|}{\tan (|\psi| + \phi_j)}$$

In terms of δ, substitute $90 - |\delta| = |\psi|$ to give

$$K_f = \frac{\tan(|\delta| - \phi_j)}{\tan|\delta|} \tag{1}$$

EQUILIBRIUM EQUATION

In the absence of body forces and in an axisymmetric problem where nothing varies with θ, the sum of the forces on the boundaries of the differential element of the following diagram must equal zero. Therefore,

$$\sigma_r(r\,d\theta) - \left(\sigma_r + \frac{d\sigma_r}{dr}\,dr\right)(r + dr)d\theta + 2\sigma_\theta \sin\frac{d\theta}{2}\,dr = 0$$

We know that $\sin d\theta/2 \approx d\theta/2$ and $dr^2 d\theta$ is negligible, thus

$$\boxed{\frac{d\sigma_r}{dr} + \frac{\sigma_r - \sigma_\theta}{r} = 0} \tag{2}$$

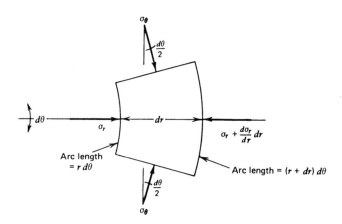

SOLUTION IN THE PLASTIC ZONE, $a < r < R$

Substituting $\sigma_\theta = \sigma_r/K_f$ in (2) yields

$$\frac{d\sigma_r}{dr} = \left(\frac{1}{K_f} - 1\right)\frac{\sigma_r}{r} \tag{3}$$

whose solution is

$$\sigma_r = Ar^Q \qquad \text{where} \qquad Q = \frac{1}{K_f} - 1 \tag{4}$$

The boundary condition is

$$\sigma_r = p_i \quad \text{when} \quad r = a$$

where p_i is the internal pressure due to supports on the inner surface of the tunnel; this gives

$$\sigma_r = p_i \left(\frac{r}{a}\right)^Q \tag{5}$$

and since $\sigma_\theta = \sigma_r / K_f$

$$\sigma_\theta = \frac{p_i}{K_f} \left(\frac{r}{a}\right)^Q \tag{6}$$

Solution for $S_j \neq 0$

Equations 5 and 6 refer to the case $S_j = 0$. If S_j is not zero, these equations apply to a τ' axis passing through $\sigma = -H = -S_j \cot \phi_j$. Thus it is necessary to add H to each pressure or stress term, giving

$$\sigma_r + S_j \cot \phi_j = (p_i + S_j \cot \phi_j) \left(\frac{r}{a}\right)^Q \tag{7}$$

$$\sigma_\theta + S_j \cot \phi_j = \frac{p_i + S_j \cot \phi_j}{K_f} \left(\frac{r}{a}\right)^Q \tag{8}$$

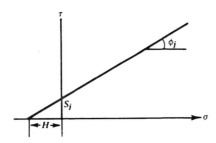

SOLUTION IN THE ELASTIC ZONE, $r \geq R$

The equilibrium equation, Equation 2, still applies. The elastic stress strain relations are now appropriate. It is most convenient to write them with stress as a function of strain (Equation 6.2). For plane strain these reduce to

$$\sigma_r = (\lambda + 2G)\varepsilon_r + \lambda\varepsilon_\theta$$

and $\hfill (9)$

$$\sigma_\theta = (\lambda + 2G)\varepsilon_\theta + \lambda\varepsilon_r$$

where

$$\lambda = \frac{E\nu}{(1 + \nu)(1 - 2\nu)}$$

The strains are related to displacement u by

$$\varepsilon_r = -\frac{du}{dr}$$

and

$$\varepsilon_\theta = -\frac{u}{r} \tag{10}$$

(Even though the problem is axisymmetric, with zero displacement v in the tangential direction perpendicular to r, there is still a tangential strain ε_θ since an arc length is stretched when it is displaced a distance u away from the origin.) Inputting (9) and (10) in (2) gives

$$r^2 \frac{d^2u}{dr^2} + r \frac{du}{dr} - u = 0 \tag{11}$$

The general solution is found by substituting $u = e^t$ and can be written

$$u = -A_1 r - \frac{B_1}{r} \tag{12}$$

Then using (10) and (9) in (12) gives

$$\sigma_r = -(\lambda + 2G)\left(-A_1 + \frac{B_1}{r^2}\right) - \lambda\left(-A_1 - \frac{B_1}{r^2}\right)$$

or

$$\sigma_r = A_2 - \frac{b}{r^2} \tag{13}$$

Similarly,

$$\sigma_\theta = A_2 + \frac{b}{r^2} \tag{14}$$

The boundary condition that $\sigma_r = \sigma_\theta = p$ as $r \to \infty$ gives

$$A_2 = p$$

$$\sigma_r = p - \frac{b}{r^2} \tag{15}$$

$$\sigma_\theta = p + \frac{b}{r^2} \tag{16}$$

THE ELASTIC-PLASTIC BOUNDARY, $r = R$

At $r = R$, both (15) and (7) apply. Eliminating σ_r between these expressions gives

$$(p_i + H)\left(\frac{R}{a}\right)^Q - H = p - \frac{b}{R^2} \tag{17}$$

The tangential stress σ_θ at $r = R$ is that required to fracture the rock by the Coulomb criterion and is also given by (16). Therefore

$$q_u + \sigma_r \tan^2\left(45 + \frac{\phi}{2}\right) = p + \frac{b}{R^2} \tag{18}$$

Adding (17) and (18) yields

$$(p_i + H)\left(\frac{R}{a}\right)^Q - H + q_u + \sigma_{r(R)}N_\phi = 2p \tag{19}$$

where

$$N_\phi = \tan^2\left(45 + \frac{\phi}{2}\right)$$

Solving (19) gives

$$R = \left(\frac{2p + H - q_u - \sigma_r N_\phi}{p_i + H}\right)^{1/Q} a \tag{20}$$

Combining (18) and (15), and solving for b results in

$$b = \left(\frac{q_u + (N_\phi - 1)p}{N_\phi + 1}\right) R^2 \tag{21}$$

Then inserting (21) in (15) and removing σ_r from (20) gives

$$R = a \left(\frac{2p + (N_\phi + 1)H - q_u}{(N_\phi + 1)(p_i + H)}\right)^{1/Q} \tag{22}$$

DISPLACEMENTS

Consider a set of joints with spacing S very small (a). The shear displacement(s) is approximated by the continuous function

$$s = D \tan \Gamma \approx D\Gamma \tag{23}$$

If the normal to joints is inclined α with x the displacements parallel to x and y are (b)

$$u = s \sin \alpha \tag{24}$$

and

$$v = -s \cos \alpha \tag{25}$$

The distance D normal to the joints from the origin to a point x, y is

$$D = x \cos \alpha + y \sin \alpha \tag{26}$$

Combining (23)–(26) gives

$$u = \Gamma(x \cos \alpha + y \sin \alpha)\sin \alpha$$

and

$$v = -\Gamma(x \cos \alpha + y \sin \alpha)\cos \alpha$$

Then

$$\left.\begin{aligned}
\varepsilon_x &= \frac{-\partial u}{\partial x} = -\Gamma \cos \alpha \sin \alpha = -\tfrac{1}{2}\Gamma \sin 2\alpha \\[2mm]
\varepsilon_y &= \frac{-\partial v}{\partial y} = \Gamma \sin \alpha \cos \alpha = +\tfrac{1}{2}\Gamma \sin 2\alpha \\[2mm]
\gamma_{xy} &= \frac{-\partial u}{\partial y} - \frac{\partial v}{\partial x} = -\Gamma \sin^2 \alpha + \Gamma \cos^2 \alpha = \Gamma \cos 2\alpha
\end{aligned}\right\} \tag{27}$$

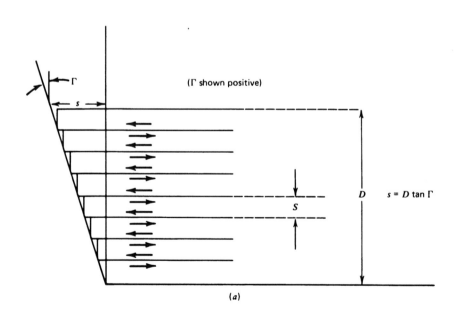

(Γ shown positive)

$s = D \tan \Gamma$

(a)

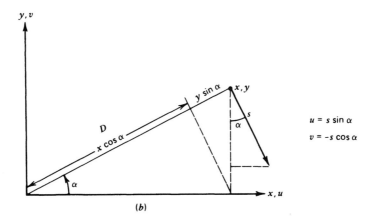

(b)

$u = s \sin \alpha$

$v = -s \cos \alpha$

Assuming a displacement u positive (in direction of positive r) (a), the strain is the sum of $\Gamma_1 = \Gamma$ on factures whose normal is inclined at $\alpha = 90 - \delta$ and (b) $\Gamma_2 = -\Gamma$ on fractures whose normal is inclined at $\alpha = 90 + \delta$ with r. The first gives

$$\varepsilon_r = -\tfrac{1}{2}\Gamma_1 \sin 2\delta = -\tfrac{1}{2}\Gamma \sin 2\delta$$
$$\varepsilon_\theta = \tfrac{1}{2}\Gamma_1 \sin 2\delta = \tfrac{1}{2}\Gamma \sin 2\delta$$
$$\gamma_{r\theta} = -\Gamma_1 \cos 2\delta = -\Gamma \cos 2\delta$$

The second gives

$$\varepsilon_r = \tfrac{1}{2}\Gamma_2 \sin 2\delta = -\tfrac{1}{2}\Gamma \sin 2\delta$$
$$\varepsilon_\theta = -\tfrac{1}{2}\Gamma_2 \sin 2\delta = \tfrac{1}{2}\Gamma \sin 2\delta$$
$$\gamma_{r\theta} = -\Gamma_2 \cos 2\delta = +\Gamma \cos 2\delta$$

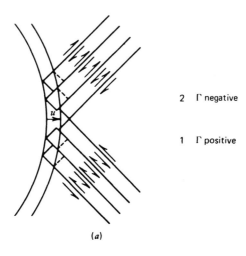

2   Γ negative

1   Γ positive

(a)

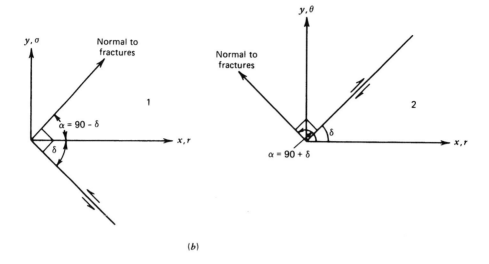

(b)

The total plastic strain due to slip of fractures is therefore

$$\varepsilon_r = -\Gamma \sin 2\delta$$
$$\varepsilon_\theta = \Gamma \sin 2\delta$$
$$\gamma_{r\theta} = 0$$

The total strain is the sum of elastic displacement and strain due to slip on the log spiral fractures:

$$\varepsilon_\theta = \Gamma \sin 2\delta + \frac{1}{E}\sigma_\theta - \frac{\nu}{E}\sigma_r$$

$$\varepsilon_r = -\Gamma \sin 2\delta + \frac{1}{E}\sigma_r - \frac{\nu}{E}\sigma_\theta \qquad (28)$$

$$\gamma_{r\theta} = 0$$

THE PLASTIC ZONE

σ_r and σ_θ are given by (7) and (8). Ignoring cohesion S_j, inserting (7) and (8) in (28) yields

$$\varepsilon_\theta = \frac{-u}{r} = \Gamma \sin 2\delta + \frac{p_i}{E}\left(\frac{1}{K_f} - \nu\right)\left(\frac{r}{a}\right)^Q$$

$$\varepsilon_r = -\frac{du}{dr} = -\Gamma \sin 2\delta + \frac{p_i}{E}\left(1 - \frac{\nu}{K_f}\right)\left(\frac{r}{a}\right)^Q \qquad (30)$$

Multiply (29) by r and differentiate with respect to r, then subtract it from (30) to give

$$\left(r\frac{d\Gamma}{dr} + 2\Gamma\right)\sin 2\delta + \frac{p_i}{E}\left(\frac{1 - \nu K_f)(Q + 1) - (K_f - \nu)}{K_f}\right)\left(\frac{r}{a}\right)^Q = 0$$

which simplifies to

$$\frac{d}{dr}r^2\Gamma\sin 2\delta = \frac{-p_i}{EK_f}\frac{[(1/K_f) - K_f]}{a^Q}r^{Q+1} \tag{31}$$

Integrating gives

$$\Gamma\sin 2\delta = \frac{-p_i}{E}\frac{[(1/K_f) - K_f]}{1 + K_f}\left(\frac{r}{a}\right)^Q + \frac{t}{r^2} \tag{32}$$

where t is a constant. Now substituting (32) in (29) and collecting terms yields

$$u = \frac{-p_i}{E}(1 - \nu)\frac{r^{Q+1}}{a^Q} - \frac{t}{r} \tag{33}$$

The displacement of a point at radial coordinate r due to raising the field stresses to the initial value p are

$$u = -r\varepsilon_\theta = -r\frac{p}{E}(1 - \nu) \tag{34}$$

This cannot be measured in the field because the displacements are referred to the state of initial stress. Therfore, u given by (34) must be subtracted from the total displacement given by (33):

$$u = -\frac{1 - \nu}{E}\left(p_i\frac{r^{Q+1}}{a^Q} - pr\right) - \frac{t}{r} \tag{35}$$

DISPLACEMENTS AT $r = R$

The elastic zone stresses are given by (15) and (16):

$$\sigma_\theta = p + \frac{b}{r^2} \quad \text{and} \quad \sigma_r = p - \frac{b}{r^2}$$

$$u = -r\varepsilon_\theta = \frac{-r}{E}\left[\left(p + \frac{b}{r^2}\right) - \nu\left(p - \frac{b}{r^2}\right)\right]$$

$$u = \frac{-r}{E}\left(p(1 - \nu) + \frac{b}{r^2}(1 + \nu)\right) \tag{36}$$

As before, subtracting (34) from (36) gives the net displacement. At $r = R$

$$u_r = -\frac{1 + \nu}{E}\frac{b}{R} \tag{37}$$

Also at $r = R$ (35) gives

$$u_R = -\frac{1-\nu}{E}\left(p_i\frac{R^{Q+1}}{a^Q} - pR\right) - \frac{t}{R} \tag{38}$$

Equating (37) and (38) yields

$$t = \frac{1-\nu}{E}R^2\left[p - p_i\left(\frac{R}{a}\right)^Q\right] + \frac{1+\nu}{E}b \tag{39}$$

At $r = a$ (35) gives

$$\boxed{u_a = \frac{1-\nu}{E}(p - p_i)a - \frac{t}{a}} \tag{40}$$

SUMMARY

The radial outward displacement at the wall $r = a$ is given by (40) where p is the initial pressure and p_i is the support pressure. The constant t is given by (39). R and b are given by (22) and (21).

Equation 8.6a

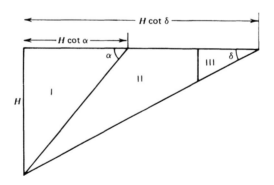

Let I represent the area of region I, etc.

$$II = (I + II + III) - I - III$$
$$II = \tfrac{1}{2}H^2 \cot \delta - \tfrac{1}{2}H^2 \cot \alpha - \tfrac{1}{2}Z^2 \tan \delta$$

Equation 8.12 **465**

Equation 8.6b

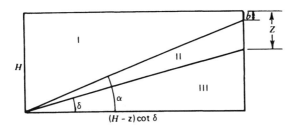

$$b = H - (H - Z)\cot \delta \tan \alpha$$
$$II = (I + II + III) - I - III$$
$$II = H(H - Z)\cot \delta - \tfrac{1}{2}\{H + [H - (H - Z)\cot \delta \tan \alpha]\}(H - Z)\cot \delta$$
$$\quad - \tfrac{1}{2}(H - Z)^2 \cot \delta$$
$$\quad = \tfrac{1}{2}(H - Z)^2 \cot \delta(\cot \delta \tan \alpha - 1)$$

Equation 8.6

The driving force $=$ the resulting force $\div F$ or

$$F(W \sin \delta + V \cos \delta) = S_j A + (W \cos \delta - U - V \sin \delta)\tan \phi \qquad (1)$$

Equation 8.6a can be written

$$W = a - b \cot \alpha \qquad (2)$$

and a and b are then as given in the text following Equation 8.6. Inserting (2) in (1) and solving for $\cot \alpha$ gives (8.6).

Equation 8.12

If the upper block (the active block) has $\delta_1 > \phi_1$, loads N_3, T_3 are transferred to the lower block (the passive block). Equilibrium in the y direction gives

$$N_3 \sin \delta_1 - T_3 \cos \delta_1 + W_1 \cos \delta_1 - N_1 = 0 \qquad (1)$$

Equilibrium in the x direction gives

$$-N_3 \cos \delta_1 - T_3 \sin \delta_1 + W_1 \sin \delta_1 - T_1 = 0 \qquad (2)$$

At the limit of equilibrium,

$$T_1 = N_1 \tan \phi_1 \tag{3}$$

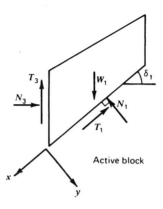

Active block

Combining (1) and (3) gives

$$T_1 = \tan \phi_1 (N_3 \sin \delta_1 - T_3 \cos \delta_1 + W_1 \cos \delta_1) \tag{4}$$

Substituting (4) in (2) gives

$$N_3(\sin \delta_1 \tan \phi_1 + \cos \delta_1) + T_3(\sin \delta_1 - \cos \delta_1 \tan \phi_1)$$
$$= W_1(\sin \delta_1 - \cos \delta_1 \tan \phi_1) \tag{5}$$

The following are applicable:

$$\sin A \tan B + \cos A = \frac{\cos(A - B)}{\cos B} \tag{6}$$

and

$$\sin A - \cos A \tan B = \frac{\sin(A - B)}{\cos B} \tag{7}$$

Let

$$T_3 = N_3 \tan \phi_3 \tag{8}$$

Combining (8) and (5) with (6) and (7) gives

$$N_3 \frac{\cos(\delta_1 - \phi_1)}{\cos \phi_1} + N_3 \tan \phi_3 \frac{\sin(\delta_1 - \phi_1)}{\cos \phi_1} = \frac{W_1 \sin(\delta_1 - \phi_1)}{\cos \phi_1}$$

Finally,

$$N_3 = \frac{W_1 \sin(\delta_1 - \phi_1) \cos \phi_3}{\cos(\delta_1 - \phi_1 - \phi_3)} \tag{9}$$

Equation 8.12 **467**

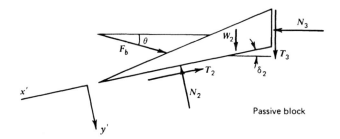

Passive block

In the passive block, similarly, equilibrium in the x' and y' directions (see figure) gives

$$-N_3 \sin \delta_2 + T_3 \cos \delta_2 + W_2 \cos \delta_2 + F_b \cos \theta \sin \delta_2$$
$$+ F_b \sin \theta \cos \delta_2 - N_2 = 0 \tag{10}$$

and

$$N_3 \cos \delta_2 + T_3 \sin \delta_2 + W_2 \sin \delta_2 - F_b \cos \theta \cos \delta_2$$
$$+ F_b \sin \theta \sin \delta_2 - T_2 = 0 \tag{11}$$

and at the limit of equilibrium

$$T_2 = N_2 \tan \phi_2 \tag{12}$$

From (10) and (12)

$$T_2 = \tan \phi_2(-N_3 \sin \delta_2 + T_3 \cos \delta_2 + W_2 \cos \delta_2$$
$$+ F_b \cos \theta \sin \delta_2 + F_b \sin \theta \cos \delta_2) \tag{13}$$

Substituting in (11) gives

$$N_3(\cos \delta_2 + \sin \delta_2 \tan \phi_2) + T_3(\sin \delta_2 - \cos \delta_2 \tan \phi_2)$$
$$= W_2(\cos \delta_2 \tan \phi_2 - \sin \delta_2) + F_b \sin \theta(\cos \delta_2 \tan \phi_2 - \sin \delta_2)$$
$$+ F_b \cos \theta(\sin \delta_2 \tan \phi_2 + \cos \delta_2) \tag{14}$$

Introducing (6)–(8) gives

$$N_3 = \frac{F_b \cos \theta \cos(\delta_2 - \phi_2) - (W_2 + F_b \sin \theta)\sin(\delta_2 - \phi_2)}{\cos(\delta_2 - \phi_2) + \tan \phi_3 \sin(\delta_2 - \phi_2)} \tag{15}$$

Equating (15) and (9) we have

$$\frac{W_1 \sin(\delta_1 - \phi_1)}{\cos(\delta_1 - \phi_1 - \phi_3)}$$
$$= \frac{F_b[\cos \theta \cos(\delta_2 - \phi_2) - \sin \theta \sin(\delta_2 - \phi_2)] - W_2 \sin(\delta_2 - \phi_2)}{\cos(\delta_2 - \phi_2 - \phi_3)} \tag{16}$$

Using

$$\cos A \cos B - \sin A \sin B = \cos(A + B)$$

simplifies (16) to Equation 8.12.

Equations 9.1, 9.2

These equations are derived in most textbooks on solid mechanics, for example, S. Timoshenko and J. N. Goodier, *Theory of Elasticity* (McGraw-Hill, New York, 1951) pp. 85–91, also Obert and Duvall and Jaeger and Cook (works cited earlier).

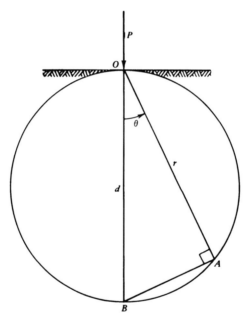

That Equation 9.1 described a circular locus of constant radial stress is seen as follows. Consider a circle with radius $d/2$ centered $d/2$ below the point of load. Any point on this circle has polar coordinates $(d \cos \theta, \theta)$. The radial stress all along the circle is then given by Equation 9.1 as

$$\sigma_r = \frac{2P \cos \theta}{\pi d \cos \theta} = \frac{2P}{\pi d} = \text{constant}$$

If r is a constant R, Equation 9.1 gives $\sigma_r = 2P \cos \theta/(\pi R)$. Note from the figures that if $0B$ is set equal to $2P/(\pi R)$, then $0A$ equals $2P/(\pi R)\cos \theta$. Therefore as stated in Chapter 9 $|0A|$ gives the magnitude of the pressure along any line $0A$ acting on a circle of radius R centered about 0.

Equation 9.8   469

Equations 9.3 to 9.5

The stresses under a line load acting perpendicular to a surface in a principal symmetry direction of a transversely isotropic material are given by A. E. Green and W. Zerna, *Theoretical Elasticity* (Oxford, Univ. Press, London, 1954), p. 332. John Bray showed that this solution also holds for an arbitrarily inclined load on a surface at any angle relative to the principal symmetry directions.

Equations 9.6, 9.7

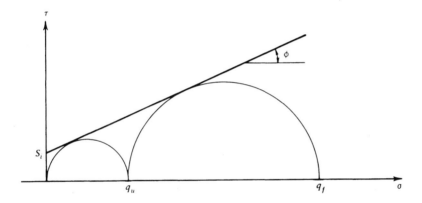

Using Equation 3.8 with the right circle in the figure gives

$$q_f = q_u \tan^2 \left(45 + \frac{\phi}{2} \right) + q_u = q_u(N_\phi + 1)$$

Equation 9.8

Assume a circular footing of diameter B centered on a circular joint block of diameter S. The problem is axisymmetric and the equation of equilibrium in the radial direction is as for the derivations of Equations 7.11 to 7.16:

$$\frac{d\sigma_r}{dr} + \frac{\sigma_r - \sigma_\theta}{r} = 0 \qquad (1)$$

If σ_r causes the rock to break, σ_θ corresponds to σ_3 and Equation 3.8 gives

$$\sigma_r = m + N_\phi \sigma_\theta \qquad (2)$$

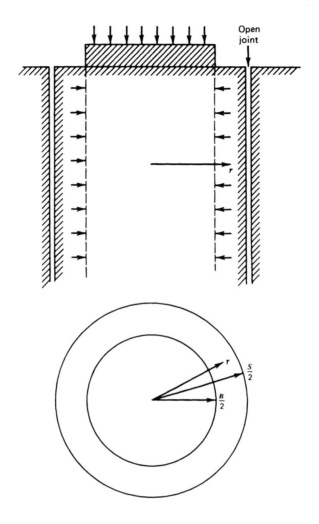

where $m = 2S_i(N_\phi)^{1/2}$ and $N_\phi = \tan^2(45 + \phi/2)$. Inserting (2) in (1) and solving gives

$$\sigma_r = \frac{N_\phi}{N_\phi - 1}\left(Ar^{(1-N_\phi)/N_\phi} - \frac{2S_i}{N_\phi^{1/2}}\right) \tag{3}$$

At $r = B/2$, let $\sigma_r = p_h$; this gives

$$A = \left(p_h \frac{N_\phi - 1}{N_\phi} + \frac{2S_i}{N_\phi^{1/2}}\right)\left(\frac{B}{2}\right)^{N_\phi/(1 - N_\phi)} \tag{4}$$

Equation 9.10   **471**

Substituting (4) in (3) gives

$$\sigma_r = \left(p_h + \frac{2S_iN_\phi^{1/2}}{N_\phi - 1}\right)\left(\frac{r}{B/2}\right)^{(1-N_\phi)/N_\phi} - \frac{2S_iN_\phi^{1/2}}{N_\phi - 1} \tag{5}$$

At $r = S/2$, let $\sigma_r = 0$ corresponding to an open joint condition. Inserting this boundary condition in (5) and solving for p_h gives

$$p_h = \frac{2S_i(N_\phi)^{1/2}}{N_\phi - 1}\left[\left(\frac{B}{S}\right)^{(1-N_\phi)/N_\phi} - 1\right] \tag{6}$$

We consider the cylinder under the footing to be a kind of large triaxial compression specimen with confining pressure p_h. Its strength is, therefore, using the notation of (2)

$$q_f = p_hN_\phi + 2S_iN_\phi^{1/2} \tag{7}$$

Substituting (6) in (7) and simplifying gives Equation 9.8.

Equation 9.10

Equilibrium in the vertical y direction gives

$$d\sigma_y\pi r^2 + \tau 2\pi r\, dy = 0 \tag{1}$$

In the pier, with $\sigma_x = \sigma_r$,

$$E_c\varepsilon_r = \sigma_r - v_c\sigma_r - v_c\sigma_y \tag{2}$$

and

$$\varepsilon_r = -du/dr \tag{3}$$

Substituting (3) in (2) and integrating from 0 to a, the outward radial displacement u at the surface of the pier is

$$u = \frac{-(1 - v_c)}{E_c}a\sigma_r + \frac{v_c}{E_c}a\sigma_y \tag{4}$$

In the rock, the radial pressure on the surface $r = a$ is similar to a uniform pressure acting on the inside wall of an infinitely thick hollow cylinder. This solution was presented as Equation 6.13 in Chapter 6 in connection with the borehole dilatometer test. Inserting u in place of Δu and σ_r in place of Δp, with properties E_r and v_r in (6.13), the outward radial displacement of the rock at the surface of the pier is

$$u = \sigma_r\frac{(1 + v_r)a}{E_r} \tag{5}$$

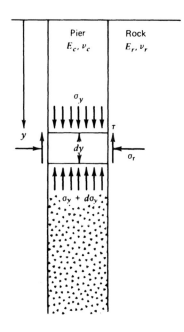

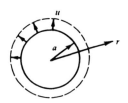

Equating (4) and (5) and solving for σ_r yields

$$\sigma_r = \left(\frac{\nu_c}{1 - \nu_c + (E_c/E_r)(1 + \nu_r)}\right)\sigma_y \tag{6}$$

If the pier/rock contact is a surface without adhesion but with friction, then coefficient $\mu = \tan \phi_j$:

$$\tau = \mu\sigma_r \tag{7}$$

Substituting (6) in (7) and then in (2) gives

$$\sigma_y = A \exp\left(\frac{-2\nu_c\mu}{1 - \nu_c + (1 + \nu_r)E_c/E_r}\frac{y}{a}\right) \tag{8}$$

At $y = 0$, $\sigma_y = p_{total}$. Therefore $A = p_{total}$, giving (9.10).

Equation 9.15   **473**

Equation 9.14

The footnote in Chapter 9 after Equation 9.14 gives the settlement of a pier, exclusive of base settlement, corresponding to the load transfer relation (9.10). With reference to the preceding figure, the vertical displacement obeys

$$-\frac{dv}{dy} = \varepsilon_y = \frac{1}{E_c}\sigma_y - \frac{2v_c}{E_c}\sigma_r$$

Substituting (6) and (8) for σ_r and σ_y in the above and integrating from $y = l$ to the surface ($y = 0$) gives the downward displacement of a pile of length l continuously embedded in rock (exclusive of end settlement). Let

$$g = \frac{-2v_c}{1 - v_c + (E_c/E_r)(1 + v_r)}$$

Then integration yields

$$v = \left(1 - \frac{v_c^2}{1 - v_c}\right)\left(\frac{P_{\text{total}}}{E_c}\frac{a}{g}e^{-[(g/r)y]}\right)_l^0$$

Simplifying and adding a term for the shortening of the unsupported length of pile above the top of rock yields the equation given in the footnote in the text.

Equation 9.15

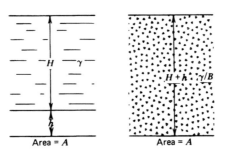

Area = A   Area = A

Consider a mine opening of height h at depth H in material of weight density γ. The weight of the material over area A at the mine floor is $W = \gamma HA$. Let the average weight density of the rock after collapse be γ/B; B represents a bulking factor. Since no new material has been added, the weight is unchanged. Therefore, $(\gamma/B)(H + h)A = \gamma HA$. This leads directly to Equation 9.15.

The Use of Stereographic Projection

Introduction

The stereographic projection simplifies graphical solutions to problems involving the relative orientations of lines and planes in space. In rock mechanics contexts, stereographic projection is appealing for analyzing the stability of excavations as shown in Chapter 8, as well as for exploring and characterizing discontinuities in rocks. Many publications in structural geology, crystallography, and rock mechanics show constructions and tricks available using stereographic projection. Especially helpful references for the purposes of rock mechanics are Phillips (1972), Hoek and Bray (1977), and Goodman (1976). For the restricted applications considered in this volume, it will suffice to explain the underlying principles and to demonstrate the most essential operations.

Figure A5.1*a* shows the stereographic projection of a plunging line. The line passes through the center of a reference sphere at 0, and pierces its surface at P in the lower hemisphere, and at $-P$ in the upper hemisphere. In all applications we will cling to the convention that the line or plane we wish to project contains the center of the reference sphere. The horizontal plane through 0 is termed the *projection plane*. A perpendicular to the projection plane pierces the top of the reference sphere at F, which will be termed the *focus for lower hemisphere projection*. The stereographic projection consists of projection of lines and points on the surface of a reference sphere from a single perspective point to corresponding points in the projection plane. To find the lower hemisphere stereographic projection of any line through 0 we find the

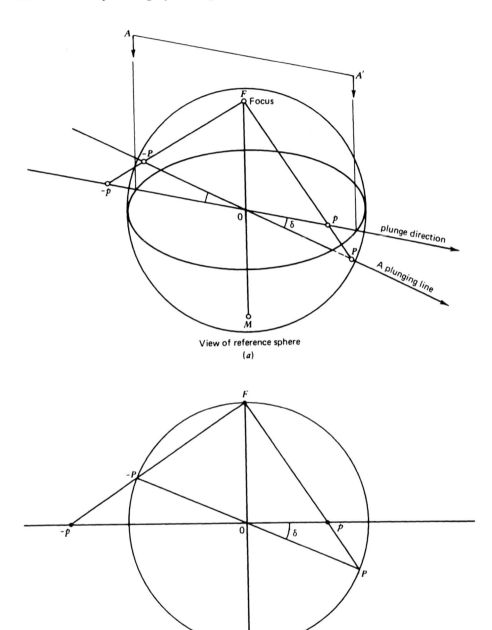

View of reference sphere
(a)

Vertical section through AA'
(b)

Figure A5.1

point where this line pierces the surface of the reference sphere, construct a straight line between the piercing point and F, and find the point where the construction line to F crosses the projection plane. For example, the line $0P$ in Figure A5.1a pierces the reference sphere at point P and the construction line PF crosses the projection plane at point p. The latter is then the correct representation of $0P$ in a lower hemisphere stereographic projection. Similarly, the opposite end of $0P$, which pierces the upper hemisphere of the reference sphere at point $-P$, projects to point $-p$ as shown. Figure A5.1b presents a vertical section of the reference sphere through the line $0P$. It is perhaps easier to visualize the spatial relations of the line and its projection in this slice. Although the construction shown in Figure A5.1b, or its mathematical equivalent, can always be invoked to locate the stereographic projection of a line, it proves most convenient to plot the stereographic projection by tracing from a stereonet, as shown later.

The stereographic projection of a plane consists of finding the locus connecting the stereographic projections of all the lines it contains. A theorem holds that any circle on the reference sphere projects as a circle in the projection plane. (This is not true for the "equal area projection," a variant of the stereographic projection.) Since any plane we wish to project must contain the center of the reference sphere, it must pierce the surface of the sphere along a great circle. In view of the above theorem, the stereographic projection of a plane must therefore project as a true circle. To find its center, it is sufficient to construct a circle through the stereographic projections of the strike line and the dip vector.

Figure A5.2 shows a horizontal plane piercing the reference sphere along great circle SMT. These points are unmoved by the projection from F. Therefore, a circle centered about 0 in the projection plane represents the stereographic projection of a horizontal plane. Points inside it, when projected from F at the top of the reference sphere, belong to the lower hemisphere; all other points belong to the upper hemisphere. This figure also slows an inclined plane passing through 0 and intersecting the reference sphere along great circle SDT. Line $0S$ and its opposite $0T$ represent the strike of the inclined plane; these project at points S and T. Line $0D$ is the dip vector of the inclined plane; it projects to point d as shown. Other lines in the plane, $0A$, $0B$, $0C$, etc. project to points a, b, c, etc. to define the circular locus TdS as shown. To plot this locus is to determine the stereographic projection of the plane. One way to do this would be to construct a circle through points T, d, and S. The center for this construction will be found along line $0V$ at a distance corresponding to the projection of a line plunging at an angle from vertical equal to twice the dip (which is measured from horizontal). Another way to construct the projection of the plane is to plot the projection of $0W$, the opposite of the dip vector, and bisect its distance to d to locate the center of the projected circle. However, the

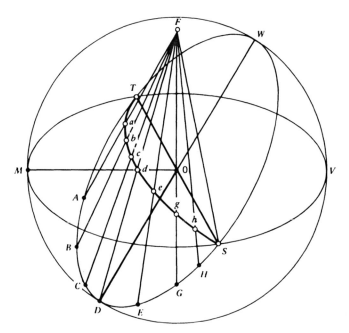

Figure A5.2

most convenient way to project the plane is by tracing it from the family of great circles provided in a stereonet.

A *stereonet* is a stereographic projection of a set of reference planes and lines within one hemisphere. Figure A5.3 is an equatorial stereonet presenting a series of great circles of planes having a common intersection and dipping in increments of 2°. These great circles resemble the lines of longitude on a map of the earth. A family of small circles are also provided, each representing the locus of lines making a constant angle with the line of intersection of the great circles. The small circles, which resemble the lines of latitude on a map of the globe, calibrate the great circles; that is, angles in any great circle are measured by counting small circles. This procedure will be illustrated in examples that follow. To follow these examples, detach the copy of Figure A5.3 that is reprinted near the back endpaper of this book, and pierce it with a thumbtack from behind that passes exactly through the center. Tracings placed on the stereonet can then be rotated about the center. An additional stereonet has been provided at the end of the book to permit continuation of the great and small circles beyond the limits of one hemisphere, for use in applications of block theory.

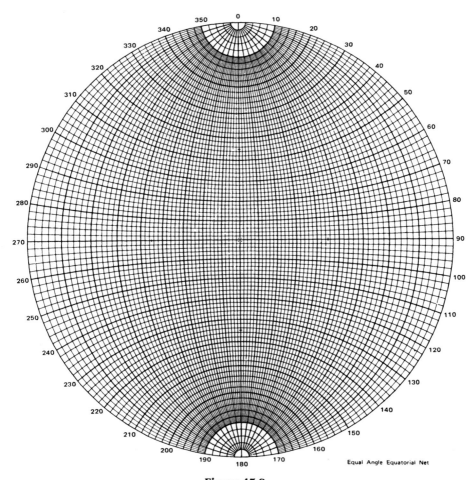

Figure A5.3

Projection of a Line

Line 1 *plunges with vertical angle* 40° *below horizontal toward the N* 30° *E; plot it on a lower hemisphere stereographic projection.* The line will be assumed to pass through the center of the reference sphere. If the focus is assumed to lie atop the reference sphere, its stereographic projection will be a point inside the "horizontal circle" (i.e., the projection of a horizontal plane). The letters L.H. (lower hemisphere) on the tracing will indicate that this is the selected option. In Figure A5.4*a*, a tracing has been superimposed on the stereonet, north has

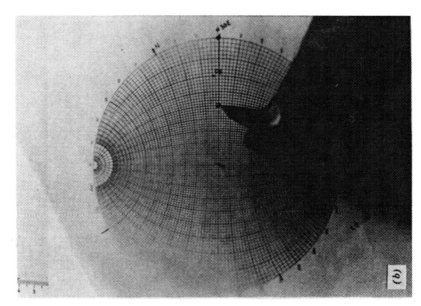

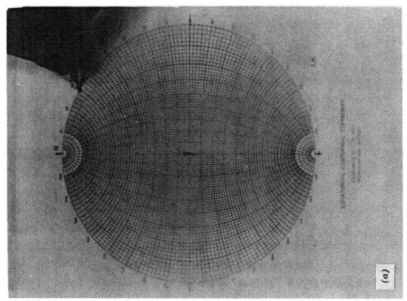

Figure A5.4

480

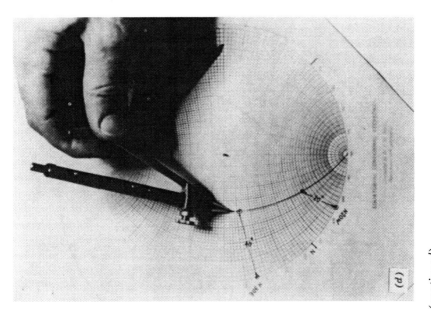

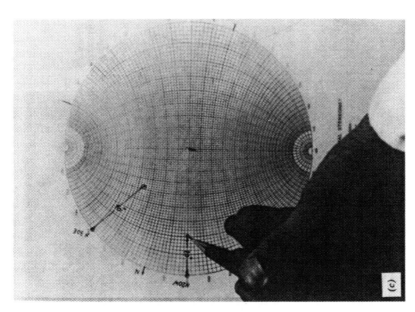

Figure A5.4—(*continued*)

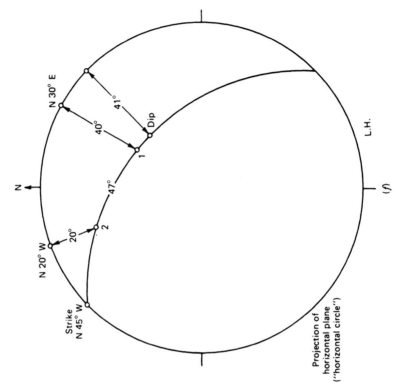

N 30° E

N 30° E

41°

Dip

40°

1

N

47°

N 20° W

20°

2

Strike
N 45° W

L.H.

(f)

Projection of
horizontal plane
("horizontal circle")

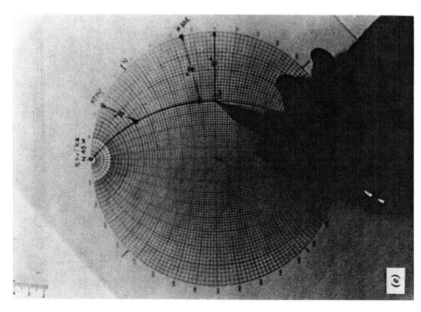

(e)

Figure A5.4—(continued)

been marked arbitrarily and then registry marks have been placed at E, S, and W positions around the horizontal circle. The horizontal line bearing N 30° E has then been projected by marking a point 30° east of north on the horizontal circle. In Figure A5.4b, the tracing has been rotated to line up the previously plotted point with a ruled diameter of the stereonet. The straight line marking the chosen diameter is the one member of the family of small circles that happens to be also a great circle; since it is a straight line, it must be the projection of a vertical plane. The 2° family of great circles calibrates the vertical plane represented by this diameter, and so the vertical angle of 40° can be laid off by counting 20 great circle intersections. The point thus marked, point 1, is the stereographic projection we require.

The Angle between Two Lines

A line (2) plunges 20° to the N 20° W. Plot this line and measure its angle with line 1 plotted previously. Line 2 is added to the tracing using the same sequence of steps as above (Figure A5.4c). To measure the angle from line 1 to line 2, it is now necessary to determine the plane common to both lines. Since each line passes through the center of the reference sphere, a common plane exists. It is found by rotating the tracing until both points fall along the same great circle (Figure A5.4d). The angle between (1) and (2) is then measured by counting the intersections with the small circles (which are spaced every 2°). The angle is 47°. The strike and dip of the plane common to (1) and (2) are indicated on Figure A5.4e, obtained by rotating the tracing so that the point of intersection of the great circle and the horizontal circle overlies the axis of the great circle family on the stereonet. Figure A5.4f shows the tracing at the end of this step.

Projection of a Plane Given its Strike and Dip

Plot the stereographic projection of a plane (1) striking N 50° E and dipping 20° to the N 40° W. On a new tracing, the strike vector, a horizontal line bearing N 50° E, is plotted as a point 50° east of north along the horizontal circle (Figure A5.5a). Next rotate the tracing to place the strike vector over the axis of the great circles and plot the dip vector along the diameter at right angles to the strike (Figure A5.5b). The dip vector is a line plunging 20° to the N 40° W, so

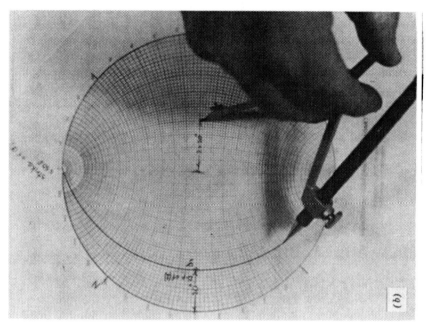

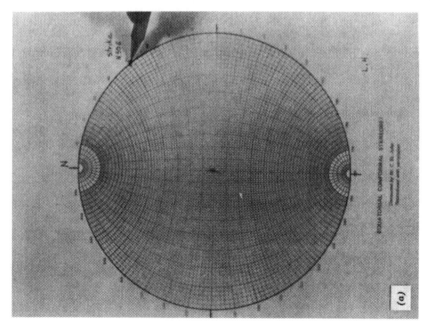

Figure A5.5

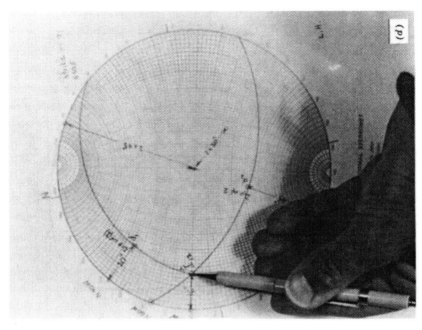

(d)

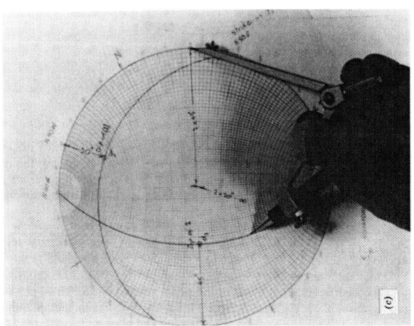

(c)

Figure A5.5—*(continued)*

485

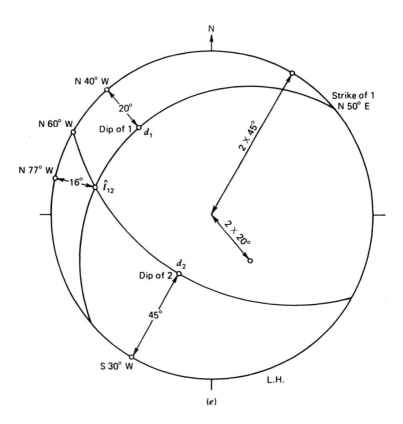

(e)

this step follows from the example discussed previously. Now trace the great circle common to the strike and dip vectors. To increase accuracy, the great circle may be constructed with a compass. Since the dip vector plunges 20°, the center of the great circle is 40° from vertical along the diameter containing the dip vector, as shown in Figure A5.5b.

Plot the stereographic projection of a plane (2) *striking N* 60° *W and dipping* 45° *to the S* 30° *W and find the bearing and plunge of the line of intersection of planes* 1 *and* 2. Emulating the steps above for plane 1, the new plane (2) yields the great circle shown in Figure A5.5c. This circle crosses the previously constructed great circle at the point marked I_{12}. Since I_{12} is a point in the projection of each plane, it represents a line that lies in each plane; it is therefore the required intersection. The bearing and plunge of I_{12} are read from the stereonet by rotating the tracing to the diameter of the net as shown in

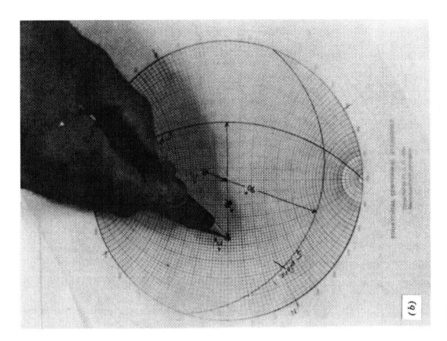

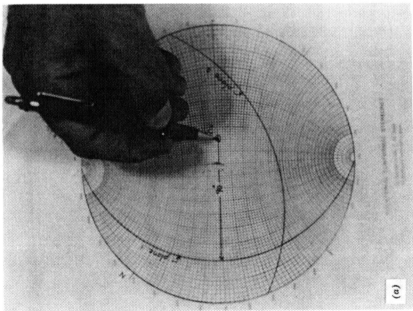

Figure A5.6

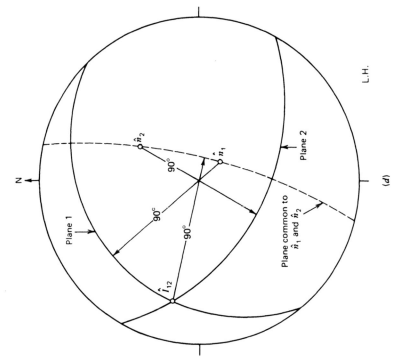

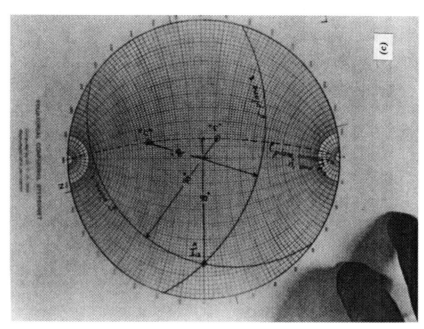

Figure A5.6—(*continued*)

Figure A5.5*d*. In this position, the vertical angle from I_{12} to horizontal (the plunge of I_{12}) can be found by counting the number of great circle intersections between I_{12} and horizontal. The intersection plunges 16° to the N 77° W. Figure A5.5*e* shows the tracing at the conclusion of this step.

There is a more convenient way to find the line of intersection of two planes when they are represented by their normals. If it is understood that a line identified as a normal to a plane is intended to represent the plane, then in place of the great circle a plane can be plotted by means of a single point. To find the intersection line of two planes plotted by their normals, n_1 and n_2, the method shown in Figure A5.6 can be used. In this figure, the projections of planes 1 and 2 found in the previous example have been traced on a clean overlay. The normal to plane 1 (n_1) is plotted in Figure A5.6*a* by lining up the dip vector of plane 1 with the diameter of the stereonet and measuring 90° along this diameter through the vertical. (A vertical line is represented by the point at the center of the projection, this being a lower hemisphere projection.) The normal to plane 2 is plotted similarly in Figure A5.6*b*. Then, in Figure A5.6*c*, the two normals, n_1 and n_2 are lined up on a common great circle by rotating the tracing appropriately. The normal to this great circle is I_{12} (Figure A5.6*c*). Figure A5.6*d* shows the tracing at the end of this step. Note that it was not necessary to draw the great circles of planes 1 and 2 to find I_{12} by this construction. They were drawn in the figure to demonstrate that the two methods of construction do in fact lead to the same result.

The Locus of Lines Equidistant from a Given Line

The locus of lines making a constant angle with a certain line is a circular cone with vertex at the center of the reference sphere. This cone projects as a small circle. By the theorem stated previously, the projection of a small circle is a true circle, that is, it may be drawn with a compass. A way to do this is shown in Figure A5.7.

Plot the locus of lines at 45° with the normal to plane 1 from the previous problem. In Figure A5.7*a*, the point n_1, traced from the Figure A5.6*d*, has been lined up with the net's diameter. Two lines on the cone are then plotted by moving away from n_1 by the required 45° along the diameter in each direction. In Figure A5.7*b*, the distance between these two points is then bisected to find the center of the small circle. Note that the center for construction does not coincide with the axis of the cone (n_1). The circle is drawn from the center using a compass as shown in Figure A5.7*c*. The tracing after this step is shown in Figure A5.7*d*.

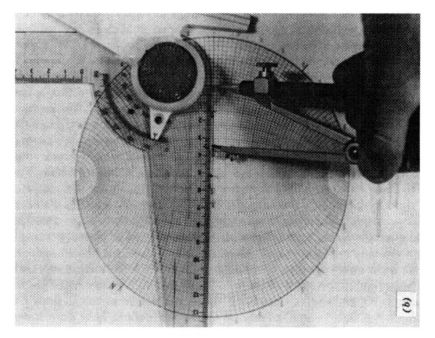

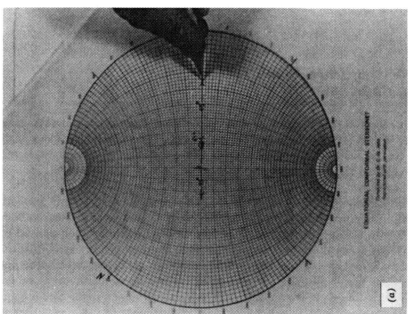

Figure A5.7

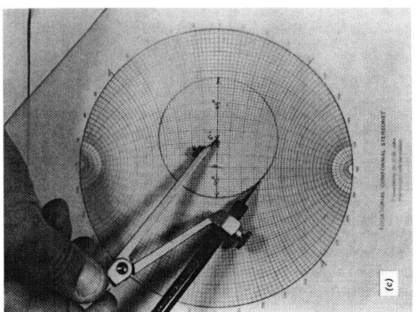

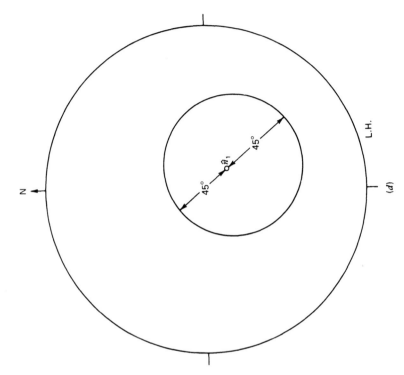

(d)

Figure A5.7—(*continued*)

Vectors

Problems in slope stability and rock foundations involve manipulations with vectors. Since the direction of a vector can be shown as a point on the stereographic projection, the preceding constructions prove applicable to stability analysis, as discussed in Chapter 8. However, there is a world of difference between the tip and the tail of a vector in this connotation; therefore one must distinguish carefully between a line and its opposite. In structural geology work, such a distinction is not usually required and a point on one hemisphere can be replaced by its opposite in the other hemisphere without penalty. In rock mechanics we must work with the whole sphere.

There is no essential difficulty in working with both hemispheres, the only requirement being either a very large piece of paper or two separate tracings, one marked L.H. to denote lower hemisphere (as in all the examples of this section) and another marked U.H. to identify the upper hemisphere. Both hemispheres can be shown on one plot but only one hemisphere is located inside the horizontal circle. Constructions and manipulations helpful for problems embracing the whole sphere are presented by Goodman (1976).

References

Donn, W. L. and Shimer, J. A. (1958) *Graphic Methods in Structural Geology*, Appleton Century Crofts, New York.

Goodman, R. E. (1976) Principles of stereographic projection and joint surveys, in *Methods of Geological Engineering in Discontinuous Rocks*, West, St. Paul, MN.

Hoek, E. and Bray, J. W. (1977) Graphical presentation of geological data, in *Rock Slope Engineering*, 2d ed., Institute of Mining and Metallurgy, London.

Phillips, F. C. (1972) *The Use of Stereographic Projection in Structural Geology*, 3d ed., Arnold, London.

Priest, S. D. (1985) *Hemispherical Projection Methods in Rock Mechanics*, George Allen & Unwin, London.

Problems

1. Write an equation for the distance from the center of the projection circle (corresponding to a unit reference sphere) to the lower hemisphere stereographic projection of a line plunging $\delta°$ below horizontal.

2. Determine the angle between lines 1 and 2 and the strike and dip of their common plane: line 1 plunges 70° to the N 30° E; line 2 plunges 15° to the N 60° E.

3. Determine the bearing of the plunge of the line of intersection of two planes given as follows: plane 1 strikes N 70° E and dips 60° S 20° E; plane 2 strikes N 20° W and dips 40° N 70° E.

4. Show how a line in the upper hemisphere is plotted in a lower hemisphere projection.

5. Given a line plunging in the lower hemisphere 30° to the north, plot its opposite in an upper hemisphere projection (i.e., a projection in which the focus is at the bottom of the reference sphere). Also, plot the line (not its opposite) in the lower hemisphere projection. Compare the two results and generalize if possible.

6. Construct the locus of lines making an angle of 35° with a line plunging 60° to the N 30° E. (Use a lower hemisphere projection.) What is the minimum angle between line 2 of Problem 2 and any point on the locus?

Answers to Problems

CHAPTER 2

1. With $G_{\text{illite}} = 2.75$, $G_{\text{chlorite}} = 2.84$, and $G_{\text{pyrite}} = 4.9$, $\overline{G} = 3.20$ and $\overline{n} = 0.24$. Then $\gamma_{\text{dry}} = 23.83$ kN/m^3 $= 151.8$ P/ft^3. Average water content $\overline{w} = 0.099$ giving γ_{wet} 26.2 kN/m^3 $= 166$ P/ft^3.

$$\sigma_v = 47.9 \text{ MPa} = 6948 \text{ psi}$$

2. $q_u = 19.0$ MPa (2760 psi); 53.1 MPa (7700 psi); 137.0 MPa (19,800 psi).

3. $\gamma_{\text{wet}} = 24$ kN/m^3; $\gamma_{\text{dry}} = 22.76$ kN/m^3; $n = 12.8\%$

4. $\gamma_{\text{wet}} = 20.11$ kN/m^3 (assuming $G = 2.70$)
   The loss in wet weight/m^3 $= 3.89$ kN; then the loss in volume of saturated solid/m^3 $= 0.233$ m^3, giving $\Delta n = 0.233$ and $n = 0.361$.

5. $V_1^* = 6440$ m/s; moderately to strongly fissured

6. $\gamma_{\text{dry}} = 0.028$ MN/m^3 $= 178$ P/ft^3

7. $w = 5.25\%$

8. See Equation 2.6.

9. $q = 9.63 \times 10^{-7}$ cm^3/s per cm^2 area

10. $\sigma_v = 38.1$ MPa $= 5520$ psi

11. 0.040 mm

12. $2k_f e = kS \rightarrow k = k_f \dfrac{2e}{S}$

13.

Joint Condition	Total Rating	Description
Rough; hard wall rock	60	Fair rock
Slightly rough; hard rock; aperture < 1 mm	55	Fair rock
Ditto above but soft wall rock	50	Fair rock
Smooth; open 1–5 mm or gouge 1–5 mm	40	Poor rock
Open more than 5 mm or gouge thicker than 5 mm	30	Poor rock

14. $\dfrac{e^3}{6S} = 55$ darcies $= 55 \times 9.8 \times 10^{-9}$ cm^2

    $e = (6 \times 50 \text{ cm} \times 55 \times 9.8 \times 10^{-9} \text{ cm}^2)^{1/3}$

    $= 0.546$ mm

15. $C = \dfrac{4.0 \dfrac{\text{gal}}{\text{min}}}{10 \text{ ft} \times 55 \text{ psi}} = 7.3 \times 10^{-3}$ gal/min/ft/psi

    $C_{\text{Lugeons}} = \dfrac{4.0 \dfrac{\text{gal}}{\text{min}} \times \dfrac{1}{0.264 \text{ gallon}} \dfrac{\text{L}}{}}{\dfrac{10 \text{ ft} \times 12 \text{ in./ft}}{39.37 \text{ in./m}} \times 55 \text{ psi} \dfrac{1}{145} \dfrac{\text{MPa}}{\text{psi}}} = 13.1$ Lugeons

16. $\gamma_1 = (1 - n)\gamma$

    (a) First 17.6 kN/m^3; then 24.8 kN/m^3

    (b) First 17.6 to 13.5 kN/m^3; then 24.8–20.3 kN/m^3

CHAPTER 3

1. $S_i = 1.17$ MPa; $\phi = 40°$

2. $P_w = 3.27$ MPa $= 474$ psi

3. $K = 0.217$

4. $v = 0.178$

5. $q_u = 3.84$ MPa; assume $T_0 = 0.05 q_u$, giving $T_0 = 0.19$ MPa

6. $\Delta p_w = 2.45$ MPa $= 356$ psi; $\Delta h_{\text{water}} = 250$ m $= 820$ ft

7. $\sigma = \dfrac{\sigma_1 - T_0}{2} - \dfrac{\sigma_1 + T_0}{2} \sin \phi$ and $\sigma_1 = -T_0 \tan^2 \left(45 + \dfrac{\phi}{2}\right) + 2S_i \tan \left(45 + \dfrac{\phi}{2}\right)$

8. (a) Linear regression gives $\sigma_{1p} = 11,980$ psi $+ 6.10 \sigma_3$; thus $\phi_p = 45.9°$, $S_{ip} = 2425$ psi

    (b) Linear regression gives $\sigma_{1r} = 1020 + 5.74 \sigma_3$; thus $\phi_r = 44.7$ and $S_{ir} = 143$ psi

    Note in (a) and (b) that the determined values of q_u are slightly different than those measured in the unconfined compression tests.

    (c) Using $q_u = 11,200$ as measured, power law regression gives $\sigma_{1p}/q_u = 1 + 5.65(\sigma_3/q_u)^{0.879}$.

    (d) $\sigma_{1p} = 40,980$ psi $+ 12.68 \sigma_3$, giving $\phi_p = 58.6°$ and $S_{ip} = 5750$ psi

    (e) $\sigma_{1r} = 3470 + 10.37 \sigma_3$, giving $\phi_r = 55.5$ and $S_{ir} = 540$ psi

    (f) Using $q_u = 41,000$ as measured rather than 40,980 as determined by linear regression in (d) gives $\sigma_{1p}/q_u = 1 + 11.91(\sigma_3/q_u)^{0.979}$.

9. The maximum moment is $M = [(P/2) \cdot (L/2)]$ where L is the length of the beam. The maximum tensile stress is $\sigma_{\max} = Mc/I$ with $c = d/2$ and $I = \pi d^4/64$, (d being the diameter of the core sample). Therefore, $T_{\text{MR}} = \sigma_{\max} = 8PL/(\pi d^3)$.

10. Consider a unit cube with edges aligned to x, y, and z axes. Its initial volume $V = 1$. After undergoing strains ε_x, ε_y, and ε_z, the lengths of the edges become $1 + \varepsilon_x$, $1 +$

ε_y, and $1 + \varepsilon_z$. The change in volume ΔV is therefore $(1 + \varepsilon_x)(1 + \varepsilon_y)(1 + \varepsilon_z) - 1 = 1 + \varepsilon_x + \varepsilon_y + \varepsilon_z + \varepsilon_x\varepsilon_y + \varepsilon_y\varepsilon_z + \varepsilon_z\varepsilon_x + \varepsilon_x\varepsilon_y\varepsilon_z - 1$. If the strains are small, their products can be neglected. Therefore $\Delta V = \varepsilon_x + \varepsilon_y + \varepsilon_z = \Delta V/V$.

11.

ψ	S_i	ϕ	q_u	σ_{1p} for $\sigma_3 = 30$
0°	60.2	33.0	221.7	323.5
30°	26.4	21.1	77.0	140.7
60°	60.2	18.6	167.6	225.7
90°	69.8	28.9	236.5	322.6

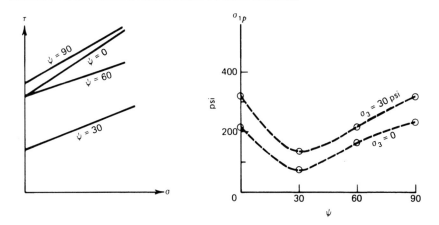

12. Assuming that $S_{i,\min}$ occurs when $\psi = 30$ and that $n = 1$, Equation 3.18 becomes $S_i = S_1 - S_2 \cos 2(\psi - 30)$. Let $S_{i,30}$ and $S_{i,75}$ be the values of S_i for $\psi = 30°$ and $75°$, respectively. Then $S_1 = S_{i,75}$ and $S_2 = S_{i,75} - S_{i,30}$. Since $q_u = 2 \tan(45 + \phi/2)S_i$, then

$$q_u = q_{u,75} - (q_{u,75} - q_{u,30})\cos 2(\psi - 30)$$

Substituting this value for q_u in Equation 3.8 yields the desired result.

13. We may assume that anisotropy arises from the presence of planar fissures. These close under high mean stress.

14.

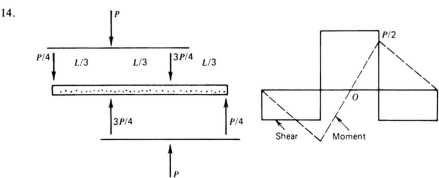

15. (a) Assume that the residual strength of an unconfined compression specimen is essentially zero. Then

$$\frac{\sigma_{1r}}{q_u} = M \left(\frac{\sigma_3}{q_u}\right)^N$$

(b) For the sandstone $\dfrac{\sigma_{1r}}{q_u} = 5.27 \left(\dfrac{\sigma_3}{q_u}\right)^{0.867}$

(c) For the norite $\dfrac{\sigma_{1r}}{q_u} = 4.21 \left(\dfrac{\sigma_3}{q_u}\right)^{0.633}$

16. (a) for $m = 0$

$$\frac{\sigma_{1p}}{q_u} = \frac{\sigma_3}{q_u} + 1 \qquad \text{by Hoek and Brown}$$

Equation 3.15 is

$$\frac{\sigma_{1p}}{q_u} = N \left(\frac{\sigma_3}{q_u}\right)^M + 1$$

So the two are identical if $N = M = 1$ when $m = 0$.

(b) Substitution in the Hoek and Brown criterion gives the following values corresponding to the three rock types:

		σ_{1p}	
σ_3	$m = 7$	$m = 17$	$m = 25$
0	100	100.0	100
10	140.4	174.3	197.1
20	174.9	229.8	264.9
40	234.9	319.3	371.7
70	312.9	429.2	500.1
100	382.8	524.3	609.8

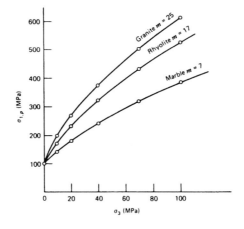

(See the Figure.) It can be appreciated that parameter m is a generalization of the Mohr-Coulomb friction angle for the case of a downward-curved failure envelope.

17. (a) Equation 2.17 gives RMR $= 9 \log Q + 44$
Solving for Q yields

$$Q = e^{\left(\frac{RMR - 44}{9}\right)}$$

Substituting this result for Q in the given expression gives

$$A = 0.0176 e^{M[(RMR-44)/9]}$$

(b) For $M = 0.65$, $A = 0.0176 e^{(0.072RMR-3.177)}$ or $A = e^{(0.0722RMR-7.217)}$. Substituting in Equation 3.15 with A in place of 1 and with $N = 5$, $q_u = 2$, and RMR $= 50$ gives

$$A = e^{-3.607}$$

and

$$\sigma_{1,p} = 2[e^{-3.607} + 5(\sigma_3/2)^{0.65}]$$

Finally

$$\sigma_{1,p} = 0.054 + 6.37(\sigma_3)^{0.65} \quad (MPa)$$

CHAPTER 4

1. $\sigma_v = 13.5$ MPa $= 1960$ psi
$\bar{\sigma}_H = 6.75$ MPa $= 979$ psi

2. Original water pressure $= 1300$ psi
$\sigma_{h,min} = 9.72$ MPa $= 1410$ psi
$T_0 = 3.45$ MPa $= 500$ psi
$\sigma_v = 24.7$ MPa $= 3580$ psi
$\sigma_{h,max} = 18.76$ MPa $= 2720$ psi if pore pressure is neglected. If pore pressure is considered using the answer to Problem 4.11 with $p_w = 13,000$ psi: $\sigma_{h,max} = 1,420$ psi $= 9.79$ MPa.

3. $\sigma_{major} = 3.10$ MPa $= 450$ psi
$\sigma_{minor} = 1.38$ MPa $= 200$ psi
$\theta_1 = -63.4°$

4. 278 m

5. $\sigma_1 = \sigma_{h,max} = 82.33$ MPa $= 11,938$ psi
$\sigma_2 = \sigma_{h,min} = ?$
$\sigma_3 = \sigma_v = 25.15$ MPa $= 3646$ psi
Figure 4.7 gives $\bar{K} = 1.91$, or $\sigma_{h,max} = 48.0$ MPa $= 6957$ psi.

6. $K = 1.11$

7.

8. Assuming that the Kirsch solution applies, with a perfectly circular opening in a homogeneous, isotropic, elastic, continuous mass, with the radius of the tunnel much larger than the width of the jack; that the value of E for loading equals the value for unloading; and that the slot and jack have the same dimension, then

$$\sigma_{horiz} = 4.48 \text{ MPa} = 650 \text{ psi}$$
$$\sigma_{vert} = 7.24 \text{ MPa} = 1050 \text{ psi}$$

Note that if the stress concentrations are taken at the center of the jack, Equations 4.13 become

$$\begin{Bmatrix} \sigma_{\theta,w} \\ \sigma_{\theta,R} \end{Bmatrix} = \begin{pmatrix} -0.635 & 2.47 \\ 2.47 & -0.635 \end{pmatrix} \begin{Bmatrix} \sigma_{horiz} \\ \sigma_{vert} \end{Bmatrix}$$

giving

$$\begin{Bmatrix} \sigma_{horiz} \\ \sigma_{vert} \end{Bmatrix} = \begin{pmatrix} 0.111 & 0.434 \\ 0.434 & 0.111 \end{pmatrix} \begin{Bmatrix} \sigma_{\theta,w} \\ \sigma_{\theta,R} \end{Bmatrix}$$

Then

$$\sigma_{horiz} = 4.61 \text{ MPa} = 669 \text{ psi}$$
$$\sigma_{vert} = 8.16 \text{ MPa} = 1185 \text{ psi}$$

9. In pounds and inches

$$\begin{Bmatrix} \sigma_x \\ \sigma_z \\ \tau_{xz} \end{Bmatrix} = 10^8 \begin{pmatrix} 182.5 & -57.5 & 0 \\ 2.5 & 122.5 & 207.85 \\ 2.5 & 122.5 & -207.85 \end{pmatrix}^{-1} \begin{Bmatrix} 0.003 \\ 0.002 \\ 0.001 \end{Bmatrix}$$

$$\begin{Bmatrix} \sigma_x \\ \sigma_z \\ \tau_{xz} \end{Bmatrix} = 10^3 \begin{pmatrix} 544.4 & 127.8 & 127.8 \\ -11.1 & 405.6 & 405.6 \\ 0 & 240.6 & -240.6 \end{pmatrix} \begin{Bmatrix} 0.003 \\ 0.002 \\ 0.001 \end{Bmatrix}$$

$$\begin{Bmatrix} \sigma_x \\ \sigma_z \\ \tau_{xz} \end{Bmatrix} = \begin{Bmatrix} 2017.0 \\ 1183.0 \\ 241.0 \end{Bmatrix} \text{ psi}$$

giving

$$\sigma_{max} = 2082 \text{ psi}$$
$$\sigma_{min} = 1118 \text{ psi}$$
$$\alpha = 15.0°$$

10. $\sigma_h = 180$ MPa $= 26,100$ psi
    The measurement gives $\sigma_h = 80$ MPa $= 11,600$ psi.
    The rock could not withstand such a high stress difference ($\sigma_v = 0$) and developed
    fractures.

11. $p_{c1} - p_w = 3\sigma_{h,min} - \sigma_{h,max} - 2p_w + T_0$ by substituting $\sigma_{h,max} - p_w$, $\sigma_{h,min} - p_w$, and
    $p_{c1} - p_w$ in place of $\sigma_{h,max}$, $\sigma_{h,min}$, and p_{c1}, respectively.

12. (a) $\sigma_{h,max} = 1333$ psi
    $\sigma_{h,final} = 1333 - 449 = 884$ psi
    (b) Glaciation, or sedimentation followed by uplift and erosion

13. $\sigma_x = 48.25$, $\sigma_z = 10.15$, $\tau_{xy} = -1.13$, $\sigma_1 = 48.28$, $\sigma_2 = 10.12$, $\alpha = -1.70$

CHAPTER 5

1. There are three sets of joints:

   (a) Strike S 38.4° E   (b) Strike S 34.3° W
       dip 36.8° NE   dip 62.2° NW
       $K_f = 557$ $K_f = 439$

   (c) Strike N 18.5° E
       dip 63.2° SE
       $K_f = 238$

2. See diagram.

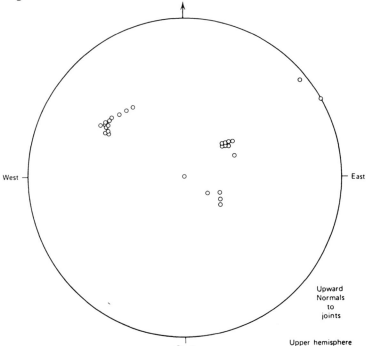

3. $\phi_\mu = 34.5°$ (the values of σ, τ on the sawed joint during sliding are (0.32, 0.22), (0.97, 0.67), (1.61, 1.11), and (3.23, 2.23)).

4. 252 MPa

5.

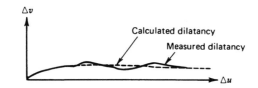

6. $\sigma_v = 16.2, \sigma_h = 10.6$ MPa

7. $p_w = 0.50$ MPa

8. $\tau_{peak} = \dfrac{6.5 \text{ kN}}{0.5 \text{ m}^2} = 13 \text{ kN/m}^2$

$\left.\begin{array}{l} \\ \\ \end{array}\right\}$ $\tau_{peak} = \tan 33°$

$\sigma = \dfrac{10 \text{ kN}}{0.5 \text{ m}^2} = 20 \text{ kN/m}^2$

$\tau_{resid} = \dfrac{5.3 \text{ kN}}{0.5 \text{ m}^2} = 10.6 \text{ kN/m}^2$ $\tau_{resid} = \tan 27.9°$

Shear stiffness at peak:

$$k_s = \frac{13 \text{ kN/m}^2}{5.2 \text{ mm}} = 2.50 \text{ MPa/m}$$

The initial shear stiffness is 4.00 MPa/m. The peak dilatancy angle is $\phi_{peak} - \phi_{resid} = 5.1°$ (assuming no wear on the joint). Using Schneider's equation (5.8), i varies as

u (mm)	5.2	7.5	9.5	11.0	≥ 12
i (°)	5.1	3.1	0.91	0.45	0

9. (a) At slip, with angle of friction ϕ_j on the joint

$$T - F_B \sin \alpha + F_B \cos \alpha \tan \phi_j \tag{1}$$

Then, to prevent slip

$$\boxed{F_B = \frac{T}{\cos \alpha \tan \phi_j + \sin \alpha}} \tag{2}$$

(b) The forces across the joint are inclined at ϕ_j with the normal at the point of slip.

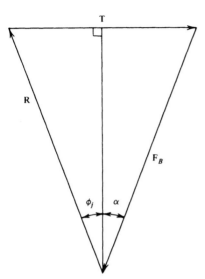

The directions of T and F_B are given. From the force triangle, F_B is minimum when

$$\phi_j + \alpha = 90° \tag{3}$$

So

$$\boxed{\alpha_{min} = 90 - \phi_j} \tag{4}$$

In this direction, (2) gives

$$F_b = T \cos \phi \tag{5}$$

(c) First, dilation would change the strength so that in place of ϕ_j one should substitute $\phi_j + i$ in (2) and (4). Second, the dilatant displacement component in the direction of the bolt would encounter resistance due to the bolt stiffness k_B causing an increment of bolt force:

$$\Delta F_B = k_B \, (u \tan i \cos \alpha + u \sin \alpha) \tag{6}$$

In place of (2), we obtain

$$F_B = \frac{T}{\cos \alpha \tan(\phi_j + i) + \sin \alpha} + k_B u (\tan i \cos \alpha + \sin \alpha) \tag{7}$$

An additional resistance would come from bolt shear stiffness when the steel hits the sides of the borehole. As this induces rock crushing and sharp curvature in the steel, with combined, nonhomogeneous stresses, the solution becomes more complex. Equation 5 can be simplified to

$$F_b = \frac{\cos(\phi_j + i)}{\sin(\alpha + \phi_j + i)} + k_B \, u \, \frac{\sin(\alpha + i)}{\cos i} \tag{8}$$

10.

	σ_3'/σ_1'	
$|\psi|$ (°)	$\phi_j = 20°$	$\phi_j = 30°$[a]
0, 180	0.000	0.000
5, 175	0.086	0.125
10, 170	0.305	0.210
15, 165	0.383	0.268
20, 160	0.434	0.305
25, 155	0.466	0.327
30, 150	0.484	0.333

35, 145	0.490	0.327
40, 140	0.484	0.305
45, 135	0.466	0.268
50, 130	0.434	0.210
55, 125	0.383	0.125
60, 120	0.305	0.000
65, 115	0.086	−0.188
70, 110	0.000	−0.484
75, 105	−0.327	−1.000
80, 100	−1.000	−2.064
85, 95	−3.063	−5.330
90	−∞	−∞

[a] See figures for $\phi_j = 20°$ and $30°$.

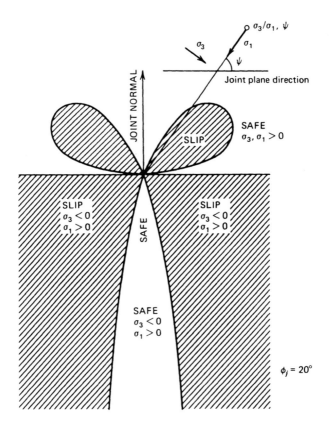

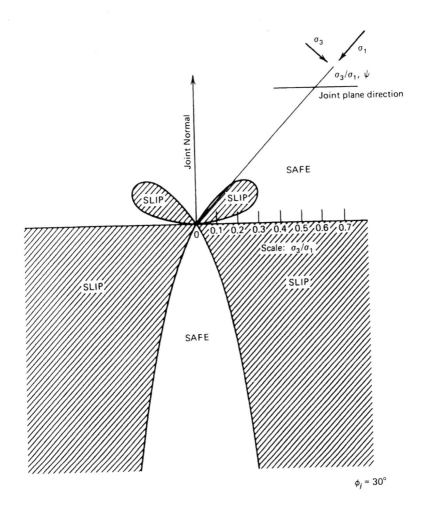

$\phi_j = 30°$

11. $\phi_j = 28.2$, $\psi = 50°$, $p_w = 10$ MPa, $\sigma_3 = 1.5$ MPa, $\sigma_1 = 4.5$ MPa

$$\frac{1.5 - p_w}{4.5 - p_w} = \frac{\tan 50}{\tan 78.2} = 0.249$$

$$1.5 - p_w = 1.121 - 0.249 p_w$$

$$p_w = \frac{0.379}{0.751} = 0.505 \text{ MPa}$$

12. $S_i = 1.0$, $\phi = 30°$

$\gamma = 0.025$ MPa/m

$\nu = 0.2$ (see figure)

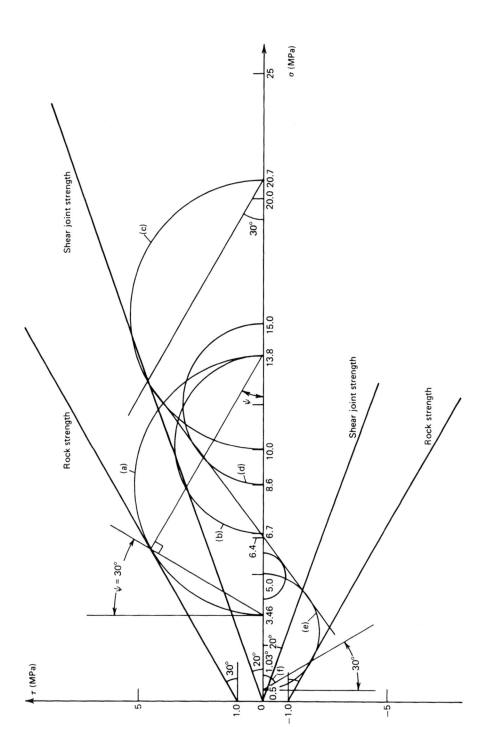

(a) $\sigma_h/\sigma_v = \dfrac{\nu}{1 - \nu} = \dfrac{0.2}{0.8} = 0.25$

Equation 3.14 → $\sigma_{1,p} = \dfrac{q_u}{1 - k\,\tan^2(45 + \phi/2)}$

$q_u = 2S_i\,\tan(45 + \phi/2) = 2(1.0)\tan 60 = 3.46$ MPa

$\sigma_v = \sigma_{1,p} = \dfrac{3.46}{1 - 0.25\,\tan^2 60} = 13.8$ MPa

Depth $Z = 13.8$ MPa/0.025 MPa/m $= 554$ m

$\sigma_h = 0.25\sigma_v = 3.46$ MPa

(b) Using Bray's equation, given in Problem 5.10, after the shear joints have formed,

$$\sigma_h/\sigma_v = \frac{\tan\psi}{\tan(\psi + \phi_j)} = \frac{\tan 30}{\tan 50} = 0.484$$

σ_h becomes $0.484 \times 13.8 = 6.7$ MPa

(c) $\sigma_v = 1.5 \times 13.8 = 20.7$ MPa

$\sigma_h = 0.484 \times 20.8 = 10.0$ MPa

$Z = 831$ m

(d)

$\Delta\sigma_h = 0.25\Delta\sigma_v$

$10.0 - \sigma_h = 0.25(20.7 - \sigma_v)$

$\sigma_h = 4.825 + 0.25\sigma_v$

σ_v	σ_h
15	8.58
10	7.32
6.43	6.43
4	5.83

and $\sigma_h = \sigma_v$ when

$\sigma_v = 4.825 + 0.25\sigma_v$

$\sigma_v = 6.43$ MPa

$Z = 257$ m

(e) The equation connecting σ_h and σ_v is

$\sigma_h = 4.825 + 0.25\sigma_v$

at rupture, with $\sigma_h = \sigma_1$ and $\sigma_v = \sigma_3$

$\sigma_h = q_u + \sigma_v\,\tan^2(45 + \phi/2)$

giving

$4.825 + 0.25\sigma_v = 3.46 + 3\sigma_v$

or

$$\sigma_v = 1.365/2.75 = 0.50 \text{ MPa}$$

and

$$\sigma_h = 4.95 \text{ MPa}$$
$$Z = 20 \text{ m}$$

(f) $\dfrac{\sigma_v}{\sigma_h} = \dfrac{\tan 30}{\tan 50} = 0.484$

Note that $\psi = 30°$ is now to be measured from horizontal, whereas previously it was measured from vertical:

$$\sigma_h = \frac{\sigma_v}{0.484} = \frac{0.50}{0.484} = 1.03$$

(g) See figure.

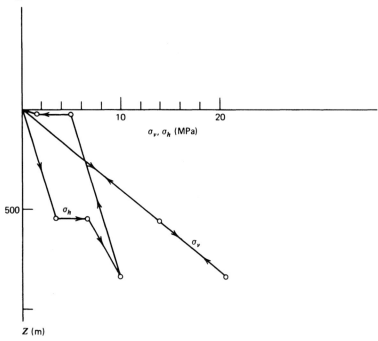

13. See figures
   (a) In direction $\perp$ fractures

$$\lambda = \frac{n}{L}$$

In direction $\pm\theta$ from normal

$$\lambda_{(\theta)} = \frac{n}{L/\cos\theta} = \lambda|\cos\theta|$$

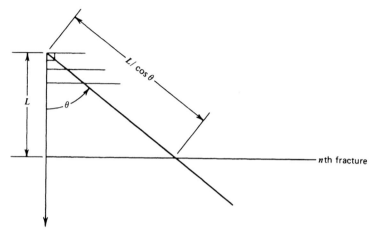

Normal to fracture set

(b) $\lambda_{(\theta)} = \lambda_1|\cos \theta_1| + \lambda_2|\cos \theta_2|$

$\quad = \lambda_1|\cos \theta_1| + \lambda_2|\sin \theta_1|$

(c) For max λ,

$$\frac{d\lambda}{d\theta} = -\lambda_1 \sin \theta_1 + \lambda_2 \cos \theta_1 = 0$$

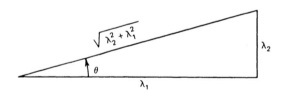

or

$$\tan \theta = \lambda_2/\lambda_1$$

then

$$\cos \theta = \frac{\lambda_1}{\sqrt{\lambda_1^2 + \lambda_2^2}}$$

and

$$\sin \theta = \frac{\lambda_2}{\sqrt{\lambda_1^2 + \lambda_2^2}}$$

giving

$$\lambda_{max} = \sqrt{\lambda_1^2 + \lambda_2^2} = 5.39$$

$$\theta_{max} = 21.8° \text{ from normal to set } 1$$

Average spacing in this direction $= (\lambda_{max})^{-1} = 0.19$.

14.! By analogy with Equation 5.8, ϕ_b is the residual friction angle of the joint and [JCR log (JCS/σ_n)] is the roughness angle of the joint surface. When σ_n is small compared with JCS, the dilatancy angle is large and when JCS/σ is 10 the roughness angle equals JCR (degrees). The roughness angle declines with increasing normal stress until σ_n = JCS after which the roughness angle is zero. (It makes no sense to discuss the shear strength, using Barton's equation, when the normal stress is higher than JCS.)

CHAPTER 6

1. Denoting deviator strain components by e and mean strain by $\bar{\varepsilon}$, as in Appendix 2,

$$e_x = \varepsilon_x - \bar{\varepsilon} = \tfrac{2}{3}\varepsilon_x - \tfrac{1}{3}\varepsilon_y - \tfrac{1}{3}\varepsilon_z \tag{1}$$

Similarly,

$$\sigma_{x,\text{dev}} = \sigma_x - \bar{\sigma} = \tfrac{2}{3}\sigma_x - \tfrac{1}{3}\sigma_y - \tfrac{1}{3}\sigma_z \tag{2}$$

where $\sigma_{x,\text{dev}}$ is deviatoric normal stress in the x direction and $\bar{\sigma}$ is the mean stress. Substituting Equation 6.1 in (1) gives

$$
\begin{aligned}
e_x &= \frac{2}{3}\left(\frac{1}{E}\sigma_x - \frac{\nu}{E}\sigma_y - \frac{\nu}{E}\sigma_z\right) - \frac{1}{3}\left(\frac{1}{E}\sigma_y - \frac{\nu}{E}\sigma_x - \frac{\nu}{E}\sigma_z\right) \\
&\quad - \frac{1}{3}\left(\frac{1}{E}\sigma_z - \frac{\nu}{E}\sigma_x - \frac{\nu}{E}\sigma_y\right) \\
&= \frac{1+\nu}{E}\frac{2}{3}\sigma_x - \frac{1+\nu}{E}\frac{1}{3}\sigma_y - \frac{1+\nu}{E}\frac{1}{3}\sigma_z = \frac{\sigma_{x,\text{dev}}}{2G}
\end{aligned}
$$

Similar expressions arise by considering e_y and e_z. Also, $e_{xy} = \tfrac{1}{2}\gamma_{xy} = \tfrac{1}{2}(\tau_{xy}/G)$ with similar expressions for e_{yz} and e_{zx}.

2. Let $\dfrac{\varepsilon_{\text{lateral}}}{\varepsilon_{\text{axial}}} = R$; then

$$\nu = \frac{R\sigma_{\text{axial}} - p}{p(2R-1) - \sigma_{\text{axial}}} \quad \text{and} \quad E = \frac{\sigma_{\text{axial}} - 2\nu p}{\varepsilon_{\text{axial}}}$$

3. During a triaxial compression test, the mean stress is $\bar{\sigma} = (\sigma_{\text{axial}} + 2p)/3$. To hold $\bar{\sigma}$ constant as σ_{axial} is increased, it is sufficient to decrease the confining pressure from its initial value $p = \bar{\sigma}$ to a value $p = (3\bar{\sigma} - \sigma_{\text{axial}})/2$. Manual feedback can achieve the control for slow rates of load; a computer driving a servofeedback system is necessary for fast load rates and for precise control near the peak. In terms of the change in p required,

$$\bar{\sigma} = \tfrac{1}{3}\sigma_{\text{axial}} + 2p = \text{constant}; \text{ then } \Delta\sigma_{\text{axial}} + 2\Delta p = 0 \text{ and } \Delta p = \tfrac{1}{2}\Delta\sigma_{\text{axial}}$$

4. The data are plotted in the diagram. Using the load cycle from 0 to 5000 N,

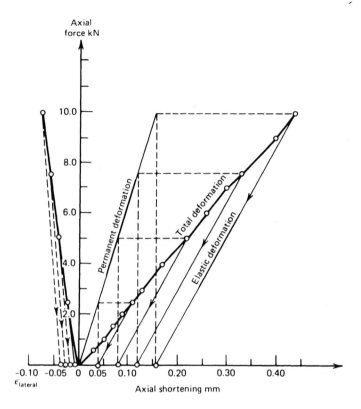

$E = 1821$ MPa (elastic deformation)

$M = 3188$ MPa (permanent deformation)

$\nu = 0.300$ (elastic deformation)

$\nu_p = 0.400$ (permanent deformation)

From the total deformation on loading,

$E_{\text{total}} = 1159$ MPa   and $\nu_{\text{total}} = 0.336$

5. $D_1 = 3K = 3 \dfrac{E}{3(1 - 2\nu)}$ and $D_2 = 2G = \dfrac{2E}{2(1 + \nu)}$

(see Problem 1). Then

$$E = \frac{3D_1 D_2}{D_2 + 2D_1} \quad \text{and} \quad \nu = \frac{D_1 - D_2}{2D_1 + D_2}$$

6. (a) $\dfrac{1}{E_{\text{total}}} = \dfrac{1}{M} + \dfrac{1}{E}$

   (b) See the diagram.

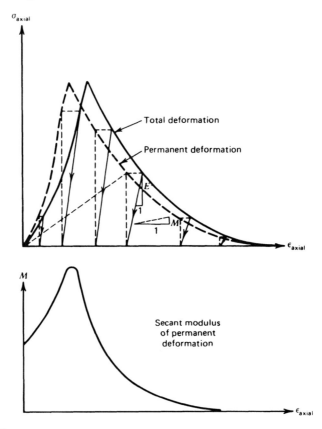

7. $\nu = 0.277$

   $E = 43{,}900$ MPa $(=43.9$ GPa$) = 6.37 \times 10^6$ psi

8. Permanent deformation arises from irreversible closure of fissures, pores, and joints. The second and flatter slope (Γ) could reflect sliding on joints.

9. (a) Joint normal deformation $= \dfrac{\sigma}{k_n}$ (1)

   Rock normal deformation $= \dfrac{\sigma}{E} S = \dfrac{\sigma}{E} 0.4$ m   (2)

   Equating 1 and 2 gives

   $$k_n = \frac{E}{S} = 2.5E \tag{3}$$

Similarly,

$$k_s = \frac{G}{S} = \frac{E}{2(1 + \nu)S} = \frac{2.5E}{2(1 + \nu)} \tag{4}$$

(b)

$$\frac{1}{E_n} = \frac{1}{E} + \frac{1}{k_n S} = \frac{1}{E} + \frac{1}{E} = 2 \times 10^{-4} \text{ (MPa)}^{-1}$$

$$\frac{1}{G_{sn}} = \frac{1}{G} + \frac{1}{k_s S} = \frac{1}{G} + \frac{1}{G} = \frac{4(1 + \nu)}{E} = 5.32 \times 10^{-4} \text{ (MPa)}^{-1}$$

$$E_s = E = 10^4 \text{ MPa}$$

$$\nu_{sn} = 0.33$$

$$\nu_{ns} = \frac{E_n}{E} \nu = \frac{0.5 \times 10^4}{10^4} (0.33) = 0.165$$

$$\nu_{st} = 0.33$$

Then the terms of the strain-stress matrix corresponding to (6.9) are

$$10^{-4} \begin{bmatrix} 2 & -0.33 & -0.33 & 0 & 0 & 0 \\ -0.33 & 1 & -0.33 & 0 & 0 & 0 \\ -0.33 & -0.33 & 1 & 0 & 0 & 0 \\ 0 & 0 & 0 & 5.32 & 0 & 0 \\ 0 & 0 & 0 & 0 & 5.32 & 0 \\ 0 & 0 & 0 & 0 & 0 & 2.66 \end{bmatrix}$$

10. Let joint set 1 have spacing S_1, etc. Define directions 1, 2, and 3 normal to sets 1, 2, and 3. Then,

$$\frac{1}{E_{n1}} = \frac{1}{E} + \frac{1}{k_{n1} S_1}$$

etc., for 2 and 3 and

$$\frac{1}{G_{12}} = \frac{1}{G} + \frac{1}{k_{s1} S_1} + \frac{1}{k_{s2} S_2}$$

etc., for 23 and 31.

$$\nu_{12} = \nu_{13} = \frac{E_{n1}}{E} \nu$$

$$\nu_{21} = \nu_{23} = \frac{E_{n2}}{E} \nu$$

$$\nu_{31} = \nu_{32} = \frac{E_{n3}}{E} \nu$$

11. We assume the rock is isotropic and that the strain/stress matrix is symmetric. Since s, and t directions are both in the plane parallel to the joint, there is no difference in deformability constants relating n to s and relating n to t directions. Thus it suffices to discuss: (1) ε_n due to application of σ_n; (2) ε_s due to application of σ_s; and (3) ε_s due to application of σ_n.

(1) Applying only σ_n causes closure Δn_j due to the joint, and Δn_r due to the rock. Then

$$\varepsilon_n = \frac{\Delta n_j}{S} + \frac{\Delta n_r}{S} = \frac{\sigma_n}{k_n S} + \frac{\sigma_n}{E} = \frac{\sigma}{E}\left(\frac{1}{k_n S} + \frac{1}{E}\right)$$

finally

$$\varepsilon_n = \frac{\sigma}{E}\left(\frac{E}{k_n S} + 1\right) = \frac{1}{E}(p)\sigma_n$$

(2) Applying only σ_s causes strain in direction s due only to the rock:

$$\varepsilon_s = \varepsilon_{s,r} = \sigma_s = \frac{1}{E}(1)\sigma_s$$

(3) Applying only σ_n causes lateral strain due only to the rock

$$\varepsilon_s = \varepsilon_{s,r} = \frac{-\nu}{E}\sigma_n = \frac{1}{E}(-\nu)\sigma_n$$

12. The slopes of the unloading, reloading ramps give as the bulk modulus k_B

$$k_B = \frac{2.4}{0.0007} = 3430 \text{ MPa}$$

$$k_B = \frac{4.8}{0.0014} = 3430 \text{ MPa}$$

$$k_B = \frac{10.3}{0.0030} = 3430 \text{ MPa}$$

Assuming the plastic deformation arises entirely due to unrecoverable (plastic) closing of the joints, then

$$\frac{\Delta V}{V(\text{plastic})} = \varepsilon_{1\text{plastic}} + \varepsilon_{2\text{plastic}} + \varepsilon_{3\text{plastic}} = 3\frac{p}{k_n S}$$

where S = the spacing of joints. Substituting $S = 5$ cm $= 0.05$ m, and solving yields

$$k_n = \frac{3}{0.05}\frac{p}{(\Delta V/V)_{\text{plastic}}}$$

p (MPa)	$\Delta V/V$	$\Delta V/V_{\text{plastic}}$	k_n (MPa/m)
2.4	0.0034	0.0027	53,000
4.8	0.0057	0.0043	67,000
10.4	0.0088	0.0058	107,000

13. $\nu_t = \dfrac{\nu M + \nu_p E}{M + E}$

CHAPTER 7

1.

Point	θ	r	σ_θ	σ_r	α	σ_n	τ_{ns}
A	61°	29.15	695	404	−31°	481	−128.5
B	30°	25.00	748	352	0	352	0
C	−1°	29.15	695	404	+31°	481	128.5

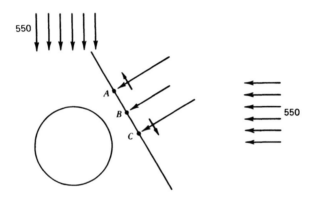

2. For $\gamma = 1.1$ psi/ft,

K	0	$\frac{1}{3}$	$\frac{2}{3}$	1	2	3
σ_θ/σ_v	2.0	1.67	2.33	4.0	9.0	14.0
$\sigma_{\theta,max(psi)}$	2200	1837	2563	4400	9900	15,400
Location	Wall	Wall	Roof	Roof	Roof	Roof

3. $\Delta q = 1.71$ psi $= 11.81$ kPa
$u_{max} = 0.136$ in. $= 3.45$ mm
$\sigma_{max,s.s} = 363$ psi $= 2.50$ MPa
$\sigma_{max,shale} = 72.7$ psi $= 0.50$ MPa

4. (a) $Q = 1.0396$; $R = 34.74a$; $b = 2.67 \times 10^{10}$
$t = 53,390$
$u = 556.5$ in. $\gg a$, meaning the tunnel fails

Elastic zone:

$$\sigma_r = 4000 \text{ psi} - 2.898 \times 10^6 \frac{1}{(r/a)^2}$$

$$\sigma_\theta = 4000 \text{ psi} + 2.898 \times 10^6 \frac{1}{(r/a)^2}$$

Plastic zone:

$$\sigma_r = 40 \left(\frac{r}{a}\right)^{1.0396}$$

$$\sigma_\theta = 2.0396\sigma_r$$

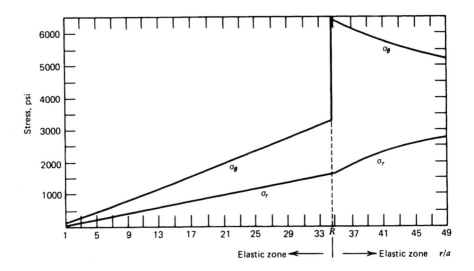

Elastic zone ◄——— | ———► Elastic zone r/a

(b) For $p_i = 400$ psi $R = 3.79a$, $u_r = -6.35$ in. in elastic zone:

$$\sigma_r = 4000 \text{ psi} - 3.453 \times 10^4 \frac{1}{(r/a)^2}$$

$$\sigma_\theta = 4000 \text{ psi} + 3.452 \times 10^4 \frac{1}{(r/a)^2}; \text{ in plastic zone: } \sigma_r = 400 \left(\frac{r}{a}\right)^{1.0396};$$

$$\sigma_\theta = 2.0396\sigma_r$$

5. (a) 3608 lb
   (b) 17.1 psi in limestone
       113 psi in sandstone

6. $p_1 = 4000$ and $p_2 = 2000$ psi

t	20 min	1 hr	12 hr	1 day	2 days	4 days
u_r (in.)	1.15	1.15	1.19	1.23	1.32	1.48

	1 week	2 weeks	8 weeks	1 yr	10 yr
	1.73	2.28	4.92	10.69	13.84

7. Anchor of bolt has coordinates 480″, 30°.
   Head of bolt has coordinates 300″, 30°.
   $E = 30 \times 10^6$ psi; area of bolt = 1.227 in.^{2}

t	u in. Anchor	Head	Δu in. $u_{head} - u_{anchor}$	Δu since $t = 12$ hr	ε	σ psi	F lb
12 hr	0.83	1.23	0.40	0	0	0	0
24 hr	0.86	1.27	0.41	0.01	5.56×10^{-5}	1,667	2,045
2 days	0.92	1.36	0.44	0.04	2.22×10^{-4}	6,667	8,182
4 days	1.04	1.55	0.51	0.11	6.11×10^{-4}	18,333	22,500
1 week	1.22	1.81	0.59	0.19	1.06×10^{-3}	31,667	38,360
2 weeks	1.63	2.41	0.78	0.38	2.11×10^{-3}	63,333 YIELD	77,710

The bolts should become plastic after about 2 weeks.

8.

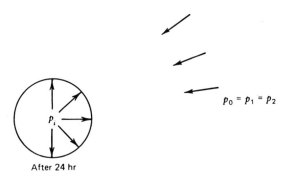

After 24 hr

$p_0 = p_1 = p_2$

For the tunnel in an axisymmetric stress field at $r = a$ in Equation 7.18, we use $A = p_0 a/2$, $B = C = 0$, giving

$$u_r(t) = \frac{p_0 a}{2} \frac{G_1 + G_2}{G_1 G_2} - \frac{p_0 a}{2G_1} e^{-(G_1 t/\eta_1)} + \frac{p_0 a}{2\eta_2} t$$

For the dilatometer, with p_1 internal pressure, Equation 6.33 is equal to

$$u_r = \frac{p_i a}{2} \frac{G_1 + G_2}{G_1 G_2} - \frac{p_i a}{2G_1} e^{-(G_1 t/\eta_1)} + \frac{p_i a}{2\eta_2} t$$

Thus the displacements are the same (but opposite in magnitude, of course). For a depth of 1000 ft, with $\gamma = 150$ P/ft^3, $p_0 = 1042$.

Time	Time Since Dilatometer Applied	$u(p_0)$ in.	$u(p_i)$	u_{total}
1 min		0.21		0.21
5 min		0.22		0.22
15 min		0.22		0.22
30 min		0.25		0.25
1 hr		0.28		0.28
3 hr		0.41		0.41
6 hr		0.59		0.59
12 hr		0.85		0.85
24 hr	0	1.16	0	1.16
36 hr	12 hr	1.31	−0.08	1.23
2 days	24 hr	1.39	−0.11	1.28
3 days	2 days	1.44	−0.13	1.32
4 days	3 days	1.46	−0.14	1.32
5 days	4 days	1.46	−0.14	1.32
6 days	5 days	1.46	−0.14	1.32

9. For RMR = 20, Figure 7.13 indicates that the following relationship exists:

Unsupported Span (m)	Stand-up Time (hr)
0.8	2
1.2	1
1.8	0.5
2.3	0.3

10. For a maximum unsupported span of 4 m, Figure 7.13 indicates the following:

Rock Mass Rating	Stand-up Time
34	5 hr
39	1 day
53	1 month
67	2 yr

11. (a) The center of gravity of block 1 lies above the edge of block 2 if $x_1 = S/2$. The center of gravity of blocks 1 and 2 considered as a unit overlies the edge of block 3 if $x_2 = S/4$. Similarly $x_3 = S/6$, $x_4 = S/8$, etc. Thus, at depth $n \cdot t$ below the crown, the tunnel cannot maintain a width greater than $w = \sum_{i=1}^{n} S/i$. This is the harmonic series, which diverges showing that the walls approach but never become vertical.

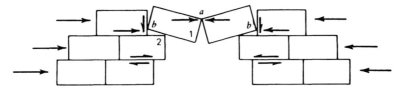

(b) Block 1 tends to fail by rotation about the upper corner of block 2, forming a hinge at its own upper right corner (*a*) (see diagram). This causes point (*b*) to swing up, sliding along the adjacent block. If there were a horizontal force on that block, the resulting friction would exert a stabilizing moment on block 1. Horizontal force on the lower rows of blocks would tend to shear them along the layer boundaries, reducing the width of the span and increasing the stability of the roof. This lateral displacement would reduce the horizontal force and the tunnel wall would then become stabilized with the maximum horizontal stress allowed by interlayer friction.

12.

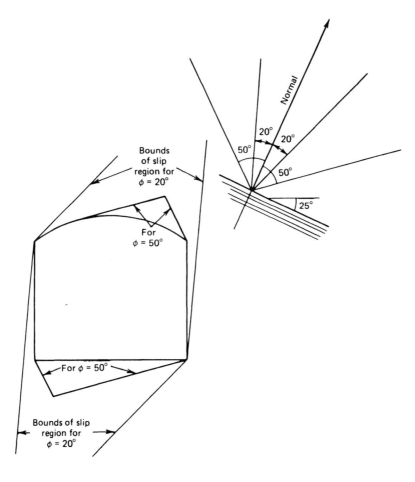

13. (a) With x, y parallel to σ_1, σ_3 directions and x' parallel to the normal to layers, α is the angle from x to x' and $\sigma_{x'}$ and $\tau_{x'y'}$ are given by Equations A1.2 (with $\tau_{xy} = 0$). By definition of ϕ_j: $|\tau_{x'y'}/\sigma_{x'}| \leq \tan \phi_j$. This lead to the limiting condition

$$\sigma_3 = \frac{\cot \phi_j - \cot \alpha}{\tan \alpha + \cot \phi_j} \sigma_1$$

Bray's formula, Prob. 10 of Chap. 5, could also be used.

(b) $\sigma_3 = 0.31$ MPa

(c) The above value of σ_3 must be supplied by the action of the supports. This can be obtained by rock bolts, for example, if the force in each rock bolt divided by the area tributary to one bolt $= \sigma_3$, providing the bolts are closely spaced.

14. For strata dipping $\delta = 45°$ to the left as shown, interlayer slip occurs from $\theta = 0°$ to 15°, from $\theta = 75°$ to 195° and from $\theta = 225°$ to 360°. $\alpha = 180 - \theta + \delta$.

θ	α	p_b (MPa)
0	225°	0.402
15°	210°	0
60°	165°	0
90°	135°	0.938
120°	105°	0.804
180°	45°	0.402

15. (a) The original volume, per unit thickness, is (see figure)

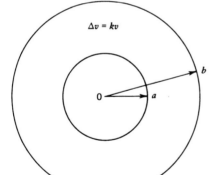

$$v = \pi(b^2 - a^2) \tag{1}$$

which bulks to a final volume

$$v_f = (1 + k_B)v \tag{2}$$

The final volume can also be expressed as

$$v_f = \pi(b^2 - (a - U_a)^2) \tag{3}$$

where U_a is the inward displacement of radius a. Assume U_b is zero; then inserting (1) and (2) and equating to (3) yields

$$U_a = a - \sqrt{a^2 - k_B(b^2 - a^2)} \tag{4}$$

(This result was first published by Labasse, in *Revue Universelle des Mines*, March 1949.)

(b) $$k_B = \frac{U_a(2a - U_a)}{b^2 - a^2} \tag{5}$$

Since U_a is always less than a, $2a - U_a > 0$ and k_B is always positive (if $U_b = 0$).

(c) $k_B = \dfrac{U_a(2a - U_a) - U_b(2b - U_b)}{b^2 - a^2}$ \qquad\qquad (6)

since $2a - U_a \approx 2a$ and $2b - U_b \approx 2b$

$$k_B \approx \frac{2(aU_a - bU_b)}{b^2 - a^2}$$ \qquad\qquad (7)

16.

Limits of Ring (m)	$\bar{r}$	$\bar{r}/a$	t (days)	k_B	Sense of Change
2.12–4.5	3.3	1.56	20	−0.0078	Compacting
			100	−0.0168	Compacting
			800	−0.0347	Compacting
4.5–7	5.75	2.71	20	0.0007	Expanding
			100	−0.0025	Compacting
			800	−0.0128	Compacting
7–9.4	8.20	3.87	20	0.0021	Expanding
			100	−0.0003	Compacting
			800	−0.0054	Compacting

These results are plotted in the answer to Problem 17.

17. (a) The results of Problem 16 are plotted as k_B versus r yielding the following:

t (days)	r_c (m)	$R = 2.7r_c$ (m)	R/a
20	5.5	14.8	7.0
100	8.6	23.2	11.0
800	10.6	28.6	13.5

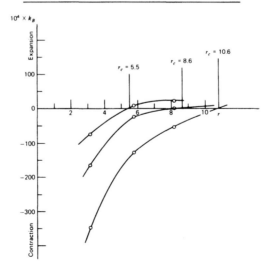

(b) $U_{elas} = \dfrac{p(1 + \nu)}{E} r$

$= \dfrac{0.4(1.2)}{5000} r$

$U_{elas} = 9.6 \times 10^{-4} r$

This relationship has been plotted with the extensometer data in $U - \log r$ coordinates. The extensometer data intersect the elastic displacement data determining R as follows:

T (days)	R (m)	R/a
20	11.7	5.5
100	17.2	8.1
800	25.7	12.1

18. (a) At limiting equilibrium, $\tau = \sigma \tan \phi$ on the vertical joints while the horizontal joint opens, becoming free of stress. Summing forces in the vertical direction gives

$$W = \gamma bh \quad \text{(per unit of thickness)}$$

$$B - W + 2h\sigma \tan \phi = 0$$

$$\frac{B}{W} = 1 - \frac{2\sigma \tan \phi_j}{\gamma b}$$

(b) If $B = 0$

$$b_{max} = \frac{2\sigma \tan \phi_j}{\gamma}$$

19. Associated with shear displacement u, each joint tends to dilate by $\Delta v = u \tan i$. If the wall rock were rigid, the normal strain increment of the block would be $\Delta \varepsilon = 2\Delta v/b$ and the normal stress increment would be $\Delta \sigma = E\Delta \varepsilon$. Thus, in view of the answer to 18(a),

$$\frac{B}{W} = 1 - \frac{4E \tan i \tan \phi_j}{\gamma_b} \frac{u}{b}$$

20. The displacement path across Figure 5.17b would be inclined α upward from horizontal. For initial normal stress equal to a, it would follow the dashed path shown (neglecting initial shear stress on the joints) (see figure). The normal stress would drop slightly; then it would increase almost to b, and then start to drop.

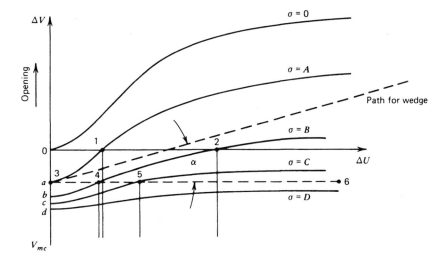

21. Before gravity is "switched on" to the wedge, the stresses tend to flow tangentially around the circular opening. Close to the periphery of the opening, the direction of the tangent makes a larger angle (α_1) with the normal to the joints than it does nearer the vertex of the wedge where the angle is α_1 as shown in the figure. Thus the upper portion of the block is effectively reinforced by shears along its slides. If the angle α_1 exceeds the friction angle for the joint, a portion of the block's weight will be transferred farther up along the joint, setting up vertical tensile stresses in the wedge. The block may therefore break into two parts, allowing the lower part to fall while the upper part is restrained.

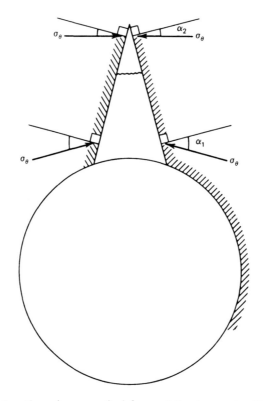

22. (a) By symmetry, there is no vertical force at the top center (0). Summing forces vertically gives

$$V = \gamma s b \tag{1}$$

Taking moments about 0, when Δy is very small

$$Hb + \gamma s b \frac{s}{2} = Vs \tag{2}$$

or with (1)

$$H = \tfrac{1}{2}\gamma s^2$$

and

$$H/V = \frac{s}{2b} \tag{3}$$

The lines of action of the resultant of H, v, W and the horizontal force through 0 intersect at a single point; this establishes the location of the reaction.

(b) The block becomes unstable if point 0 falls lower than H, which can happen for blocks in which $b \ll s$. The critical case, is shown in the figure. The block has rotated by an amount

$$\theta = \tan^{-1}(b/s) \tag{4}$$

$$\Delta x = s \cos \theta + b \sin \theta - s = s(\cos \theta - 1) + b \sin \theta \qquad (5)$$

Using (4), $\Delta x = \sqrt{s^2 + b^2} - s$ (This result can be seen directly in the figure).

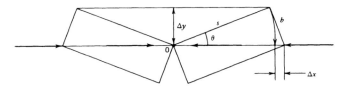

23. (a) See figure. $S = l/2 \rightarrow \dfrac{l}{S} = 2$

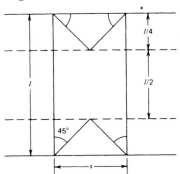

(b) See figure.

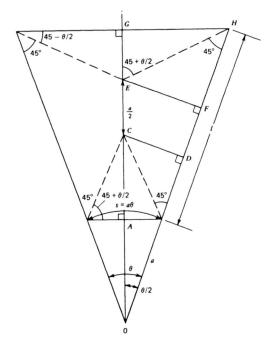

$$l = BD + DF + FH \tag{1}$$

$$BD = CB \cos 45 \; = \; \frac{AB \cos 45}{\cos(45 + \theta/2)} = \frac{a \sin \theta/2 \cos 45}{\cos(45 + \theta/2)} \tag{2}$$

$$DF = CE \cos \theta/2 \qquad\qquad = \frac{a}{2} \cos \theta/2 \tag{3}$$

$$FH = EH \cos 45 \; = \; \frac{GH \cos 45}{\sin(45 + \theta/2)} = \frac{(a + l)\sin \theta/2 \cos 45}{\sin(45 + \theta/2)} \tag{4}$$

$$\sin(45 + \theta/2) = \sin 45 \cos \theta/2 + \cos 45 \sin \theta/2 = \frac{\sin \theta/2 + \cos \theta/2}{\sqrt{2}} \tag{5}$$

$$\cos(45 + \theta/2) = \cos 45 \cos \theta/2 - \sin 45 \sin \theta/2 = \frac{\cos \theta/2 - \sin \theta/2}{\sqrt{2}} \tag{6}$$

Substituting (2), (3), and (4) with (5) and (6) in (1) and solving for l yields

$$l \left(\frac{\cos \theta/2}{\sin \theta/2 + \cos \theta/2} \right) = \frac{2a \sin \theta/2 \cos \theta/2}{\cos^2\theta/2 - \sin^2\theta/2} + \frac{a}{2} \cos \theta/2 \tag{7}$$

Finally,

$$l = a(1 + \tan \theta/2)(\tfrac{1}{2} \cos \theta/2 + \tan \theta) \tag{8}$$

Letting bolt spacing be determined by arc length, then

$$s = a\theta \tag{9}$$

and

$$\boxed{\frac{l}{s} = \frac{1}{\theta} (1 + \tan \theta/2)(\tfrac{1}{2} \cos \theta/2 + \tan \theta)} \tag{10}$$

For $\theta = 40°$, $= 0.698$ radians, as drawn, $\dfrac{l}{s} = 2.56a$.

24. The angle bolts defend against diagonal tensile and shear failure above the haunches. Figure 7.6c shows an opening diagonal tension crack above the left haunch. Its growth releases a complete block in the roof, which is falling in Figure 7.6d.

25. See figure for parts (a) and (b):
    (a) 100
    (b) 011

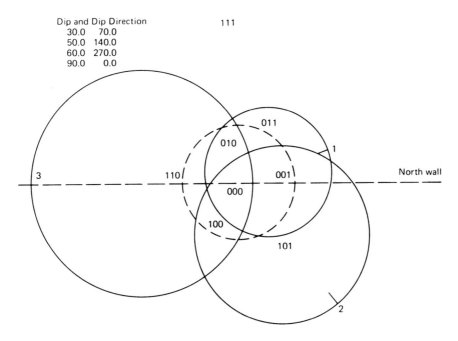

Dip and Dip Direction
30.0	70.0
50.0	140.0
60.0	270.0
90.0	0.0

See figure for part (c)

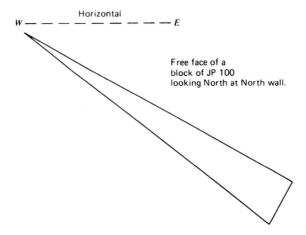

Free face of a
block of JP 100
looking North at North wall.

26. See figure

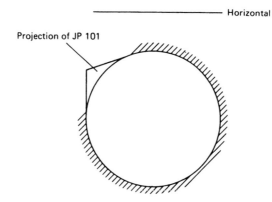

Horizontal

Projection of JP 101

CHAPTER 8

1.

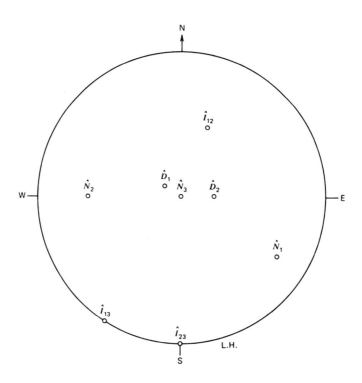

2.

Azimuth of Radius to Point	Strike of Cutting at That Point	Cut Slope Dip Direction	Steepest Safe Slope	Governing Failure Mode
0	E	S	90°	None
15	S 75° E	S 15° W	90°	None
30	S 60° E	S 30° W	90°	None
45	S 45° E	S 45° W	87°	Sliding on 1
60	S 30° E	S 60° W	53°	Toppling on 2
75	S 15° E	S 75° W	51°	Toppling on 2
90	S	W	50°	Toppling on 2
105	S 15° W	N 75° W	51°	Toppling on 2
120	S 30° W	N 60° W	53°	Toppling on 2
135	S 45° W	N 45° W	60°	Wedge (I_{12})
150	S 60° W	N 30° W	50°	Wedge (I_{12})
165	S 75° W	N 15° W	42°	Wedge (I_{12})
180	W	N	38°	Wedge (I_{12})
195	N 75° W	N 15° E	37°	Wedge (I_{12})
210	N 60° W	N 30° E	37°	Wedge (I_{12})
225	N 45° W	N 45° E	39°	Wedge (I_{12})
240	N 30° W	N 60° E	44°	Wedge (I_{12})
255	N 15° W	N 75° E	51°	Wedge (I_{12})
270	N	E	43°	Toppling on 1
285	N 15° E	S 75° E	41°	Toppling on 1
300	N 30° E	S 60° E	40°	Toppling on 1
315	N 45° E	S 45° E	40°	Toppling on 1
330	N 60° E	S 30° E	42°	Toppling on 1
345	N 75° E	S 15° E	83°	Sliding on 2

The best orientation for a highway cut through a ridge in this rock mass would be the one that produces the steepest safe slope on both sides of the highway. A cut striking east can have a slope of 90° on one side but only 38° on the other, thus it is not optimum.

Strike of Cut	Maximum Slopes from Kinematic Analysis	
E	90°	38°
S 75° E	90°	37°
S 60° E	90°	37°
S 45° E	87°	39°
S 30° E	53°	44°

S 15° E	51°	51°
S	50°	43°
S 15° W	51°	41°
S 30° W	53°	40°
S 45° W	60°	40°
S 60° W	50°	42°
S 75° W	42°	90°

The optimum is the one that minimizes the excavation and can be determined graphically if the topographic profile is drawn.

3. (a) (See diagram.) The minimum bolt force for a factor of safety of 1.0 is the minimum force that when added to 400 tons vertically will incline the resultant 20° from the vertical. The magnitude of this force is 137 tons and it is applied in a direction rising 20° above the horizontal to the S 60° W. For a factor of safety of 1.5, tan ϕ_{req} = tan $\phi_j/1.5$ giving ϕ_{req} = 21°. Therefore the minimum bolt force rises 29° above horizontal to S 60° W with magnitude 194 tons.

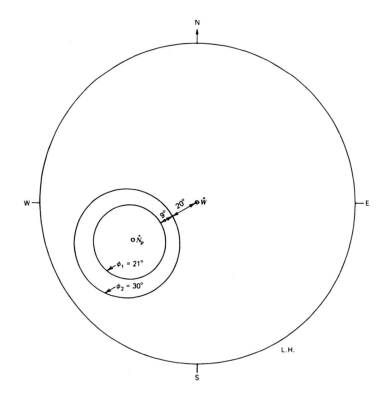

(b) (See diagram.) From point C, representing the tip of the resultant with $F = 1.5$, a force $CD = 112$ tons directed opposite to $\hat{N}$ will incline the new resultant 30° from $\hat{N}$, and is therefore the force to initiate slip. The pressure is 112 tons/200 m² = 0.56 tons/m².

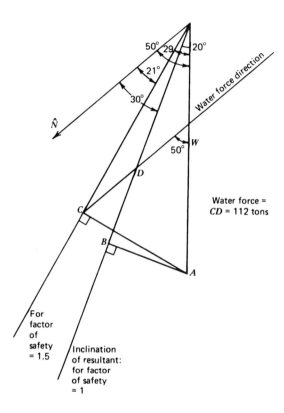

Water force =
$CD = 112$ tons

(c) The minimum force direction is not the direction for shortest bolts. The latter is parallel to $\hat{N}$. The optimum direction depends on the relative costs of steel and drill holes and lies somewhere between these two extremes.

4. (See diagram.) The friction circle for $\phi_j = 33°$ lies partly in the upper hemisphere. The diameter of the circle is from C to A'. A' is found by first marking A'' at the outer circle on line CN. A'' is 30° from N. Lay off the additional 3° along the outer circle to locate point A. A' is at the intersection of lines $0A$ and CA'' where 0 is at north. (The rationale for this construction is discussed in Goodman (1976) *Methods of Geological Engineering*.)

(a) The minimum rock bolt force $\mathbf{B}_{\min}$ must incline the resultant 42° from $\hat{W}$. It is 200 sin 42° = 134 MN, 42° above horizontal to the east.

(b) With the bolts in direction $\hat{b}$, as shown in the diagram, the required rotation from $\hat{W}$ is 46° (to point D). The angle between $\hat{W}$ and $\hat{b}$ is 80° giving $B = 255$ MPa.

(c) The inertia force $\mathbf{F}_I = (Kg)m = (Kg)W/g = KW$. The angle between the direction of F_I and point D is $80°$ and the required rotation for the orientation of the resultant $\mathbf{W} + \mathbf{B}$ is $20°$ to point E. The force triangle on the figure determines $F_I = 135$ MN. $K = F_I/W = 0.68$. Thus the block slips when the acceleration reaches $0.68g$.

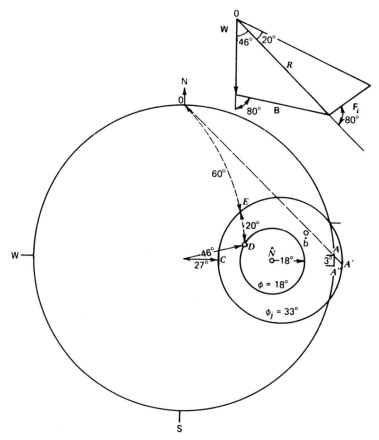

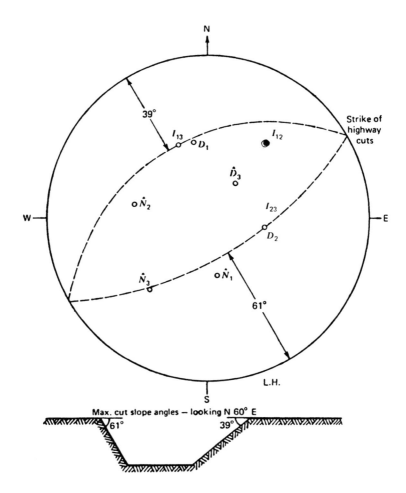

6. (a) Slides when $\delta = \phi_j$.

(b) Overturns when $\delta = \tan^{-1}(t/h)$.

The governing condition is the one that yields the smaller value of δ. This then depends on ϕ_j, and the dimensions of the block.

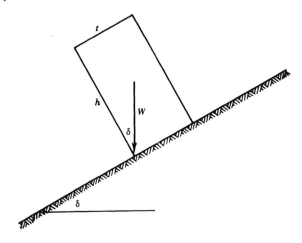

7. (a) If $\delta > \phi_j$, both blocks slide.

(b) If $t_1/h_1 > \tan \delta$ and $t_2/h_2 > \tan \delta$ and $\delta < \phi_j$, the system is stable.

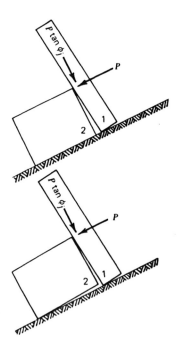

(c) If $t_1/h_1 > \tan \delta$ and $t_2/h_2 < \tan \delta$ and $\delta < \phi_j$ only the lower block rotates. In the drawing, however, $t_2/h_2 > t_1/h_1$ so this could not happen.

(d) If $\delta < \phi_j$, $t_2/h_2 > \tan \delta$ and $t_1/h_1 < \tan \delta$, the upper block (1) tends to rotate. The lower block (2) may tend to slide or to rotate and both conditions must be checked. In either case, the force P transferred to the lower block is

$$P = \frac{W_1(h_1 \sin \delta - t_1 \cos \delta)}{2h_2}$$

If the lower block slides, the limiting condition is

$$\frac{W_1(h_1 \sin \delta - t_1 \cos \delta)}{2h_2} = \frac{W_2(\cos \delta \tan \phi_j - \sin \delta)}{1 - \tan^2 \phi_j}$$

If the lower block rotates, the limiting condition is

$$\frac{W_1(h_1 \sin \delta - t_1 \cos \delta)}{2h_2} = \frac{W_2(t_2 \cos \delta - h_2 \sin \delta)}{2(h_2 - t_2 \tan \phi_j)}$$

8. According to the theory discussed in most books on strength of materials, buckling occurs when σ_1, the stress parallel to the axis of a column, reaches Euler's critical stress for buckling σ_E:

$$\sigma_E = \frac{\pi^2 E t^2}{3L^2}$$

(compare with discussion above Equation 7.5). The free-body diagram defines the force polygon, which yields

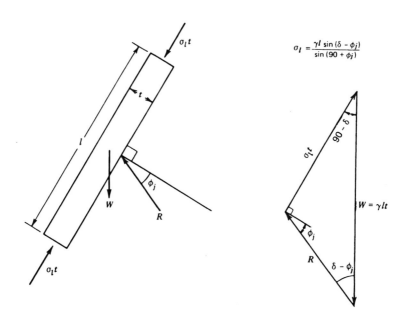

$$\sigma_1 = \frac{\gamma l \sin (\delta - \phi_j)}{\sin (90 + \phi_j)}$$

$$W = \gamma l t$$

The condition for initiation of failure is met if $\sigma_1 = \sigma_E$, giving

(a) $l_{max} = \dfrac{\pi^2 E t^2}{3L^2} \dfrac{\sin(90 + \phi_j)}{\gamma \sin(\delta - \phi_j)}$

(b) $l_{max} = 59.9$ m

9. The following values are found by trial, using Equation 8.12 (with $\theta = 0$):

ϕ_j	F_b (MN)
30	77.94
35	16.54
38	-19.31
36	4.53
37	-7.42
36.4	$-0.26 \rightarrow$
36.3	0.94

Since $F_b = 0$, $\phi_{available} = 36.4°$.

(b) Repeating a similar procedure with W_1 reduced to 6000 m³ yields $\phi_{required} = 33.3°$. Thus the factor of safety is now tan 36.4/tan 33.3 = 1.12.

ϕ_j	F_b
30	33.53
32	13.28
34	-6.77
33.5	-1.77
33.3	0.23

(c) With $W_1 = 10,000$ as originally given, and $\phi_j = 33.3$, the required anchorage force is 37.1 MN (8.3 × 10⁶ lb ≈ 4200 tons). This means approximately 42 anchors in the passive region for each meter of slide width.

10. If ϕ_1 is fixed at a value corresponding to a defined factor of safety on plane 1, Equation 8.12 can be solved for σ_2 required at the limit of equilibrium and therefore the factor of safety on plane 2. Thus there is an infinite combination of values ϕ_1, ϕ_2 corresponding to limiting equilibrium. An appreciation of the sensitivity of the system to changes in either ϕ_1 or ϕ_2 is gained by plotting the limiting values of tan ϕ_1 versus tan ϕ_2.

11. $\cot \alpha = \dfrac{d \cot \delta(F \sin \delta - \cos \delta \tan \phi)}{d(F \sin \delta - \cos \delta \tan \phi) - U \tan \phi - V(\sin \delta \tan \phi + F \cos \delta) + S_j A}$

where $d = \frac{1}{2}\gamma H^2(1 - Z/H)^2 \cot \delta$.

12. (a) Summing forces down the dip of $P3$, at limiting equilibrium gives

$$W \sin \delta - B \cos \delta - \tau_1 A_1 - \tau_2 A_2 - \tau_3 A_3 = 0 \qquad (1)$$

$$W = \gamma \tfrac{1}{2} lh \frac{h}{\tan \delta} \tag{2}$$

$$A_1 = A_2 = \tfrac{1}{2} \frac{h^2}{\tan \delta} = \frac{W}{\gamma l} \tag{3}$$

At the limit of equilibrium

$$\tau_1 A = \tau_2 A_2 = \sigma_j \tan \phi_j \frac{W}{\gamma l} \tag{4}$$

and

$$\tau_3 A_3 = W \cos \delta \tan \phi_3 \tag{5}$$

Inserting all the above in (1) gives

$$B \cos \delta = W \sin \delta - 2\sigma_j \frac{W}{\gamma l} \tan \phi_j - W \cos \delta \tan \phi_3 \tag{6}$$

Dividing by $W \cos \delta$ yields

$$\boxed{\frac{B}{W} = \tan \delta - \tan \phi_3 - 2 \frac{\sigma_j \tan \phi_j}{\gamma l \cos \delta}} \tag{7}$$

(b) $B/W = 0 = \tan 60 - \tan 30 - \dfrac{2 \, \sigma_j \tan 30}{25 \, l \cos 60}$

$\sigma_j/l = \dfrac{\tan 60 - \tan 30)(25 \cos 60)}{2 \tan 30}$

$\sigma_j = 12.5 l \text{ kN/m}$

l	σ_j Required	
(m)	(kPa)	(psi)
1.	12.5	1.81
5.	62.5	9.06
10.	125.	18.1
20.	250.	36.3

Note that only a small side stress can stabilize a large block.

13. (a) If the abutment rocks are rigid, all dilatant displacement can be expressed as a tendency toward normal strains, $\Delta \varepsilon = \Delta l / l$ in the block. Then

$$\Delta l = 2u \tan i$$

and

$$\Delta \sigma_j = E \, \Delta l / l$$

Combining these equations gives

$$\Delta \sigma_j = 2E \tan i \, u / l$$

$$\frac{B}{W} = \tan \delta - \tan \phi_3 - \left(\frac{4E \tan i \tan \varphi_j}{\gamma l^2 \cos \delta} \right) u$$

(b) In Problem 12(b), $\sigma_j = 12.5l$ with σ expressed in kPa and l in meters. Here

$$\Delta\sigma_j = (2 \times 10^7 \text{ (kPa)})(2 \tan 10\ u/l)$$

or

$$7.1 \times 10^6 u/l = 12.5l$$
$$u = 1.77 \times 10^{-6} l^2 \quad (u \text{ and } l \text{ in meters})$$

For equilibrium

l (m)	u (mm)
1	1.77×10^{-3}
2	7.1×10^{-2}
5	0.044
10	0.177
20	0.71

14. (a) $W = 100$ tons. Let $B =$ the bolt force, determined from the triangle of forces (see figure) as follows:

$$100 \text{ tons}/\sin 75° = B/\sin 5°$$

giving

$$B = 9.02 \text{ tons}$$

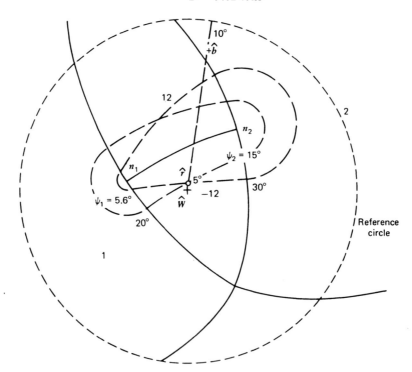

(b) 5.6°, or approximately 6°, as shown on the stereographic projection in the figure. This is an upper focal point ("lower hemisphere") projection.

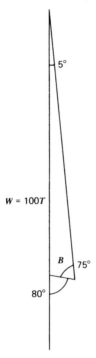

15. See the figure, constructed with a lower focal point ("upper hemisphere") projection.

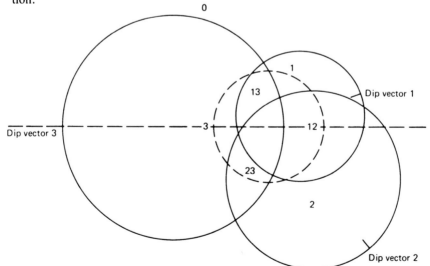

16. See the figure (in two parts):

$$W = 50 \text{ tons} = 50{,}000 \text{ kg} = 0.49 \text{ MN}$$

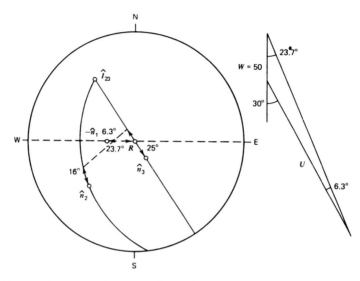

The required water force is U determined from the triangle of forces as follows:

$$50/\sin 6.3° = U/\sin 23.7°$$

giving

$$U = 183.2 \text{ tons} = 1.795 \text{ MN}$$

The water pressure (averaged over the face of plane 1) is

$$P_W = U/7.5 \text{ m}^2 = 0.239 \text{ MPa} = 34.7 \text{ psi}$$

17. (a) See the figure for the JP analysis. The only JP lacking any area inside the dashed circle is 011. This is therefore the only JP defining removable blocks.

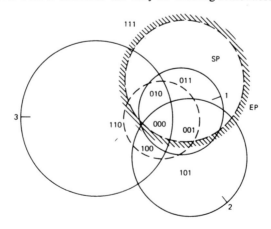

(b) See the figure for the stability analysis. For a factor of safety of 2.0 on each plane

$$\tan \phi_{\text{req'd}} = \tan 35°/2.0$$

giving

$$\phi_{\text{req'd}} = 19.3°$$

From the stereographic projection, the required rotation of **R** from **W** is 13.3°, giving $B = 90 \tan 13.3° = 21.3$ tons.

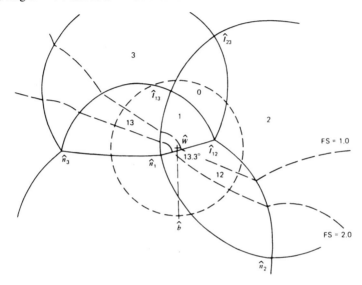

CHAPTER 9

1. For the left circle in the diagram,

$$p_h = 2S_p \tan \left(45 + \frac{\phi_p}{2} \right) = q_u$$

Then Equation 3.8 gives

$$q_f = q_u \tan^2 \left(45 + \frac{\phi_r}{2} \right) + 2S_r \tan \left(45 + \frac{\phi_r}{2} \right)$$

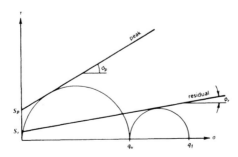

2. In addition to the forces considered in regard to Equation 8.2, we now have an additional vertical force $P \sin \beta$ and an additional horizontal force $P \cos \beta$. Thus the result for a force P bearing on the slide block follows from the following substitutions in (8.2): in place of W, input $W + P \sin \beta$; in place of V, input $V + P \cos \beta$.

3. The block slides if the resultant of P and W is inclined ϕ_j with the normal to the plane.

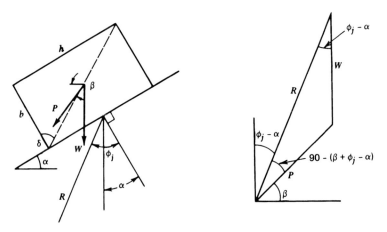

For sliding, from the law of sines applied to the force triangle we have

$$\frac{W}{\sin[90 - (\beta + \phi_j - \alpha)]} = \frac{P}{\sin(\phi_j - \alpha)} \tag{1}$$

or for limiting equilibrium in sliding

$$P_{\text{slide}} = \frac{\sin(\phi_j - \alpha)}{\cos(\beta + \phi_j - \alpha)} W \tag{2}$$

For case a: The block overturns when the resultant of P and W is oriented $\delta = \tan^{-1} b/h$ with the normal to the block. Therefore, for toppling, replace ϕ_j in (2) by δ giving

$$P_{\text{topple}} = \frac{\sin(\delta - \alpha)}{\cos(\beta + \delta - \alpha)} W \tag{3}$$

For case b: The block topples when P = one-half of the value given by (3).

4. Following closely the derivation in the appendix, with an added force P gives in place of Equation 9

$$N_3 = \frac{W_1 \sin(\delta_1 - \phi_1)\cos \phi_3}{\cos(\delta_1 - \phi_1 - \phi_3)} + \frac{P \cos(\beta - \delta_1 + \phi_1)\cos \phi_3}{\cos(\delta_1 - \phi_1 - \phi_3)} \tag{9a}$$

Equating this to (15) and solving for F_b gives as a final result

$$F_b = \frac{[W_1 \sin(\delta_1 - \phi_1) + P \cos(\beta - \delta_1 + \phi_1)]\cos(\delta_2 - \phi_2 - \phi_3) + W_2 \sin(\delta_2 - \phi_2)\cos(\delta_1 - \phi_1 - \phi_3)}{\cos(\delta_2 - \phi_2 + \theta)\cos(\delta_1 - \phi_1 - \phi_3)}$$

5. For $\nu_r = \nu_c = 0.26$ and embedment depths of a, $2a$, $3a$, and $4a$, the results of Osterberg and Gill for $E_c/E_r = \frac{1}{4}$ are fitted closely by Equation 9.10 with $\mu = \tan 59°$, as shown in the following table ($p_{end} = \sigma_y$):

	p_{end}/p_{total}	
y/a	Osterberg and Gill	Equation 9.10
1	0.44	0.44
2	0.16	0.19
3	0.08	0.09
4	0.03	0.04

However, 40 mm of settlement will most probably rupture the bond, reducing μ to a value less than unity. Initially, the Osterberg and Gill results should apply. Subsequently, the load transfer will shift to the distribution given by (9.10) with $\mu < 1$.

6. There is no rational procedure, short of full-scale load tests, that can determine the allowable bearing pressure. However, it can be estimated in several ways. First, since a 2-m diameter is about seven times the joint spacing, the q_u given by the small-lab tests should be reduced by a scale factor of, say, 5. Moreover, a factor of safety is warranted. Assume $\phi_P = 20°$, and $(q_u)_{field} = 18$ MPa/5 $= 3.6$ MPa. With a factor of safety F equal to 3, the allowable bearing pressure for a surface footing is

$$q_{allow,footing} = \left(\frac{(\tan^2 55) + 1}{3}\right) 3.6 = 3.65 \text{ MPa}$$

For a pier at depth, this value may be multiplied by 2, giving

$$q_{allow,pier} \approx 7.3 \text{ MPa}$$

The bond strength is not governed by the same size-scaling factor as the end bearing since the shear is confined to the contact. Assume $\tau_{bond} = 0.05 q_u$ with $q_u =$ one-half the lab value. Then $\tau_{bond} = 0.45$ MPa. With a factor of safety of 2, this gives

$$\tau_{allow} = 0.22 \text{ MPa}$$

7. Using radius $a = 1$ m, $l_{max} = 31.8$ m. With the Osterberg and Gill results, any length approaching this would transfer all load to the sides and we will require $l = l_{max}$. Such a long pier (a "pile" in fact) is not the most economical solution. For $a = 2$ m, $l_{max} = 15.9$ m; then with $l_1 = 6$ as a trial, $p_{end}/p_{total} = 0.07$ giving $p_{end} = \tau = 0.27$ MPa, which is much too small for the former and much too large for the latter.

As an alternative, if the bond is broken or prevented, the load transfer will occur at a lesser rate. Using Equation 9.10a with $\mu = \tan 40°$, and ignoring shear stress in the concrete since it is confined, the required lengths to satisfy the bearing capacity of 2 MPa are shown in the following table. The compressive stress in the pier above the socket is also given.

a (m)	l (m)	Socket Volume (m³)	p_{end}/p_{total}	p_{end} (MPa)	$\bar{\tau}$ (MPa)	$\sigma_{max,concrete}$ (MPa)
1.0	8.97	28.2	0.31	2.00	0.24	6.37
0.9	9.54	24.3	0.255	2.00	0.28	7.86
0.8	9.94	20.0	0.201	2.00	0.32	9.95
0.7	10.15	15.6	0.154	2.00	0.38	12.99
0.6	10.13	11.5	0.113	2.00	0.46	17.68
0.5	9.85	7.74	0.079	2.00	0.60	25.46
0.4	9.27	4.66	0.050	2.00	0.82	39.79
0.3	8.63	2.95	0.024	2.00	1.20	70.74

If the concrete has compressive strength equal to 20 MPa, and it is desired to achieve a factor of safety of 2, the solution with minimum volume socket is a pier with radius 0.8 m and length 10 m. These results depend markedly on the choice for μ and E_c/E_r, and, to a lesser extent, on v_r and v_c.

Another solution is to use a pier seated without a socket on the surface of the rock, or, if that surface is weathered or inclined, seated inside a socket of enlarged diameter. The required pier radius is 1.78 m. The most economical choice between the alternatives depends on the volume of the pier passing through the soil.

8. Consider the sandstone roof as a continuous clamped beam. The most critical condition is tensile stress at the ends on the upper surface. Using (7.5) with $\sigma_h = 0$, $\gamma = 150$ lb/ft³, and $T_0 = 2$MPa gives $L = 334$ ft ≈ 100 m. This is increased if $\sigma_h \neq 0$. However, a beam 200 ft thick with $L = 334$ ft is too thick for thin-beam theory. (A finite element analysis would be useful in a particular case.)

9. $H = \dfrac{2h}{B - 1}$

10. (a) Summing forces in the y (vertical) direction acting on the differential element gives

$$s^2 \, d\sigma_v + 4\tau \, s \, dy = 0 \tag{1}$$

Substituting in (1)

$$\tau = \sigma_h \tan \phi \tag{2}$$

and

$$\sigma_h = k \, \sigma_v \tag{3}$$

yields

$$\frac{d\sigma_v}{\sigma_v} = -\frac{4}{s} k \tan \phi \, dy \tag{4}$$

Solving gives

$$\sigma_v = A \, e^{-4k \tan \phi \, y/s} \tag{5}$$

when $y = 0$, $\sigma_v = q$, giving $A = q$.

The support pressure is the value of σ_v when $y = t$, which gives

$$\boxed{p_b = q\, e^{-4k \tan \phi\ t/s}} \tag{6}$$

(b) $k = \cot^2 (45 + \phi/2) = 0.406$

$t/s = 0.67$, $q = 21$ kPa

Then (6) gives

$$p_b = 21\, e^{-0.507} = 12.65 \text{ kPa } (=1.83 \text{ psi})$$

If $s = 1.5$ m, the force per support is $T = s^2 (12.65) = 28.5$ kN (≈ 6400 lb).

11. (a) With self-weight, the free-body equilibrium gives

$$s^2\, d\sigma_v + 4\tau s\, dy = \gamma s^2\, dy \tag{1}$$

Substituting as in the answer to Problem 10(a) gives

$$d\sigma_v = \left(\gamma - \frac{4k \tan \phi}{s} \sigma_v \right) dy \tag{2}$$

Let

$$z = \gamma - \frac{4k \tan \phi}{s} \sigma_v \tag{3}$$

then

$$dz = -\frac{4k \tan \phi}{s}\, d\sigma_v \tag{4}$$

and (2) becomes

$$\frac{-s}{4k \tan \phi} \frac{dz}{z}\, dy \tag{5}$$

whose solution is

$$z = Ae^{-4k \tan \phi\ y/s} \tag{6}$$

Finally, resubstituting (3) in (6) gives

$$\gamma - \frac{4k \tan \phi}{s} \sigma_v = Ae^{-4k \tan \phi\ y/s} \tag{7}$$

Since $\sigma_v(y = 0) = q$

$$A = \gamma - \frac{4k \tan \phi}{s} q \tag{8}$$

Finally, $P_b = \sigma_v(y = t)$, giving

$$P_b = \frac{s}{4k \tan \phi} \left(\gamma - \left(\gamma - \frac{4k \tan \phi}{s} q \right) e^{-4k \tan \phi\ y/s} \right) \tag{9}$$

Simplifying yields

$$\boxed{P_b = \frac{s \gamma}{4k \tan \phi} (1 - e^{-4k \tan \phi\ t/s}) + q\, e^{-4k \tan \phi\ t/s}}$$

(b) $P_b = \dfrac{(1.5)(27)}{(4)(0.406)(0.47)} (1 - e^{-0.507}) + 21 (e^{-0.507})$

   $= 33.75$ kPa $(=4.89$ psi$)$

(c) If rock bolts were being installed, then an additional force would need to be added to the equilibrium equation to account for the action of the anchor end of the bolt. (This is discussed by Lang, Bischoff, and Wagner (1979).)

APPENDIX 1

1. (a) 1. $\sigma_{x'} = 27.7$ $\tau_{x'y'} = -18.7$
      2. $\sigma_{x'} = 20.0$ $\tau_{x'y'} = -10.0$
      3. $\sigma_{x'} = 30.0$ $\tau_{x'y'} = 20.0$
      4. $\sigma_{x'} = 50.0$ $\tau_{x'y'} = -20.0$
   (b) 1. $\sigma_{x'} = 52.7$ $\tau_{x'y'} = -7.3$
      2. $\sigma_{x'} = 52.7$ $\tau_{x'y'} = 7.3$
      3. $\sigma_{x'} = 72.7$ $\tau_{x'y'} = 27.3$
      4. $\sigma_{x'} = 108.3$ $\tau_{x'y'} = 0.0$

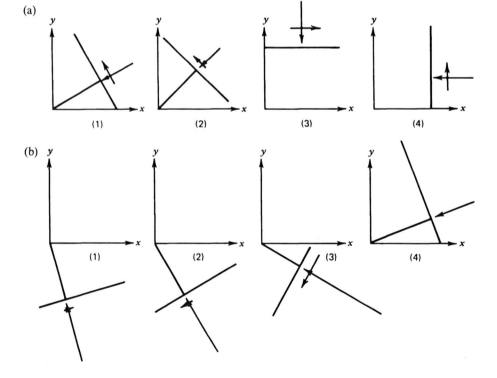

2. (a)

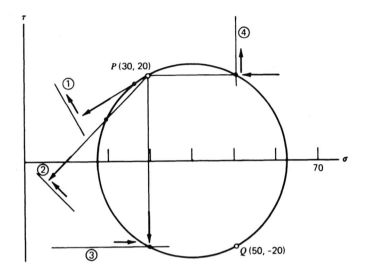

(b)

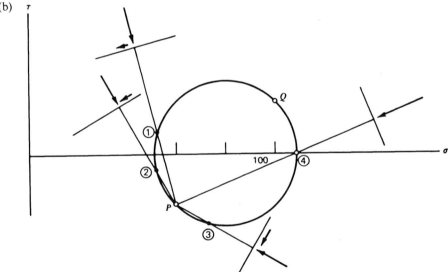

3. For Problem 1a:

$$\alpha = -31.7, 58.3$$
$$\sigma_1 = 62.4, \sigma_2 = 17.6$$

For Problem 1b:

$$\alpha = 22.5, 112.5°$$
$$\sigma_1 = 108.3$$
$$\sigma_2 = 51.7$$

4. $\sigma_{y'} = (\sin^2\alpha \quad \cos^2\alpha \quad -\sin 2\alpha) \begin{Bmatrix} \sigma_x \\ \sigma_y \\ \tau_{xy} \end{Bmatrix}$

5. $\sigma_{x'} + \sigma_{y'} = (\sigma_x \cos^2\alpha + \sigma_y \sin^2\alpha)$
$$+ (\sigma_x \sin^2\alpha + \sigma_y \cos^2\alpha)$$
$$= \sigma_x + \sigma_y$$

6. (a) $\alpha = 67.50$
$\sigma_1 = 108.28$
$\sigma_2 = 51.72$

(b) $\alpha = -67.50$
$\sigma_1 = 108.28$
$\sigma_2 = 51.72$

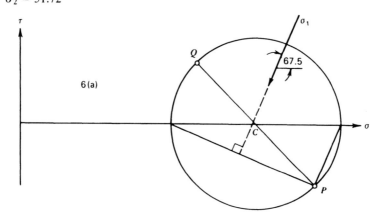

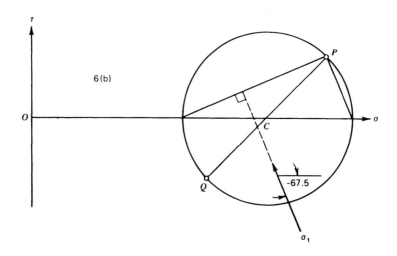

7.

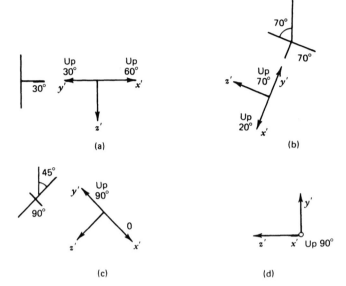

	Line	Bearing	β	δ	l	m	n
(a)	x'	East	0	60	0.50	0.00	0.866
	y'	West	180	30	−0.866	0.00	0.50
	z'	South	−90	0	0.00	−1.00	0.00
(b)	x'	S 20° W	−110	20	−0.321	−0.883	0.342
	y'	N 20° E	70	70	0.117	0.321	0.940
	z'	N 70° W	160	0	−0.940	0.342	0.00
(c)	x'	S 45° E	−45	0	0.707	−0.707	0.00
	y'	N 45° W	135	90	0.00	0.00	1.00
	z'	S 45° W	−135	0	−0.707	−0.707	0.00
(d)	x'	—	—	90	0.00	0.00	1.00
	y'	North	90	0	0.00	1.00	0.00
	z'	West	180	0	−1.00	0.00	0.00

8.

	$\sigma_{x'}$	$\tau_{x'y'}$	$\tau_{x'z'}$	$\lvert\tau_{x'\mathrm{max}}\rvert$
(a)	593.30	234.81	−25.00	236.13
(b)	265.55	141.03	−10.28	141.40
(c)	100.00	35.36	50.00	61.24
(d)	700.00	0	−50.00	50.00

9.

	$P_{x'x}$	$P_{x'y}$	$P_{x'z}$	$|\tau_{x'\text{max}}|$
(a)	93.30	25.00	631.22	236.13
(b)	−59.19	−192.67	223.34	141.40
(c)	35.56	−106.06	35.36	61.24
(d)	50.00	0	700.00	50.00

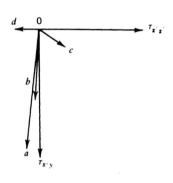

10.

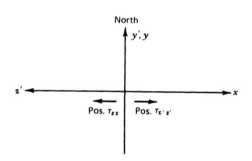

$\tau_{zx} = 50$ (given); acts to left

$\tau_{x'z'} = -50$ (calculated); acts to left

11. $I_1 = 1000 \ F/L^2$

$I_2 = 225 \times 10^3 \ (F/L^2)^2$

$I_3 = 11.75 \times 10^6 \ (F/L^2)^3$

12.

	σ_1	σ_2	σ_3	$\sigma_{x'}$	$\tau_{x'y'}$	$\tau_{x'z'}$
(a)	150	0	0	50.0	70.7	0
(b)	100	50	0	50.0	35.4	20.4
(c)	100	25	25	50.0	35.4	0
(d)	50	50	50	50.0	0	0
(e)	75	75	0	50.0	17.7	30.6
(f)	200	0	−50	50.0	106.07	20.4

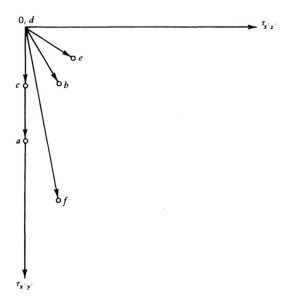

APPENDIX 2

1. Equations A1.2 together with the answer to Problem 4 in Appendix 1 for stress transformation in two dimensions are

$$\left\{ \begin{matrix} \sigma_{x'} \\ \sigma_{y'} \\ \tau_{x'y'} \end{matrix} \right\} = \left(\begin{matrix} \cos^2 \alpha & \cos^2 \alpha & \sin 2\alpha \\ \sin^2 \alpha & \cos^2 \alpha & -\sin 2\alpha \\ -\tfrac{1}{2} \sin 2\alpha & \tfrac{1}{2} \sin 2\alpha & \cos 2\alpha \end{matrix} \right) \left\{ \begin{matrix} \sigma_x \\ \sigma_y \\ \tau_{xy} \end{matrix} \right\}$$

Replacing τ by $\gamma/2$ and σ by ε, with appropriate subscripts, yields:

$$\left\{ \begin{matrix} \varepsilon_{x'} \\ \varepsilon_{y'} \\ \gamma_{x'y'} \end{matrix} \right\} = \left(\begin{matrix} \cos^2 \alpha & \sin^2 \alpha & \tfrac{1}{2} \sin 2\alpha \\ \sin^2 \alpha & \cos^2 \alpha & -\tfrac{1}{2} \sin 2\alpha \\ -\sin 2\alpha & \sin 2\alpha & \cos 2\alpha \end{matrix} \right) \left\{ \begin{matrix} \varepsilon_x \\ \varepsilon_y \\ \gamma_{xy} \end{matrix} \right\}$$

2. For a gage with $\alpha_A = 0$, $\alpha_B = 60$, $\alpha_C = 90$, the coefficient matrix in (A2.3) becomes

$$\begin{pmatrix} 1 & 0 & 0 \\ 0.25 & 0.75 & 0.433 \\ 0 & 1 & 0 \end{pmatrix}$$

The inverse of the above is

$$\begin{pmatrix} 1 & 0 & 0 \\ 0 & 0 & 1 \\ -0.577 & 2.309 & -1.732 \end{pmatrix}$$

3. $(\varepsilon_x \varepsilon_y \gamma_{xy}) =$
    (a) $(1.0 \times 10^{-3}, 0, 5.774 \times 10^{-4})$
    (b) $(1.0 \times 10^{-2}, 3.0 \times 10^{-2}, -1.155 \times 10^{-2})$
    (c) $(2.0 \times 10^{-4}, 5.33 \times 10^{-4}, -1.61 \times 10^{-4})$

4. (a) $\alpha = 15.0°$, $\varepsilon_1 = 1.077 \times 10^{-3}$, $\varepsilon_2 = -7.736 \times 10^{-5}$
    (b) $\alpha = -75.0°$, $\varepsilon_1 = 3.155 \times 10^{-2}$, $\varepsilon_2 = 8.452 \times 10^{-3}$
    (c) $\alpha = -78.29°$, $\varepsilon_1 = 5.344 \times 10^{-4}$, $\varepsilon_2 = 1.856 \times 10^{-4}$

APPENDIX 5

1. $x = \tan(45 - \delta/2)$.

2. The angle between (1) and (2) is 59°. Their common plane strikes N 64° E and dips 78° N 26° W.

3. $\bar{I}_{12}$ plunges 37° to the S 84° E.

4. The answer is given in Figure A5.1. The line from 0 to $-P$ is directed into the upper hemisphere. It plots outside of the horizontal circle at position $-p$ as shown.

5. Let the position of line $0Q$ in a lower hemisphere projection be point q. Then the position of the opposite to $0Q$ when plotted in an upper hemisphere projection is obtained by rotating the tracing 180°. What was *north* must be relabeled as *south*.

6. 15°.

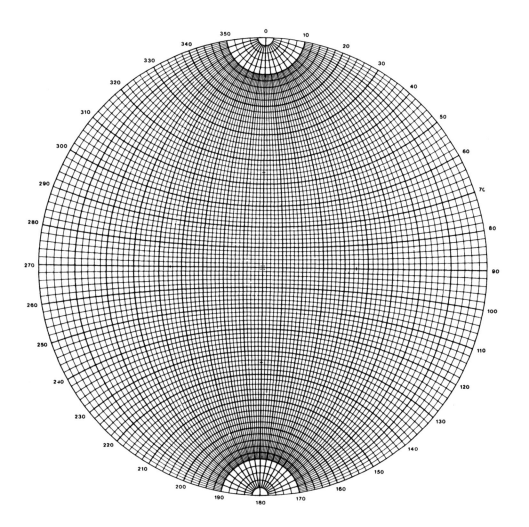

EQUATORIAL CONFORMAL STEREONET

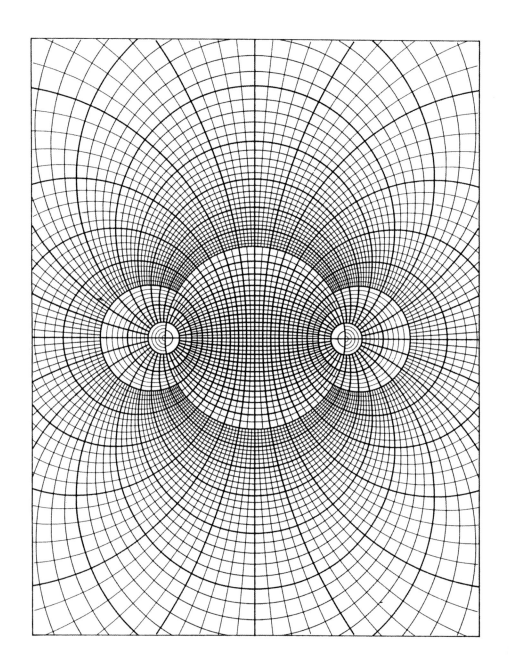

Index

Aastrup, A., 419
Adhesion, *see* Bond strength
Age of a rock:
 effect on porosity, 29
 terminology, 424
Alexander, L. G., 122
Alto Rabagao dam, 369
Alvarez, L. G., 122
Anchor piles, 382
Anhydrite:
 specific gravity, 31
 Mohr–Coulomb parameters, 83
Anisotropy, 13
 bulbs of pressure, effect on, 355–361
 in elasticity, 182, 183
 in rock fabric or structure, 420–423
 in strength, 93–95
 in tunnel support needs, 274–280
Archambault, G., 166–171

Barr, M. V., 378
Barton, N., 42, 166, 177
Basalt, 20
 compressive and tensile strength, 61
 conductivity, 35
 density, 33
 Hoek–Brown parameters, 99
 modulus ratio and Poisson's ratio, 186
 Mohr–Coulomb parameters, 83
 point load strength, 37
 porosity, 83
 sound velocity, ideal, 41
Beams in mine roof, 233–237, 289
Bearing capacity:
 analysis, 361–364
 codes, 348–350
Beatty, R. A., 132
Bedding, 13, 144
Benson, R. P., 223
Bernaix, J., 34
Bernede, J., 60, 116
Bibliographies and indexes, 15
Bieniawski, Z. T., 75
 geomechanics classification, 43–46

modulus in-situ, 198
size effect, tests, 90, 91, 231, 233
standards for compression tests, 60
stand-up time in tunnels, 250
Bischoff, J. A., 546
Bishnoi, B. W., 364
Block sliding (plane sliding), 295, 296
 kinematic analysis, 301–302
 stability analysis, 310–319
Blocks:
 criteria for a key block, 258, 259
 numbers of joint faces, 322
 removability, 259, 260
 types, 259
Block theory:
 introduction, 257–262
 for rock slopes, 320–325
 for tunnels, 270–274
 for underground chambers, 262–270
Bond strength, concrete to rock, 372, 374, 376
Borehole breakouts, 114, 115
Borehole deepening method, 116
Borehole jack test, 190, 191
Boyle, W., 258
Brace, W. F., 35, 69, 71
Bray, J. W.:
 bulbs of pressure in anisotropic rock, 358–361
 Mohr's circle construction, 163, 392
 plastic zone around tunnels, 243–250, 454–464
 slip on joints, 176, 455
 toppling, 296
Brazilian test, 60, 65
Broch, I., 37
Brown, E. T., 99, 108, 111, 378
Buckling:
 of rock slopes, 337
 of roof beams, 233
Building codes, 349, 350
Bulb of pressure, 355–361
Bulking of rock, 53, 287
Bulk modulus, 68, 182, 441

Breinigsville, PA USA
08 August 2010
243208BV00005B/22/A